THE FINE HARDWOOD VENEER INDUSTRY IN THE UNITED STATES: 1838-1990

THE FINE HARDWOOD VENEER INDUSTRY IN THE UNITED STATES: 1838-1990

John C. Callahan

NATIONAL WOODLANDS PUBLISHING COMPANY

Lake Ann, Michigan

This book was produced on an Apple Macintosh™ SE computer with Quark**XPress**™ desktop publishing software.

Text was set in Times Roman and heads in Helvetica.

Editor and Production Supervisor: John D. Schultz

Printer: Village Press, Inc., Traverse City, Michigan

Binder: John H. Dekker and Sons, Inc., Grand Rapids, Michigan

THE FINE HARDWOOD VENEER INDUSTRY IN THE UNITED STATES: 1838 - 1990

This book was published with funds provided by Hardwood Forestry Educational Fund, Inc., (Fed. I.D. #31-1070486), in cooperation with the copyright owner.

First Edition: October 1990

Library of Congress Cataloging-in-Publication Data

Callahan, John C. (John Conrad), 1923 –
The fine hardwood veneer industry in the United States: 1838-1990.

xvi + 368 p.
Bibliography: 14 p.
Includes 2 indexes, 77 figures, 19 tables.

Library of Congress Catalog Number 90-63376

International Standard Book Number 0-9628075-0-8

90 91 92 93 94 10 9 8 7 6 5 4 3 2 1

DEDICATION

James F. Burrell, a long-time editor of veneer publications, has often referred to the patriarchs of the hardwood face-veneer industry as "The Men who Loved Wood." It is to them that this book is dedicated.

FOREWORD

This book is about a particularly interesting segment of the forest products economy, the fine hardwood veneer industry, which manufactures veneer used decoratively in the production of quality furniture and panel products. The content is derived from a 15-year search for information on how this small industry of never more than 50 firms began, evolved, and survived the many economic and social vicissitudes which have occurred since 1838. The author's interest in the industry was initially awakened when he began to gather materials for a publication titled *Indiana Forest Products: The One Billion Dollar Industry* published in 1975 as Research Bulletin No. 923 of the Purdue University Agricultural Experiment Station.

Among the findings, Professor Callahan discovered that the Indiana face-veneer firms, along with related hardwood and panel manufacturers, were not only important economic contributors to the state but also ranked first in the United States when compared with similar hardwood firms in other states as measured by "value added by manufacture." When softwood veneer and plywood sectors were included, Indiana ranked fourth in the nation at that time among all states when the same measurement technique was used.

To find out more about the Indiana veneer firms and their entrepreneurs, retired Professor Roy C. Brundage was asked to make a historical survey of the Indiana veneer industry. His report concluded that it would be a difficult task to undertake since relevant portions of the history lay outside the political boundaries of Indiana and that much of the history had been lost because of a lack of retained veneer-mill records, family documents, and the fading memories of present owners/managers regarding prior generations.

Intrigued, nevertheless, by the limited data Brundage had collected, and encouraged by Mason C. Carter, Chairman of Purdue University's Department of Forestry and Natural Resources, John Callahan took up the quest for the historical essence of the industry. The study area quickly expanded to those companies which existed or had existed in the eastern hardwood region. Many of these had historical ties to Indiana.

Data were systematically obtained by interviewing retired individuals who were active in the business before 1925; managers and owners currently involved in manufacturing and selling veneer; from library collections of trade journals; and from documents provided by veneer-trade associations and equipment manufacturers. Governmental publications were used to better understand the structural changes which have occurred in the industry and to provide quantitative statistical data for analyses.

This study was begun and the research largely carried out while Professor Callahan was associated with Purdue University as a forestry economics and policy professor. After retiring from the University in 1986, he completed his research and drafted this book written in a style which will provide a pleasurable reading experience for those who may be interested in the history of this unique and dynamic industry and its individual firms. The information is fully referenced for those who may seek source materials for additional studies.

Michael O. Hunt, Director
Forest Products Laboratory
Purdue University
West Lafayette, Indiana
September 20, 1990

ACKNOWLEDGMENTS

The author is deeply indebted to the many people who consented to be interviewed, provided encouragement, and gave substantive advice and criticism. All have left their mark. However, none should be held responsible for the form or content of the final product.

Among those who deserve special recognition is Larry R. Frye, executive director of the Fine Hardwood Veneer Association. He provided me with many essential documents so I could better understand the nature of the industry and its periodic problems and triumphs. He also introduced me to James F. Burrell who acquainted me with the rich treasury of historical information available in hardwood veneer and plywood publications.

Jack Koss, president of the Capital Machine Company, was most helpful in making available company records, in supplying needed counsel and encouragement and in reviewing portions of the drafted manuscript. His friendly hospitality, known to many in the veneer industry, and his genuine interest in this project are sincerely appreciated.

One of the most interesting and cerebral contributors to the study was Percy Simmons who arrived in Indianapolis from London in 1938 at the age of 32, having already spent 18 of these years in the hardwood veneer and lumber trade in England. Involved in a number of veneer-related activities, he eventually settled down to importing mahogany logs for midwestern mills and custom cutting veneers at selected Indiana mills. When the tide turned and Europeans looked to the United States for hardwood logs, Simmons responded to these market forces and was still actively exporting logs in 1989.

I fondly remember the occasions when I met with him for tea and conversation in his living room. A wine expert, French Consul in Indianapolis for 27 years, and a former chairman of the Indiana World Trade Council, he shared with me his experiences and provided me with a working knowledge of the mahogany trade centered in New York, Norfolk, and West Africa. Percy's enthusiasm was contagious, causing me to continue to probe whenever it appeared that the information I was seeking was beyond reach and no longer available.

One of the most dignified elder statesmen of the hardwood veneer industry I met was Ross Kreamer. He had been born to a pioneering Colorado family in 1903. Kreamer migrated eastward to attend Northwestern University. When I interviewed him in Memphis, Tennessee, he was still a veneer merchant. Earlier in his career, he had been a veneer manufacturer in Kentucky and clearly remembered the early days, before there were many roads in Kentucky, when he bought logs along the

Cumberland River and barged them back to Burnside, Kentucky, where the mill was located. As a seller of veneer, he had periodically visited all of the major midwestern furniture firms. As an important member of veneer trade associations, he had also known most of the veneer-mill owners and operators of the last 60 years. He generously shared his memories and documents with me.

Technological aspects of the hardwood face-veneer industry were willingly given to me by a number of other individuals, including Craig J. Davis, Robert E. Hollowell, Jr., and William L. Koss. Tom Hale provided his philosophy about the past and current status of the industry as well as strategies the industry should adopt in the future. As an engineer and a grandson of John N. Roberts who bought slicers No. 1 and No. 4 from the Capital Machine Company in the early years of this century, Tom learned the veneer business from the stump through the manufacturing processes. These he improved with slicing technology and veneer dryers which he conceptualized, patented and built.

All 80 of those who were interviewed deserve much credit for their contributions to this book. Many are featured and quoted at length in the context of events important to their companies and the history of the industry. Their names, interview dates and business affiliations are listed in Appendix A.

I would also like to express my deep appreciation to Dr. John D. Schultz, president of National Woodlands Publishing Company, who agreed to edit a voluminous manuscript into its present form and to provide technical assistance associated with its publication. I also wish to thank the editors of the Journal of the Forest Products Research Society and the Journal of the Forest History Society for permission to use material I had originally published in their journals.

The financial assistance of the Hardwood Forestry Educational Fund, Inc., and its contributors is gratefully acknowledged. The foundation's support was essential in completing and publishing this history and analysis of the fine hardwood veneer industry.

Lastly, I would like to thank my wife, Jacqueline, for her patience while I spent long hours pondering and writing in my loft and for her help in compiling the hundreds of documents collected and reproduced at the Library of Congress. Linda Hampton, my faithful secretary for many years at Purdue University, "word-processed" the several early drafts of the text. I am grateful to her for this and for helping me maintain a peaceful co-existence with departmental administrators, peers, and students during my active years as a professor at Purdue University.

John C. Callahan
Bradenton, Florida
September 24, 1990

CONTENTS

INTRODUCTORY PERSPECTIVES

The history of the fine hardwood veneer industry, in large measure, parallels the cultural and industrial development of the United States. In the early Colonial period, much of the furniture was primitive in the sense that it was crafted by village carpenters using local materials. With the cessation of the constant struggle to survive during the first century of occupation came a time of leisure for some and an interest in self-expression in artistic ways. Veneers and inlays were employed in the creation of fine furniture. That indicated an appreciation for the decorative values associated with a number of domestic and exotic hardwoods. Manufacture of the veneers themselves was a laborious process virtually unchanged from the days of the Roman Empire.

Veneer was not produced efficiently and uniformly until the Industrial Revolution in England provided the engineering means to do so. This specialized equipment became generally available in the United States in the early 1800s at a time when an expanding, more affluent population was beginning to be supplied with mass-produced products.

The two most important types of face-veneer machines invented in these early years were the veneer saw and the veneer slicer. Both utilized longitudinal sections of logs (flitches) that had been slabbed and "opened up" so as to enhance the characteristics of the veneer to be produced (Fig. I-1). A third machine, the rotary veneer lathe (Fig. I-2), also was developed to manufacture veneer from cylindrical log sections (bolts).

The federal government's 1906 report on veneer production stated that "veneers are of three general classes—sawed, sliced, and rotary cut. Sawed veneers, which have been longest used, are of highest grade. Sliced veneers rank next in order of quality." The amount of sawn veneer decreased over time because of higher relative production costs and the loss of usable veneer through the sawing process. The technique was abandoned some time after World War II. Today, nearly all of the hardwood veneer used decoratively is sliced.

Sliced veneers can be described as fine, face, figured, or decorative special veneers and are commonly used in furniture, architectural panels, and other higher-valued, veneer-covered products. In contrast, rotary-cut hardwood veneer is most often used in the production of containers, stock panels, core stock, door skins, and less expensive decorative wall panels.

"Staylog" lathes are used in addition to veneer slicers in most hardwood face-veneer plants to produce a product similar to sliced veneer. The staylog flitch is attached to a shaft which rotates the flitch past a fixed knife. For equal-sized flitches, a somewhat wider sheet of veneer results as the veneer is cut from an arc rather than the vertical section (Fig. I-3). Individual veneer sheets cut this way have grain and figure patterns representative of both sliced and rotary-cut veneers. Veneers cut by either machine are dried and reassembled into bundles of sheets laid together in the sequence in which they were cut from the flitch.

In the earlier days of commercial veneer production, the manufacturer gave little attention to lumber production potentials. The attitude of the sawmillers was similar. The veneer industry was regarded as a separate entity entirely and not closely related to saw-milling or the lumber business. By the early 1900s, there was a growing acceptance of the concept that it was financially advantageous to produce both lumber and veneer in the same facility. This was often true for enterprises near the timber source where better utilization could be obtained from purchased lots of logs. This concept is not universally accepted today, particularly among export-oriented mills and those where transportation costs are a factor in securing a supply of logs.

Traditionally, hardwood face-veneer has been produced in the United States by single entrepreneurs or entrepreneurial partnerships made up of close associates and family members, reflecting the craft nature of the business. Mergers with larger corporations have usually failed. Business failures have also occurred because of the many business cycles and other exogenous events to which the industry has been subjected.

Figure I-1

Cross-section of a veneer slicer. The flitch reciprocates past a fixed knife. Flat sliced and quarter-sliced veneer can be cut to obtain desired figured patterns.

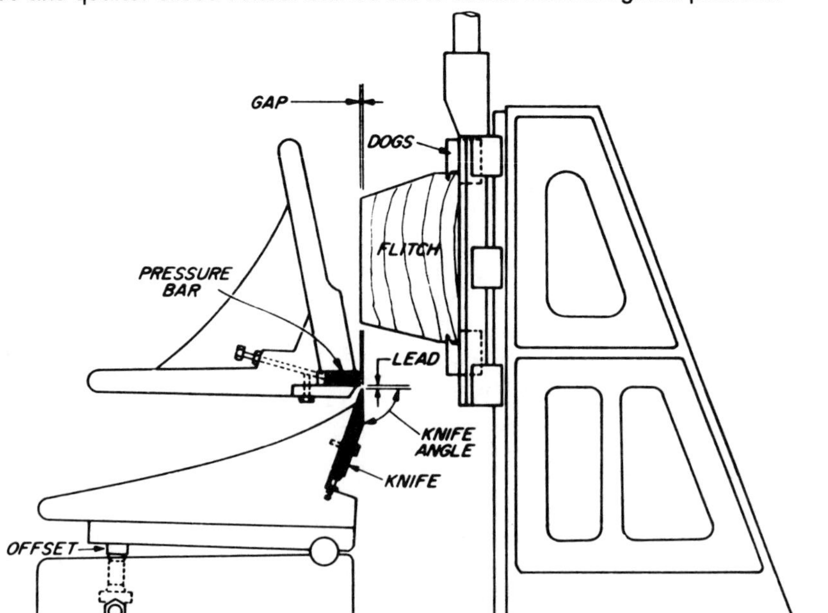

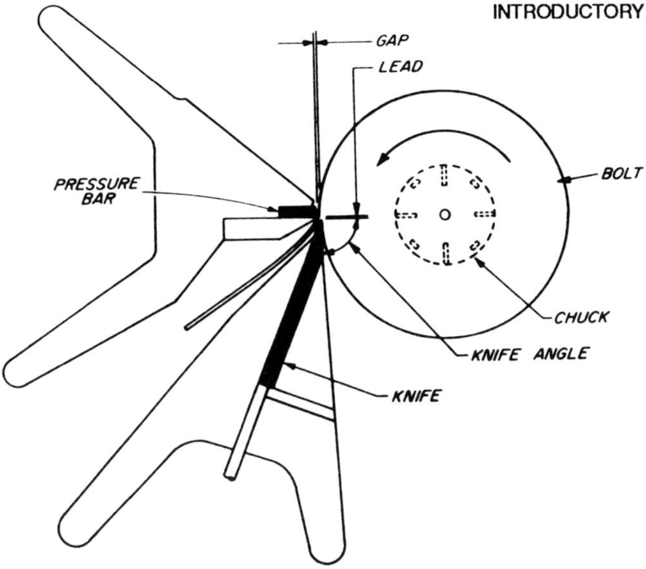

Figure I-2
Cross-section of a rotary veneer lathe. The veneer produced is often boldly figured by the annular growth rings as the bolt is "unwound."

Figure I-3
Schematic drawing of a staylog lathe. The staylog, with attached flitch, rotates past a fixed knife. Rift-cut and half-round sliced veneer can be produced with this machine.

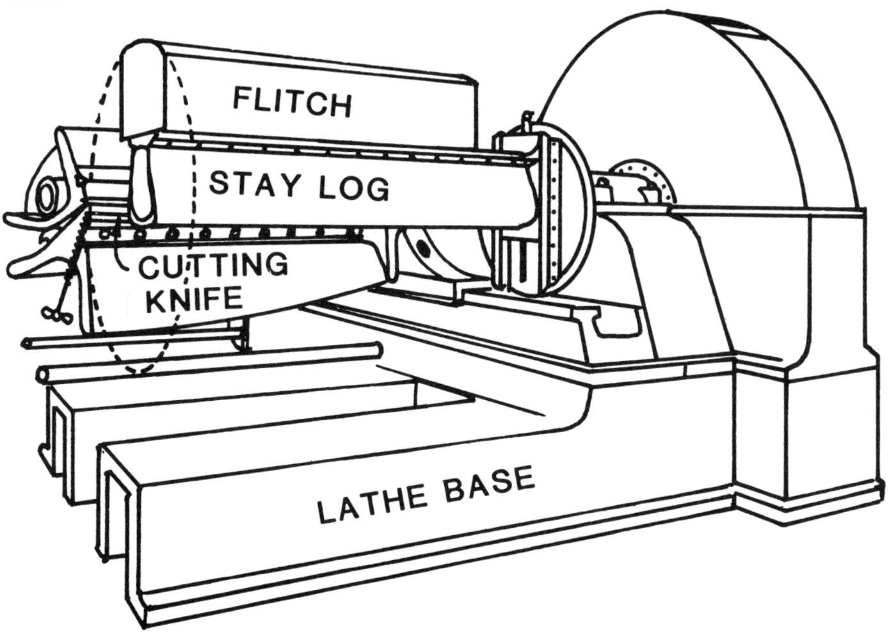

Over time, the relative volumes of the different species cut for face-veneer reflect both changing consumer preferences and the availability of domestic and foreign timber resources. Changes in the timber resources available to European veneer manufacturers have also had a significant impact on the activities of domestic producers in the United States. With the loss of their colonies after World War II and the subsequent and continuing political instability in the newly independent countries, European veneer companies have increasingly relied on North America to supply their requirements for high-quality veneer logs and manufactured face-veneers.

New mills and the adoption of the latest technology have been responsible for significant increases in production. Computers, for example, are now able to track a unit of veneer from the time a log or tree is purchased until the time it is shipped. Computers also assist in regulating the steaming of flitches before they are sliced. Other computers are used to control automated processes involved in slicing, measuring, packaging, and costing units of veneer.

The chapters which follow trace the dynamic history of the veneer industry, largely through the perspectives of those men, past and present, who were intimately involved with it. Taped interviews and trade journal literature were used extensively in an attempt to recreate the environment in which these entrepreneurs performed. To present a coherent interpretation of this industrial history from the mass of collected data was a problem I struggled with for many months. The format selected, probably, is imperfect in the eyes of some critics since the history is revealed through the chronicles of individual firms in states and regions and through the dichotomy between domestic and foreign-financed companies.

The book is organized into three parts of unequal length. The first describes the evolution of the machines and early development of the industry on the East Coast. It ends with the veneer mills of the East and Gulf Coasts during the Age of Mahogany. In the second part, I examine the emergence and growth of the domestic industry located in or near the hardwood heartland, centered in Indiana. The last chapter of the section reports on those mills purchased or built as a result of foreign investments. The final part of the book is devoted to several diverse macro-aspects of the industry.

References to materials cited are usually placed at the ends of paragraphs for the sake of reading ease. The full references are provided in the Bibliography, following Chapter 12, where they are listed alphabetically by author under their respective chapter headings. Contributors who were interviewed and then quoted in the text are listed in Appendix A rather than being listed in the Bibliography.

Please note that discussion about mills which are partially or entirely owned by foreign investors occurs in Chapter 8. Domestically-owned mills are considered according to their geographic location.

Chapter 1

EVOLUTION OF THE VENEERING PROCESS

15th CENTURY B.C. TO 1805 A.D.

Our knowledge of ancient veneering tools and techniques is derived principally from furniture which has survived from Egyptian and Roman times. One of the oldest examples of veneering craftsmanship is a 3,500 year-old Egyptian chair finished with veneered ebony. A mural decoration in the tomb of Rekmara at Thebes was dated to about the year 1490 B.C. It provides the only ancient evidence of the Egyptian art of cutting and applying veneer. The evidence suggests that each sheet was laboriously produced (Fig. 1-1). It was first sawn to separate a thin board from a larger block of wood. "The saw was about as broad as the blade of a table knife and not much longer than the knife. The teeth were merely notches..." with the teeth "set" as in modern handsaws (Wilkinson, 1883). Given the size of some of the surviving veneered furniture sections, the sawyer must have used a great deal of patience and ingenious methods in separating the thin plank from the "flitch." To smooth the surface and to work it down to the approximate thickness desired, a small adz was used.

Figure 1-1
The making and use of veneer is an activity which has existed for at least 3,500 years. This sketch of an Egyptian tomb painting illustrates the laborious nature of the process which existed until the Industrial Revolution.

> The blade of the Egyptian adz was about an inch wide. The handle was adjusted about in the same angle as the handle of the modern adz, but it was only a few inches long. The board...to be dressed was stood upright and held with one hand while the small adz was manipulated with the other.
>
> Samples [of furniture]...in the Field Museum, Chicago, show traces of smoothing tools for polishing surfaces, rubbing down high places and reducing inequalities. The scratches left on the surface [provide a] reason to believe that much of the rubbing was done with stones (Anon., 1919).

The veneered furniture produced by Egyptian craftsmen influenced the early Greeks and other Mediterranean travelers. Although no actual specimens of Greek furniture survived into the modern age, beautiful furniture is depicted on countless ceramic vessels and wall decorations found in tombs at many European and Asian locations. These pictures and paintings clearly indicate that the Greeks knew the art of veneering table tops, beds, and even the arms of chairs.

The veneering knowledge of the Greeks and the utilization of veneers in furniture making were "civilizing" influences carried back to Rome by its conquering forces. The furniture forms of both Greece and Egypt were combined in a distinct Roman furniture style well suited to the luxurious way of life of wealthy Romans.

> Roman furniture was magnificent. Roman taste despised solid block pieces as being essentially barbarian and crude. Pliny, Rome's greatest historian, even went so far in his admiration of the veneerer's art as to denounce the over-use of gold and paint "because it covered up the beauty of the veneer" (Anon., 1924)

The Roman veneer saws were improved versions of Egyptian tools made possible by improved metallurgy and design. Planes and chisels were crude but similar to those of today. Production was slow and the product correspondingly costly. From references made by writers in the time of the Romans, it is believed that the workers actually employed in making the expensive veneered tables were Greeks who carried the art of veneering from Greece to Rome. The Romans called the process of laminating veneer to the core stock "xilotarsia" and the resultant finished veneered piece "sectiles laminae." Unfortunately, none of the well-crafted furniture found at Pompeii has contained any of the fine veneers that Pliny wrote about.

Veneering became a lost art after the Roman Empire was finally destroyed in 476 A.D. Objects having elegance of design and beauty were replaced by the "coarse and gross...in this dismal period of cultural reaction to the primitive and barbaric. It was the cubistic school of the semi-civilized. But time cures all things and out of this slough of despond came the brilliant dawn of the Renaissance" (Anon., 1924).

This period which may have had its early beginnings in the late 1200s reached its peak in Italy in the 15th and 16th centuries as the powers of the church weakened while city states developed. The growth of urban populations and trade brought about wealth accumulation and a more worldly point of view. Princes and merchants sought a society in the classical model of ancient Greece and Rome. In their protest against the darkness and constraints of the Middle Ages, art, design, and a cultural preference for beautiful objects became predominant. The classical design of furniture was rediscovered and coveted. Artisans accordingly developed the necessary skills and tools required to provide their patrons with furniture pieces suitable for palaces and mansions. Tools remained primitive by modern standards and veneering was restricted for the most part to inlays and other forms of marquetry.

This cultural revolution slowly diffused through Europe and continued well into the 18th century and the beginning of the Modern Era. The climax of marquetry skill was the creation of the Bureau du Roi, now in the Louvre museum in Paris. This desk was begun in 1760 for Louis XV by Oeben and finished and signed by Jean Henri Riesener in 1769.

> It is a cylinder desk...prototype of the "roll-tops" of our later days. Its decorations are done in magnificent marquetry of flowers and trophies of poetry and war. [This piece of furniture cost the] King 500,000 francs and required nine years in the making before it was ready for use (Anon., 1924).

A number of other examples of French and British artistic triumphs are preserved in European and American museums.

In the decades prior to 1800 veneers were cut by pit sawyers. The saw blade used was thinner than the more common pit frame saw but otherwise similar to it. These were used by cabinet makers who would cut seven or eight veneers to the inch. However, in the late years of the 18th century mechanical machines were being developed which would forever change the technique for producing veneer and provide the opportunity for making veneer available for a wider range of potential applications and markets.

Machines driven by waterpower was an idea developed into a reality by the Dutch. However, it was the English who borrowed the machine concept and paved the way for the Industrial Revolution. An epochal invention in wood working machinery was that of Sir Samuel Bentham who patented a large power-drawn plane in 1793. His brother, Jeremy Bentham, a famous political economist, was in charge of several industrial prisons where prisoners by English custom were required to produce products to be sold as commercial goods. Bentham's inmates were seemingly incapable of learning a manual trade although wood-working was considered the chief possibility. Bentham appealed to his brother, Sir Samuel, believing that even his prisoners could be taught to run a machine. Sir Samuel proceeded to develop a series of machines and specialized tools. Among these was a large power-drawn plane to produce shavings that could be laminated into an early form of plywood. "Thus under the stress of prison economy, was launched the most ambitious aggregation of wood-working machines that the world had known up to that time." His recognized ingenuity eventually led to Sir Samuel's appointment as

Inspector-General of the naval works in England (Anon., 1957).

The Bentham machine seems to have led to the invention of what was known as the scale board or scaffard plane for cutting off wide, thin boards used in the early 1800s for making hat boxes.

> A machine of this type was used in the works of Esdalle & Margrave, in London, in the very early days of saw-milling. It was more in the form of a huge metal plane, working in a slide and inclined horizontally about 40-deg., and the wood to be cut by it into thin sheets was first boiled to make it softer and more elastic. It is easy enough to understand how from a machine of this kind the modern slice-cutting machine was developed. Also it is possible that from this idea was developed what we know as the rotary machine, because the rotary machine is essentially a knife held in a carriage while a block rotates against it and the knife and carriage suggesting the throat and chipbreaker of the plane. The first invention and use of the rotary machine, as such, is really lost in obscurity. Some claim that it originated in Russia, some claim it was developed in England, but none of the available records are very clear (Taylor, 1917).

1805 TO 1870

A most significant contribution to the veneer industry occurred in 1805 when Sir Mark Ishambart Brunel invented the first power veneer saw (Fig. 1-2). Brunel, born near Rouen, France, in 1769, was a well-recognized civil and nautical engineer who later resided in England. The saw was first used at the Chatham dockyard and subsequently at his wood-working mill at Battersea. It is probable that the saw blade was not segmented as a report published in England at about this time describes a veneer saw "...as consisting of a circular saw made thick at the middle and tapering to a very thin edge at the periphery" (Bardin, 1922a).

Eventually, Brunel equipped his shop with two veneer saws 18 feet in diameter and two that were 9 feet running at about 65 revolutions per minute. These saws "...were used in cutting rosewood and mahogany veneer as thin as 1/16 inch with great precision. It was considered a wonderful feat at that time [and] we are impressed with the idea that Brunel is entitled to the great historical honor in connection with the veneer industry" (Taylor, 1917).

In 1806, Brunel patented another slicing device which was described as "...a method of cutting veneer by means of horizontal knife exceeding in length the block to be cut." It was powered by a steam engine, a novelty at the time. However, there appears to be no evidence that the machine was extensively used. Although the slicer was reasonably successful with straight-grained wood, veneer had a tendency to curl up and split when cut from irregularly-grained and brittle woods.

Just when veneer was first made in the United States will never be known. "Here and there we find evidence of [veneer manufacture] following soon after its extensive introduction into England. For example,

Figure 1-2
This 1905 Coe segmented veneer sawing machine was manufactured in several models equipped with saws ranging from 58 to 76 inches in diameter. All were direct descendants of the veneer saw which Brunel invented in 1805.

it was only about twenty years after Brunel had his veneer mill going in good shape in London until we find the origin of a pioneer veneer concern at New York City coming closely in the footsteps of the early piano manufacturers in this country" (Anon., 1908). Even though the early records concerning the introduction of veneer are obscure and there are no specific data as to when the first sheet was cut on this side of the Atlantic, veneer cutting technology lagged but was definitely patterned after developments in England. Therefore, the earliest commercial veneer was probably produced in the United States by utilizing the veneer sawing method no later than the early 1830s. Other methods were soon to follow.

A crude form of horizontal slicer was first patented (U.S. Patent 387) on September 21, 1837, by Joseph Skinner of Stockbridge, Massachusetts. It was noted in the patent application that "Hard timber may be softened by steam or otherwise heated to facilitate the cutting of veneers and to render more pliant and yielding." The first authentic record of an operational slicing machine in this country was one built by John Strong and L. and J. Cutting in 1848. The three-foot machine, operated by horse power, was installed in the Strong Veneer Company plant in Gerry, New York. After making veneer with the slicer, Strong and his partner built a three-foot rotary machine which was also operated with horse power.

> Usually, it took two horses to run the 3-ft. rotary. It was rigged up with a long sweep, one end extending each side, and he attached a horse to each end on the opposite sides of the driving power. It was a part of the duty of

> B. E. Strong, who was but a mere boy at that time, to drive the horses and lash them up a little with a whip when they got lazy and didn't pull; and occasionally when he would get sleepy and not look after his task properly, the veneer cutter would go to dragging and as a consequence his father would have to stir him up by calling to him (Anon., 1908).

The earliest patent on a rotary veneer machine was granted to John Dresser also of Stockbridge, Massachusetts in 1840. This, too, was a very crude and simple machine. Since these early beginnings, scores of improved machines and many attachments for these machines have been invented as part of the evolutionary process leading to the high-speed, computer driven machines of today.

Notable achievements include the first attempt using a concept that became the modern pressure bar. Its inventor, a man by the name of Humphrey, utilized the idea in a very early rotary slicer. The pressure roller acted against the wood in advance of the knife and removed the veneer. The distance between the roller and the knife regulated the thickness of the veneer. Parker's and Sleeper's original vertical slicer used a pressure bar. Although the machine was not "heavily" constructed, all of the basic elements of the modern vertical slicer were present (Bardin, 1922a).

"Light" construction of veneer cutting machines was to persist until almost 1880. The lighter built machines were limited in cutting veneer thicker than 3/16-inch even in the softer hardwoods such as yellow poplar even when the flitches had been well steamed. Machine vibrations were also a problem. One of the reasons for this type of construction was the lack of equipment manufacturers. Veneer producers built their machines in their own establishments or with the assistance of local machine shops. For example, Land and Bodley of Cincinnati ran a general machine shop and manufactured veneer saw equipment on order as did R. Hoe & Company in New York City. Hoe's main business was the manufacturer of printing presses. The Farrel Foundry and Machine Co. of Ansonia, Connecticut, was also among the first to build some of the early veneer machines.

1870 TO PRESENT

Beginning about 1876, machine manufacturing specialization began to occur in firms such as "...Coe and Wilkes at Painesville, Ohio; the Merritt Machine Co., Lockport, New York; the Capital Machine Co., at Indianapolis; the Blakeslee Manufacturing Co., Birmingham, Alabama; the St. Joseph Machine Co., St. Joseph, Michigan; and a party by the name of Titus, at Petersburg, Virginia, and even today [1922] machines bearing some of the above names are foremost in the veneer industry. With specialization there came heavier and better machines...many new features were added and types changed" (Bardin, 1922b). The two most prominent hardwood veneer machine manufacturing firms which have survived to the present are the Capital Machine Company, Inc., and the Coe Manufacturing Company.

The Capital Machine Company

There has been no manufacturer of veneer equipment in the world more important to the hardwood, face-veneer industry in the United States than the Capital Machine Company of Indianapolis, Indiana. Louis Frederick Koss (Fig. 1-3) was the scion of an industrious family "of good German stock." His father designed and built the large pumps required for the fountains at the base of the Soldiers and Sailors monument on Monument Circle at the hub of activities in downtown Indianapolis. An uncle was in the stone construction contracting business and built viaducts and large bridges in Indianapolis (Esarey, 1924).

The Capital Machine Company was loosely organized by Louis Koss as the Capital Machine Works in 1887. The business office was "...located [at 502-504 South Pennsylvania Street] in its own building, a commodious brick structure. All the latest and most improved machinery and appliances [were] employed and from seven to 10 skilled workman [were] given steady employment, their products being sold in all parts of the United States" (Hyman, 1902). The machine shop, itself, was located at 60-100 East Garden Street.

Figure 1-3
Louis Frederick Koss, founder of Capital Machine Company and president until his death in 1947.

The first two staylog lathes were built in 1887 for the Singer Sewing Machine Company of South Bend, Indiana. Between 1887 and 1897, the adjustable pressure bar was developed and perfected. It remains a standard in the industry. Veneer slicer No. 1 was sold to the Indianapolis firm of Lilly and Roberts Veneer Company on December 17, 1900. This horizontal slicer was built from a design engineered by John N. Roberts who later applied for the patent on May 13, 1901 which was subsequently granted in 1902. The completed machine weighed about 75,000 pounds. At least two of these horizontal slicers were eventually built. One of the machines was still in use in 1989 after nearly 90 years and could be found operating on a daily basis at the Lane furniture plant in Alta Vista, Virginia. (The other sank with the ship carrying it to Brazil in 1917.)

However, Koss was not satisfied with the operational efficiency of the machine. After building and selling several early models he applied for a patent on a vertical slicing machine in 1904. U. S. Patent 793,306 was granted on June 27, 1905. His comment was supposed to have been: "The horizontal slicer isn't the right one for the customer." (This and many other details about the company were found in Capital Machine Company Minutes Books or came from conversations with L. Jack Koss.) The first two vertical slicing machines were sold in 1903 to the Talge Mahogany Company and the Indiana Lumber and Veneer Company, both Indianapolis firms. Sales to out-of-state companies began in 1904 to veneer companies in Illinois and New York. Expanding equipment sales after 1904 dictated a move to larger facilities which were constructed on vacant property at 2801 Roosevelt Avenue where the company continues to maintain business and machine shop facilities.

In 1908, the Koss company received a 50-year charter from the State of Indiana to do business as the Capital Machine Company and to issue $50,000 of Capital stock. The original stockholders were: Louis F. Koss, C. W. Talge, Edward Resener, and George R. Thoms. Koss was granted a yearly salary of $2,000 for his position as general manager.

In 1913, Louis Edward (L. Edward) Koss, a first cousin of Louis F. Koss, was appointed to the board of directors. L. Edward was to assume a managerial role in the company's business affairs until 1939. (In 1924, he became the company's general manager). Although the *History of Indiana* suggested that: "The firm's enviable reputation in manufacturing circles is due in large measure to L. Edward's executive ability and keen business judgment...," independent observers were less kind (Esarey, 1924).

Louis F. Koss was elected by the board of directors in 1921 to be both president and treasurer, positions he held until 1940. Thereafter, until his death in 1947, he retained the title of president while major responsibilities were delegated to others. Condolence letters written to Louis F. Koss, Jr., at the time of his father's death in 1947, from former business colleagues and customers, praised Louis, Sr., for his stature in the industry and his personal traits. Typical sentiments expressed can be summarized by the following passages from two of the many letters received: "...you have reasons to be proud for he was really the father of the entire veneer industry." The second letter speaks of his compassionate nature. "When I became acquainted with him in 1919, we were starting a veneer mill in Des Moines and I knew nothing whatsoever about the veneer business. He was very patient with me and gave me many pointers. I shall always remember him with a very kindly feeling." These traits of dedication to the industry and his humane qualities have been passed down from generation to generation.

Louis F. Koss, Jr., was born in 1900 and began working for the Capital Machine Company part-time and in vacation periods when he was 18 years old (Fig. 1-4). For the next four years he was a full-time mechanical engineering student at Purdue. After being granted his engineering degree in 1922, he immediately went to work in the machine shop to further perfect the Capital veneering machines and equipment.

Figure 1-4
Louis Frederick Koss, Jr., became vice-president and secretary-treasurer of Capital Machine Company in 1940. He subsequently served as president until his death in 1963.

He was made a director in 1929 and on July 11, 1940, became vice-president and secretary-treasurer after his purchase of a substantial interest in the company. The position of Louis, Jr., in the company was further strengthened in 1946 when he also assumed the duties of general manager. In 1947, after distribution of Capital Machine Company stock from the estate of Louis F. Koss, Sr., his three children (Louis F., Jr., Faye and Laura) held 80-percent of the company's outstanding stock.

One share of stock was given to Louis Jack Koss in 1951 by his father. This permitted Jack to be elected a director of the company in that same year and a vice-president in 1962. The remaining 251 shares controlled by Louis F. Koss, Jr., were placed in a family trust prior to his death in 1963.

His contributions to the company were many, including U.S. Patents for "veneer machine knife clamps" (1932), for a "variable speed drive mechanism" for the veneer slicer (1933), for a "knife rocking mechanism for veneer cutting machines," (1942) and for "the hydraulic control of veneer slicing machines" (1952).

Percy Simmons, veneer merchant and log exporter based in Indianapolis, has been an astute observer of the industry and the Capital Machine Company for about 50 years and remembers Louis F. Koss, Jr., as:

> ...an engineer par excellante. The word "professionalism" is not enough to describe Louis. He was devoted to his work. He was a perfectionist. A machine had to be bloody all right or it wouldn't leave the shop. His devotion to perfection enabled the American veneer industry to have the machinery and equipment that produced some of the finest veneers that were made anywhere. It was his devotion to perfection that could not be duplicated, that enabled this industry to have the best veneering equipment in the world. Only a person like Louis Koss could have done it.
>
> In the eyes of the veneer manufacturer, he was a wonderful guy, because he never charged enough for his excellent products. Never. As to whether this was because of the depression years or not, I don't know. Like most of the rest of the American industry, he kept his prices too low to the point that the Capital Machinery Company was not making any money either.
>
> Prior to the war the European veneer manufacturers had a low regard for the Capital slicer which was a vertical slicer whereas the European slicer was a horizontal slicer with few exceptions and cut only 12 to 14 strokes per minute. After World War II the whole philosophy of European veneer manufacturing changed when they became convinced that you could cut at 40 strokes per minute and still produce a first class product. So the Capital slicer became both famous and in great demand. Louis still was the perfectionist, and one had to wait sometimes 6 to 12 months to get delivery on one of these machines.
>
> As a personality, he was on good terms with all the veneer manufacturers. They needed him more than he needed them. He never took advantage of them. Never. His personality was outgoing, kind, and decent. He was an extrovert, an absolute extrovert and loved to entertain people.

Robert Mathers, a long-time Indiana veneer manufacturer, was a friend of Louis Koss and had similar impressions of him.

> Louis was a very outgoing individual, he tried to be a friend to everybody in the social world as well as and, especially, in the business world. He really worked to serve you and sell you equipment that you needed. In engineering you couldn't beat him. He was very sharp. Here's an example: The veneer slicers used to have old style bearings. You used to get hot bearings if you speeded up the machine. The early machines had streetcar controls. With a streetcar, if you wanted it to go faster,...it just had a bunch of brass dots...you set the throttle here, then there, like a rheostat. After you get the slicer going you could jump a couple of dots which would increase the speed of the slicer. There is a lot of weight there and you had to start out slow to get it to go up and down. After you got two strokes in, then you could speed it pretty fast. But if the machine wasn't "honed-down" after you got 28 slices per minute, you had to be careful because it would heat up if you went to 32 or 34. It's optimum movement was 28 until Lou made all those changes. Louis had good ideas. No one was jumping on him to fix the bearings. Until then, most were satisfied with 28 as labor was pretty reasonable.

During 15 of the 23 years that Louis Koss, Jr., was company president, he was putting his son, Louis Jack Koss, through a rigorous and practical apprentice training program except for those periods when Jack was in the U. S. Navy or away at school (Figure 1-5). Upon returning to

Figure 1-5
Louis Jack Koss. Currently president of Capital Machine Company, Inc., he is grandson of Louis F. Koss, founder of the company.

the family business in 1949, Jack was able to blend skillfully the knowledge he had acquired about wood technology and engineering at Purdue University with the mechanical potentials of the Capital Machine Company veneer slicer. Upon the death of his father, Jack was the most competent person available to assume the leadership role in the company as president, treasurer, and general manager.

One of the things he immediately discovered in 1963 was that the 50-year business charter granted to the company in 1908 had expired five years earlier. New papers were filed with the State of Indiana to incorporate, thus making the life of the corporation unlimited. After transferring the assets of the Capital Machine Company to the Capital Machine Company, Inc., new corporate officers were elected. L. Jack Koss remained as president and treasurer; Marlow Kluter and Edward Thoms, vice-presidents; and Mary Joan Koss Rothenberger, secretary.

The traditions of innovative engineering, dedicated service, and social responsibility continued under the leadership of Jack Koss who justifiably gained the respect and admiration of veneer manufacturers throughout the world. One of the major unsolved problems attacked by him was the staining of veneer as it was being sliced.

> Blue stain usually forms on the slicer blade at the beginning of daily operations. The flitch, still hot from its bath in a hot vat, is placed in the slicer, and the machine starts. The steel blade is cold. When the hot flitch and cold blade make contact, there is a chemical reaction between the tannic acids and other chemicals [exuding from the hot wood] and the cold steel. The result is a complex of ferrous salts which are a dingy blue. They stain the slices of veneer and make them worthless for ordinary purposes.
>
> In the early efforts, the senior Koss [Louis F. Koss, Jr.] tried using fans that blew warm air against the blade, but results were not very satisfactory. [Jack] earmarked this as one of his problems and worked on it until it was solved..." (Anon., 1971).

The company had already provided the industry with Stellite, stain resisting, pressure bars in 1930 and with the first stainless steel knives in 1963. The next step was to introduce extruded magnetic aluminum knife guards in 1967 with knife heating capability when the knives were stored. Two years later a method was developed to heat and cool the pressure cap (bar), knife bar holder and knife so as to maintain the same selected constant temperature in these slicer elements (U.S. Patent 3,441,069). Most of the modern slicers and staylog lathes today use a heater to warm the water pumped through the castings of both the cap (which holds the pressure bar) and the knife bed. By setting the thermostat, a constant machine temperature can be maintained which approximates the temperature of the wood being sliced. Cold water is injected when the combination of flitch temperature and knife-cutting friction unduly raises the temperature of the cap or knife bar.

In the years that followed, a number of other improvements were made to veneer slicers and staylog lathes. Using the engineering talent

Figure 1-6
The modern Capital Machine Company veneer slicer is essentially the same machine as was patented in 1905 by Louis F. Koss; however, the newer machines have available such added technological improvements as computerized controls and automated direct feed of cut veneer into the dryer.

of Tom Hale, an operator controlled, automatic remote-controlled extension dogging system was developed to assist in selective flitch positioning, a first in the industry. Also first was the development of a practical vacuum flitch table. This resource saving and revenue enhancing device enables the operator to slice veneer until only 3mm of the flitch is left as a backing board.

All of the technological improvements were engineered so that they could be retrofitted to any and all previously built Capital veneer slicers (Fig. 1-6). This engineering feature includes a 1986 patented, computer-controlled system designed by Chief Engineer Robert Brand to increase product quality and productivity by limiting operator error and reducing the time spent manually adjusting the several machine components. Although Capital Machine Company continues to be recognized world-wide for its superior veneer slicers and staylog lathes, the company also manufactures state-of-the-art veneer clippers, veneer hogs, and knife grinders.

As the grandson of the company founder, L. Jack Koss exhibits many of the qualities that have made the company successful for over 100 years. His personality and character were summed up this way by Robert Mathers.

> Jack is a very modest person. He is always backing the other guy up. He never says anything about himself. He is a wonderful fellow. We feel very close to Jack and I think the ones that have his machinery also feel the same. There is a closeness there. If you break down, you can call Jack and he will drop everything to get you going. That applies to everyone using his equipment. Jack really has done so much for his business and the industry.

Following the traditional route, Louis Frederick Koss' great-grandson, William Louis (Bill) Koss, joined the firm in 1979 after graduating from Purdue University. An Industrial Engineering and Business Management student, Bill Koss had also acquired much experience in the machine shop and, by 1989, had installed Capital slicers and lathes in Bhutan, China, Tasmania, Spain and a number of other locations. He also assisted his father as executive vice-president.

Jack Koss is justifiably proud of the Capital Machine Company and expressed his feelings this way:

> We've come a long way since our founding in 1887. But some things haven't changed at all...like our emphasis on quality. From the beginning, our philosophy has been to build the best...solid, dependable equipment, to provide the latest technology to the extent that such equipment can be seviced in remote regions, and to insure that our dependable equipment will do the job right through years of service. And we have succeeded. Today, Capital machines as much as 90 years old are still in operation - all day, every day. Our commitment to quality continues. In veneer making machinery, we build it to last.

The Coe Manufacturing Company

Although established 35 years before the Capital Machine Company and a much larger corporation, The Coe Manufacturing Company's principal manufacturing targets in the decades since World war II have been in markets associated with softwood plywood, particle board, and gypsum wallboard. Following the War, new housing and building construction created the dramatic increases in demand for these products. Coe's contributions to the supply of these materials was to expand on the company's experience and build appropriately-designed veneer lathes, clippers, dryers, and related products. For the most part, these large-scale markets were in the Western and Southern states. In contrast, the hardwood face veneer manufacturing sector, while expanding, remained small in comparison to these other sectors. Nonetheless, the Coe Manufacturing Company was important in the early development of the hardwood veneer industry by creating machines that became prototypes of the modern versions.

The company was founded in 1852 in Painesville, Ohio, to build steam engines and sawmills. The founders, Harold H. Coe and Leonard Anderson, located their plant near the "downtown" section of the small village. The business prospered by selling to regional buyers and by specializing in the manufacture of the "muley" saw (Anon., 1903). Leaving

Anderson in charge, Coe joined the Union forces in 1861 at the outbreak of the Civil War. Upon Coe's return, Anderson sold his interest to Fred Wilkes, "an engineer, inventor and practical business man" (Anon., 1951).

Wilkes had operated an iron furnace at Painesville using local iron ore. He is credited with successful iron smelting research and the use of soft coal in the process. For this reason, the site of the first Coe plant became known as "Furnace Hill." For the next 20 years, Wilkes contributed his not insignificant talents to the Coe & Wilkes Company through the improvement in design and manufacture of its machines.

The first veneer machine was not built until 1871 when Painesville businessmen approached the company to develop a machine to fabricate thin veneer. Gas street lights were then coming into widespread use and the need for long tapers to light them was apparent since this would speed the nightly task of lamp lighting. Thin veneer, it was also reasoned, could serve as the raw material in manufacturing toothpicks. A second plant was built on a site later known as "Toothpick Hill."

Even by 1903 standards, the first machine was "...somewhat crude in design compared with the finished product of later years..." Nevertheless, it served as a model for future rotary veneer machines because of its design features and weight. The making of sawmills was gradually discontinued as the sales volume of veneer machines increased.

In 1891, Coe purchased Wilkes' interest in the business and was joined by his son, Harry P. Coe, in incorporating as the Coe Manufacturing Company. By 1903, the company was making only veneer-related equipment. The new plant at the edge of town was producing one machine per day, with over 1,500 in service in the United States and foreign countries (Fig. 1-7). Coe also claimed that 75 percent of the veneer cut in the United States at that time was cut on his machines. Sixty different sizes and styles were offered "...with knives varying from two to ten feet in length, and with such variations and modifications that timber conditions in any locality of the country can be fully met." The physical layout of the company's production facilities was impressive:

> The new plant of the Coe Manufacturing Company at Painesville takes high rank among the modern and perfectly equipped manufacturing plants of the country. It contains 275,000 square feet of floor space.
>
> The machine shop is of brick and glass construction. It is thoroughly equipped with new and improved machinery and all modern appliances necessary in turning out work systematically and with dispatch. It is also equipped with one large traveling crane and six smaller cranes for handling the heavy castings used in the building and assembling of the veneer cutting machines, which weigh from 26,000 to 50,000 pounds each.
>
> The foundry, like the other buildings, is of modern construction. In this foundry are made all castings used in the construction of the various machines, and it is of sufficient size for working fifty molders. Power is taken into the foundry for driving the fan and for operating the conveying and hoisting appliances, sand mixer, etc.

Figure 1-7
The Coe Manufacturing Company facilities in Painesville, Ohio, about 1900.

> The power house is an all-brick building containing boilers of 300-horse power and a 125-horse power engine. The latest construction of rope drive transmits the power to the main machine and foundry buildings. In this power house is located the electric lighting apparatus, with a capacity of 20 arc and 100 incandescent lights.
>
> The increase in business has been such as to make necessary a material enlargement of the plant. Three buildings are now in course of construction, while a fourth building, to be devoted solely to the making of the veneer dryer, will be 250 x 80 feet in size and will be completed before spring (Anon., 1903).

Work on the development of the Coe roller dryer had been completed in 1902 (Figure 1-8). This innovative dryer moved the veneer along between drive rollers as it was dried and delivered to the stackers in a continuous operation. The dryer promised distinct advantages over the labor-intensive loft and yard drying practices being employed at the time.

These early years of the 20th century marked a zenith period in the fortunes of the company and many pieces of equipment were manufactured (Figs. 1-9, 1-10).

H. P. Coe succeeded his father as company president at the time of H. H. Coe's death in 1908. The younger Coe was apparently inept and unlucky. "Lacking the inventive ability, inspiration, drive and judgment [of his father] and confronted by poor economic conditions, the company suffered badly during the next decade..." (Anon., 1951). Coe gradually lost control of the firm to other investors. During World War I, Coe facilities were entirely devoted to the production of shell-turning lathes and other similar war-related products.

The plant that had been an engineering and technological triumph in 1901 was out-of-date and in a state of disrepair in 1918 following its wartime contributions. When the company resumed business as manufacturers of veneer equipment, the machines produced were patterned

Figure 1-8
The original Coe veneer dryer, developed in 1902.

Figure 1-9
By 1920, the Coe dryer was said to be "...the most efficient and economical equipment for the uniform drying of veneers..."

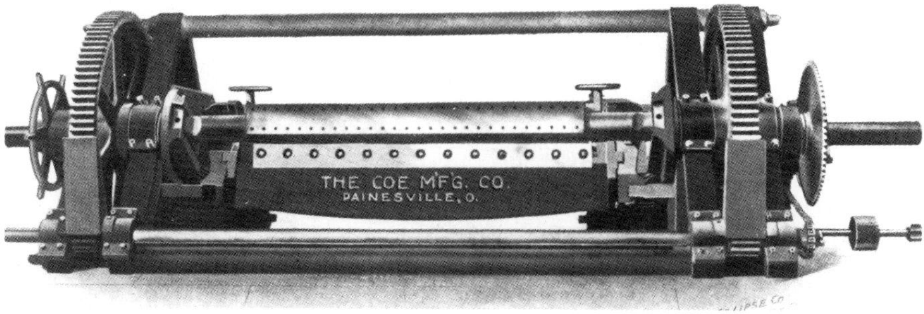

Figure 1-10
The Coe staylog rotary lathe of 1905.

after prewar designs which had not been improved during the intervening years. The downward spiral was ended, somewhat fortuitively, when an engineer acquainted with the Coe Company facilities during the war mentioned the company's plight to Frank W. Milbourn, an engineering graduate of the University of Kentucky with exceptional managerial ability and industrial experience at the Southern Engine & Boiler Works located in Jackson, Tennessee. Intrigued, Milbourn visited the Painesville plant and its facilities and immediately recognized its potential. After a period of contemplative deliberation, Milbourn made the commitment in 1919 to purchase the assets and to assume the considerable liabilities of the Coe Manufacturing Company.

Although the problems facing the company were formidable, the turn around was dramatic. Arthur B. Vance, another University of Kentucky engineer, was employed to head engineering and development. He was also responsible for sales activities. Sales engineers were recruited to provide field service and to act as a two-way conduit between the users of Coe machines and the development department. Milbourn carried the responsibility for finance and corporate restructuring and administration. The redesigned and improved machines found ready acceptance with veneer producers and fabricators. Machines for manufacturing and processing other types of building materials were developed for the softwood plywood, fibreboard and gypsum-wallboard industries in the 1920s.

During World War II, The Coe Manufacturing Company was able to contribute substantially to the war effort by assisting in the development and manufacture of synthetic-rubber dryers. This was based on the company's experience in building dryers for veneer and gypsum products. The company continued to manufacture machinery used to produce the veneers laminated into plywood for boats, airplanes, and other war-related requirements.

The post-World War II industrial and housing expansion directly provided many sales for Coe machines as company executives built new plants or refurbished old ones to meet growing market demands. While retaining the 1901 plant location and main office, the Coe Company modernized and added new production facilities in Painesville as well as building an office and plant in Portland, Oregon, in 1959 to better serve the West Coast market. Coe also established an office and warehouse in Atlanta, Georgia, in 1967 for customers in the southeastern States. From humble beginnings in 1852, with eight employees, the Coe Manufacturing Company has grown to a multimillion dollar firm with national and international stature (Coe Manufacturing Company, 1977).

Company leadership passed to Frank W. Milbourn, Jr., in 1957 upon the death of his father. Young Milbourn had joined the company in 1934 after graduating from Princeton with a degree in engineering. He became general manager in 1945, a position he held until 1957. As president, he served the company for 20 years before retiring in 1977 and being replaced by Fred W. Fields. Fields had been associated with the firm for 30 years and for the most recent 15 years had been manager of Western operations. By 1982, Fields had been succeeded by George B. Milbourn.

Other Contemporary Companies

There are few other veneer machinery manufacturing firms in the World. Two German firms are RFR (Ritter-Fleck-Roller) and Keller Hildebrand Gmbh. Valette and Garrou represent France in this machinery category. None of these have been active in selling veneer slicing equipment in the United States.

Marunaka Tekkosho, Inc. of Shizuoka, Japan, is one of the leading woodworking machine manufacturers in that country. Several of its unique veneering systems have been sold in the United States. The system slices high quality veneer from lumber in a length-wise direction. The knife reciprocates rapidly as the lumber is driven across the support table horizontally and the lumber passes the knife horizontally. The basic design was patented by Saiji Komaki on March 12, 1974 (U. S. Pat. 3,796,243). Unlike most veneer cutting practices, the wood being sliced is dry and, thus, saves time by eliminating the usual soaking and drying phases of production. The principal difficulty with the system is its low rate of production when compared to veneer slicers and staylog lathes (Miller, 1985).

The firm of Angelo Cremona & Figlio of Monza, Italy, has been much more successful in introducing its machines into the North American market. In 1984, Cremona veneering equipment was installed in veneer plants in Coldwater, Michigan; Fairfield, Ohio; St. Joseph, Missouri; Williamsport, Pennsylvania; and Kitchener, Ontario. The machines have been marketed with very favorable credit terms which has made them attractive to both new and expanding well-established firms.

Between 1973 and 1979, Cremona received eight U. S. patents on his veneering machines and added devices for the machines. His horizontal slicer machine was patented on August 7, 1973 (Pat. 3,750,725) and a

Figure 1-11
The Cremona slicer (foreground) is illustrated with a device that transports the sliced veneer directly to the dryer in the background.

vertical slicer (wood shearing machine) was patented on April 11, 1978 (Pat. 4,083,391) and again on May 16, 1978 (Pat. 4,089,354). One feature of the Cremona machine is that the veneer is sliced on the "up" stroke as opposed to the conventional vertical slicer which cuts on the "down" stroke (Fig. 1-11).

Besides criticisms about the "subsidized" credit available through the Italian company, critics point out the light-weight steel construction of the Cremona slicer as compared to American machines built with heavy cast iron components. The Italian machine, purportedly, requires skilled technical personel to maintain its more complex components.

Of alternative veneer slicing machines manufactured in the United States since 1900 only two will be mentioned. In 1954, Charles Ralph McFall was granted a patent on a vertical slicer (Pat. 2,676,627). The patent was assigned to the Johnson City (Tenn.) Foundry and Machine works. This veneer slicer was promoted as an improved, safer version of the Capital Machine Company's slicer. One or more may still be operating in the United States. In the veneer industry, the machine is referred to as the Johnson City slicer.

The Sitton Veneer Slicer was patented in 1907 by B.J. Sitton. At least one of the vertical slicing machines was built by its inventor for the Kincaid Veneer and Lumber Company of Salisbury, North Carolina. The machine was described as being "...light, compact and simple in construction. The machine now in use takes a flitch 20 by 20 inches by 10 feet long and weighs complete about 20,000 pounds..." The remarkable thing about the machine was that it ran at 40 strokes per minute and the claim was made that it could be run at a higher rate of speed. This was at a time when other slicers were operating at about one-half this speed. Apparently, the test of time eventually overwhelmed the Sitton slicer and it became a dead end, like others, in the evolution of the American veneer slicer (Anon., 1907).

Chapter 2

THE PIONEERING EAST COAST VENEER MILLS: MASSACHUSETTS AND NEW YORK

Mahogany lumber and veneer mills flourished in port cities along the Atlantic and Gulf coasts of the United States beginning about 1850. Prominent among the dynastic families that typified the industry's ownership were the Palmer-Parker families of Massachusetts and the Williams family of New York.

The reason for the prominence of mahogany in their operations was aptly summarized in 1794 by Thomas Sheraton, an English cabinet maker and furniture designer, who wrote:

> Of all woods, Mahogany is the best suited to furniture where strength is demanded. It works up easily, has a beautiful figure and polishes so well that it is an ornament to any room in which it may be placed. Other woods, formerly used for cabinet work, are quite laid by since the introduction of Mahogany (Lamb, 1948).

The immediate advantage in using mahogany was to lighten structural furniture elements. Its strength permitted more delicate and curved legs to be designed for tables and chairs. This potential was exploited by French and English cabinet makers in designing furniture for the more affluent and leisure-oriented European societies. Social customs such as afternoon tea drinking, required, for example, decorative tables and side chairs which could be drawn up closely for conversation and the tea "ceremony."

In Colonial America prior to 1700 much of the furniture was crafted by local carpenters from suitable woods available in their localities. As a rich merchant class developed in coastal cities between Philadelphia and Boston and as the landed gentry became wealthy in Virginia, they began to import fine furniture from Europe. American craftsmen ultimately began to develop similar cabinet making skills and were joined by foreign artisans who had learned their trade in some of the finest woodworking shops in their native countries. Although generally following the English style of furniture, they frequently produced interesting and beautiful variations in traditional design.

It is somewhat surprising that mahogany was not in more general use at an early date in the colonies. A large merchant fleet had quickly developed in the colonies to carry cargoes to and from the West Indies,

Great Britain, and the continental ports of Europe. One practical reason why few sailing ships returned to Colonial American ports with West Indies mahogany before 1700 was explained this way:

> [It] lies in the greater practicality of using the local woods which abounded everywhere, rather than the mahogany whose weight rendered it an expensive material when transported in the valuable cargo space. Rum and molasses paid much better, for both were consumed in large quantities...[and] as long as the call for these commodities was greater than for cabinet-woods, rum and molasses were imported. [After] fashion had set the stamp of its approval upon the use of mahogany, its import leaped to a high place on the shipping lists (Payson, 1926).

Some idea of mahogany's initial use in America can be obtained from inventories of objects belonging to deceased individuals which were then recorded at local town halls. This followed an English custom of inventorying a dead man's personal possessions room by room and supplying the information to local authorities. Payson continued:

> As early as 1708 in Philadelphia, planks of the wood are enumerated in the estate of a joiner or cabinet-maker, while before 1722 in one inventory alone are mentioned nineteen pieces of mahogany furniture including chests of drawers, tables of various sizes and shapes and several clothes presses. A mahogany four-post bedstead is found in an inventory of 1719. It is probable, therefore, that considerably before 1708 mahogany had taken its place as a cabinet wood and that a quantity of it had been imported into the colonies.

By 1750, the master craftsmen of Salem, Newport, New York and Philadelphia were working almost entirely with mahogany and making furniture that has never been surpassed. The increasing demands for mahogany lumber created the need for local producers and merchants to handle the specific requirements of cabinet makers situated in the coastal population centers. From rather modest businesses evolved the larger more permanent establishments owned and managed by entrepreneurs located at sea ports who, in the process, left their imprints in the historical records of their communities.

PALMER & PARKER COMPANY, INC.

The earliest known veneer mills in New England were located in the tide-water area of Massachusetts. Other mills followed as veneer saw technology "allegedly a French invention not protected by patents in this country," was adopted (Anon.,1933). The first with a known recorded history was established on February 28, 1833, by Harrison Parker, a native of Reading, Massachusetts. This was the beginning of Palmer & Parker Company, Inc., which remains the oldest mahogany merchant in the United States. Parker entered the venture with a broad base of experience which belied the fact that he was scarcely 20 years of age. In 1832, for example, Parker had leased water power privileges at Woburn where he built a water-powered sawmill. Before this, he had been an apprentice cabinet maker and had acquired an extensive knowledge of and appreciation for furniture woods (Burrell, 1979).

The romantic development of the business is chronicled by Parker in a hand-written document in the files of the Winchester Historical and Genealogical Society of Winchester, Massachusetts. "When that business was founded the 23rd Congress was in session and Andrew Jackson was the Chief executive; the population of Boston was 61,000;...and prayers were offered in churches for members about to set out for New York City." He goes on to relate that Parker's brother, Loea, joined him as a partner in the L and H Parker Company in 1834 when they leased sawmills in the center of South Woburn (now Winchester). New machinery was acquired for sawing veneers from local hardwoods and the business was expanded to include mahogany lumber and veneer. When the lease expired in 1842 Loea retired and Harrison Parker assumed sole control of the business. Manufacturing at the original mill continued for 24 years.

Looking for expansion potentials in the Boston market, Parker opened a Boston "store" in 1849 which was used for storage and sales. In 1856, he formed a partnership with Jonathan C. Sleeper, H. Brooks Richardson, and Joel M. Howard. The firm was known as Parker & Sleeper & Co. The firm subsequently moved to Boston's Charlestown district where it was located at 14 Charlestown Street. Veneer slicing capabilities were introduced in 1860. Parker claimed that he could cut veneer so thin that "550 of them could be packed together to make the thickness of a mass only one inch in thickness." It was a machine that he had "designed, patented and installed [and was] the first of all veneer slicers, built on principles that still obtain" (Anon.,1935). Speed was not a virtue of this machine. The lifting of the knife was first accomplished utilizing a turntable device rotated by a horse. The horse was later replaced by men who labored at this monotonous job until more modern equipment was acquired. On a good day 10,000 square feet of veneer could be produced. The two segment saws continued to operate until about 1945 when they were destroyed in a mill fire. This equipment was used to saw one-sixteenth to one-quarter inch veneers. However, most of the veneer cut was three-sixteenths of an inch in thickness and was produced for local piano manufacturers. In 1867, the business moved to a four-story building at Portland and Travers Streets in Boston. The firm became H. Parker & Co. upon Sleeper's retirement in 1871.

Harrison Parker II, and Irving S. Palmer, Harrison Parker's son-in-law, entered the business in 1875. Palmer rapidly assumed company leadership in the firm which became Palmer, Parker & Co. The company soon outgrew successive locations in Boston and in 1883 moved back to the Charlestown district at 103 Medford Street. The move had been also prompted by planned public improvements for downtown Boston which included properties owned by the company. Mills were built at Keys wharf on the Mystic river to facilitate the handling and processing of mahogany and other exotic woods received from overseas sources. Among the new employees hired was Frank Dutton Sawyer, a young boy, 14 years of age. He was to play an important role in the future of the company as the decades passed. The firm, itself, continued to grow and prosper as it adopted new technology, increased outputs, and expanded into markets outside the Boston area.

In 1893, The National Mahogany and Cedar Company was formed as a consortium of the Palmer-Parker company, Owen Bearse & Son, and Wm. R. Uptegrove of New York. The financial Panic of 1893 wiped out this fledgling company; however, Palmer, Parker & Co. was able to survive and resumed its previous business identity. It was incorporated in 1906 as the Palmer & Parker Company, Inc., with a capital of $240,000. Besides Harrison Parker II, president, and Irving S. Palmer, treasurer, the incorporating officers were Frank D. Sawyer, vice-president, who had acquired a one-third interest in the business, and William I. Palmer, assistant treasurer. At the time of Sawyer's death on July 20, 1936, he was general manager and president of the company (Anon., 1967).

The firm had always been strongly identified with the production of mahogany lumber and veneer after it located in the Boston area and was one of the "mahogany mills" situated adjacent to the principal ports along the East Coast. Mahogany logs, at first, were imported from Florida and the Caribbean islands. As these sources became depleted, logs were obtained from Central and South America (Fig. 2-1). Logs were also obtained from the West Coast of Africa beginning about 1900, usually via auctions on the docks of London and Liverpool. This was the era when Africa was still the "dark continent" and adventurous Europeans were beginning to push inland to exploit the timber resources. A West African branch was established by Palmer and Parker in 1913 for the direct importation of mahogany logs (Fig. 2-2). Company records show that it sometimes took several years of ceremonial confab with native Chiefs before their agents were permitted to buy the timber they sought. The first full cargo of two million board feet arrived at the company's Mystic River wharf the following year (Brady, 1956).

Following the Great Depression of the early 1930s, steamships again began arriving in Boston with West African logs. The Hardwood Record noted the arrival of the "...O. A. Brodin, the first full cargo of African mahogany to arrive in Boston in the past eight years; 985 logs, 2,560 tons, approximately 1,750,000 board feet were involved" (Anon., 1939a). The largest single consignment of mahogany ever imported into the United States arrived in 1955 at Palmer & Parker's Mystic River docks. "Three ships were engaged...to bring the 9,500 tons of mahogany logs from French Equatorial Africa's Ivory Coast" (Anon., 1955). This was about the "last hurrah" for mahogany in the United States because

Figure 2-1

Mahogany logging in Central and South America and in Africa were quite similar. Logs were accumulated beside streams to await the flood season. Once in the larger rivers, logs could be rafted to cargo steamers offshore. These pictures were taken in Nicaragua.

Figure 2-2
Prior to World War I, it was common practice in Africa to square each log with hand tools before it was shipped to the European log markets. The Ivory Coast was an important source for these logs and Grand Bassam was the principal port.

of radical changes in furniture styles and consumer preferences that were about to occur. However, believing that mahogany would continue to enjoy a market resurgence, the company "...joined with a large French and Belgium [holding] company, Society des Financieres de Caoutchouc" in 1956. Within the group was a subsidiary, La Forestiere Equatoriale, which operated very large tropical hardwood concessions in the Ivory Coast of Africa. The joining of hands of these two powerful interests gave the Palmer & Parker Company the assurance of receiving a continuous supply of timber. Not only did it "secure the well-known brand of 'E.O.' Grand Bassam Mahogany, but also numerous other species of hardwoods." In the transaction the foreign-based company acquired a controlling interest by purchasing 51 percent of the outstanding shares of Palmer and Parker. This corporate arrangement continued until 1965 when the holding company now known as Terre Rouge liquidated its investment (Anon., 1956).

This block of stock had represented the Palmer interest in the firm. An attempt to purchase it had been made by Frank Sawyer and Gordon Parker after the death of William Palmer in 1929. Sawyer's interest in the stock was probably enhanced after his grandson, Bernard O. Gerrish, joined the firm in 1920 and had been made company treasurer in 1929. Sawyer was eventually able to purchase Gordon Parker's minority interest in 1941 and assumed the presidency of the company. Gerrish became president when his grandfather died and retained the joint title of president and treasurer until he died in 1967. He was succeeded as president by his son, Robert S. Gerrish, whose son, Peter T. Gerrish, joined the firm in 1980 as vice-president and comptroller.

The Boston Redevelopment Agency (BRA) helped shape the future of the company when the BRA began to develop the Charlestown District of Boston. Palmer & Parker were forced again to relocate in 1967. After carefully analyzing market trends and world developments, Robert Gerrish shifted company objectives away from the production of lumber and veneer to the marketing of purchased lumber and veneer or of materials custom-cut to rigid specifications in the mills of former competitors. In doing so, the company was strategically relocated in Tewksbury, Massachusetts, where the company maintains a large inventory of foreign and domestic lumber and veneer in close proximity to the interstate highway system. It remains the oldest mahogany merchant in the country while providing an array of exotic and domestic woods for potential customers.

When Palmer & Parker celebrated its first 100 years in business in 1933, one observer wrote a commentary which is as apropos now as it was then:

> Thus reads the history of one of the most respected business houses in the United States concerned with the conversion of the decorative products of the forest to the cultivated and aesthetic taste of American families in finishing and furnishing their home surroundings. Looking to the past, the Palmer and Parker Company is seen to have contributed a fine history

> of business success and of solid tradition to the genealogy of American business institutions; consistently its contemporary activities have sustained a reputation for being always progressive (Anon.,1933).

As an old New England concern the company plans an orderly expansion in the future. They have not let their strong traditions stop them from keeping up with modern improvements including a complex computer system and state-of-the-art equipment used in processing and warehousing lumber and veneer.

ICHABOD T. WILLIAMS & SONS

One of Palmer & Parker's contemporaries was Thomas Williams. He and his descendants operated at sites on Manhattan Island and later at other New York area locations. Their business history is similar to a number of family-owned mahogany firms: a pattern of youthful growth, a vibrant maturity, later senescence, often followed by the corporate death of once proud industrial organizations.

The Williams family began its association with the sale and manufacture of hardwood lumber and veneer in 1838, a relationship that existed for the next 138 years. The company eventually became known "...as the oldest and largest firm in the world dealing in mahogany and other imported cabinet woods" (Sullivan, 1928). The original economic incentive for the Williams family and others to specialize in mahogany was the intense demand for this fine furniture wood. "The use of mahogany... was attributable not only to its superb qualities but also to the fact that it was probably more easily obtained than walnut, which did not grow in large quantities along the coastal plains. Importing hardwood logs... often cost less than hauling domestic logs from forests even thirty miles inland" (Wisdom and DelCoro, 1983).

The six generations of the family, who provided entrepreneurial direction to the company, descended in a direct line from Thomas Williams, born in 1800 in Elizabeth(town), New Jersey (Fig. 2-3). His father, Ichabod, was a farmer and a maker of cabinets and wooden cases for clocks. An example of his craftsmanship can be found in the American Museum in Bath, England.

> Thomas was brought up as a farmer, but was not very keen about it. He was much more interested in the cabinet shop. When Ichabod died in 1817, his will provided that Thomas could continue to use the cabinet shop and live in the house as long as he wished to. Thomas, however, found life a little too constricting in Elizabeth and decided to seek his fortune in New York City. He moved there, as nearly as I can judge, before 1820 (Williams, 1983).

The New York City directory of 1823 reports that he was engaged in business at Broad and Beaver Streets. The 1824 edition reported him to be a member of the firm of Williams and Dawson Cabinet Makers at 51

Figure 2-3
Thomas Williams. He and Isaac E. Smith established a lumber yard in 1838 near where the New York Stock Exchange is presently located.

Broad Street. His financial success as a cabinet maker can be inferred from his fashionable marriage to Sarah Waldron, the great granddaughter of "perhaps the most noted of the Harlem Patentees" and by family tradition which relates that he retired at the age of 37. After a few months of idleness, and at the urging of concerned friends, he realized that he should do something more constructive than pursuing pleasure. On May 1, 1838, he and a young partner, Isaac E. Smith, established the Williams and Smith lumber yard at the corner of Wall and Broad Streets near where the New York Stock Exchange is now located. This event was only coincidental with the arrival in the port of New York the same year of the side-wheeler, "Sirius," from England. This ship inaugurated transoceanic steam navigation and a new tempting opportunity for entrepreneurs of the day. "To the users of wood it meant that New York was in a position to take its pick of all the available woods in the world. And down on Broad

Street were Duncan Phyfe and other famed furniture craftsmen" (Anon.,1933). As a skilled cabinet maker, Thomas Williams knew the needs of these potential customers. Their business venture was a success from the beginning, growing in volume and importance with the city.

As the needs for space increased, the firm moved to Desbrosses Street and the North River. The location was ideal as both softwood and hardwood lumber could be brought from the Great Lakes region to Buffalo by the firm's fleet of lumber schooners and then shipped by canal and the Hudson River to New York City. Their waterfront property, which was extended into the river, also gave the firm excellent facilities for handling cargoes of exotic woods brought by ocean steamers and lumber schooners from the Orient, the West Indies, Europe, and South America. Ultimately, wherever there was a wood species demanded by designers or craftsman, there was a Williams' agent ready to send it to New York. The adventure and peril of acquiring mahogany logs in the jungles of South America and Africa are described in Appendix B.

In 1851, the company moved to a site on the Hudson River between 25th and 26th Streets (Fig. 2-4). The purchase of the Hudson River site had been prompted by the return of Ichabod T. Williams, Thomas' son, from the California gold fields in December of 1850 and the inability to expand facilities when the Pennsylvania Railroad built a ferry terminal

Figure 2-4
The Smith & Williams Company was established in 1851 at 11th Avenue, between 25th and 26th Streets. The partnership existed until 1870 when it was dissolved.

adjacent to the Williams property on Desbrosses Street (Fig. 2-5). With Ichabod in command of the business, Thomas was finally able to retire at age 50 and to live the next 25 years of his life at leisure. The company name was changed at this point to Smith and Williams, a name which was continued until the partnership was dissolved in 1870. After 1870, the firm operated as the Ichabod T. Williams Company until 1882 when "& Sons" was added. Ichabod T. Williams' sons were Thomas Williams II, admitted as a partner in 1880, and Henry K. S. Williams who was admitted as a partner in 1882 (Fig. 2-6).

It was H. K. S. Williams who introduced Cuban mahogany to his father through a coincidence arising from a vacation trip to Cuba. H. K. S. spotted some exceptionally fine mahogany logs awaiting shipment to France. By offering "a little more money," he convinced the shipper to send some of them to New York. Upon his return to New York, H. K. S. was told by his father to be prepared to go back down there "... in the fall and buy some more." H. K. S. eventually spent ten falls and Christmases in Santiago, Cuba, as the log buying agent for his father. A charming and genial man, he had started to work for the firm in 1875 and left in 1911 to pursue other business and public service interests. Ichabod's two other sons never became integral parts of the firm. Lloyd, a lawyer, and Waldron were apparently associated with the firm for only a short period of time and left the company prior to 1918.

In 1885, the City of New York condemned the firm's Hudson River property to build docks and piers and to establish 12th Avenue. However, Ichabod T. Williams & Sons continued their retail yard business and general offices on the East side of 11th Avenue. In 1899, a seven-story brick structure was erected at the corner of 11th Avenue and 25th Street for veneer storage, show-rooms, and business offices. The building served these uses until 1973 (Williams,1983).

Forced to seek a new waterfront property in 1885, a half mile of frontage was purchased on the main ship channel of New York Bay on Staten Island. "Here, with wonderful facilities of deep water and rail connections, sawmills were built and the largest Mahogany and Hardwood Yards in New York City created" (Anon.,1935). The sawmills were the first that the firm had built in New York City.

Before these production facilities were constructed, Ichabod T. Williams & Sons had imported mahogany lumber and logs by way of Liverpool, London, or Hamburg (Fig. 2-7). This pattern of trade developed because West African timber exports were largely controlled by the British and Germans through their colonial administrative and marketing infrastructures. Logs arriving in New York were custom cut at two sawmills situated along the East River in the vicinity of what is now East 8th Street to 6th Street. The Mulgrew mill was one of these and the other was owned by William Marshall. Logs were also purchased at auctions organized by importers such as Willard Hawes and Company when their

Figure 2-5
Ichabod T. Williams took over as head of the Smith and Williams Company in 1851. In 1870, the firm was renamed the Ichabod T. Williams Company.

Figure 2-6
Thomas Williams II became a partner in the business in 1880. After his brother, Henry K. S. Williams, was admitted as a partner in 1882, the company was renamed Ichabod T. Williams & Sons.

Figure 2-7
A London agent is shown inspecting a figured mahogany log purchased by Ichabod T. Williams & Sons for $10,000 in 1910.

schooners arrived from Africa laden with mahogany logs.

The increasing demand for mahogany provided the necessary incentive for the firm to own and charter its own boats to bring logs directly from Africa's Ivory Coast and Central America to Staten Island. This eliminated European handling costs and the costs of the circuitous routing from the country of origin to the United States. A cribwork, built out into the open water of the harbor, was constructed so that ships arriving with cargoes of logs could tie up and discharge logs directly into the protected water inside the cribs where they remained until processed (Fig. 2-8).

During these decades of expansion Ichabod T. Williams & Sons established an office and lumber yards in London and had agents in France, Belgium, and Germany. The company also owned and operated a sawmill in Kentucky and bought lumber from other hardwood mills. The scope of this segment of the business can be judged by a comment given by Ross Kreamer. "Perhaps, the largest single shipment of hardwoods ever made was a solid trainload of 38 cars, over 5,000,000 board feet, of FAS (First and Second) band sawn, quartered white oak, all shipped from Memphis to New York and then to Brussels, Belgium, right from the train cars onto the ship."

The firm also maintained offices and resident managers in British Honduras, Nicaragua, Mexico, Columbia, and the Ivory Coast. The Ivory Coast operation was organized in 1915 at Grand Bassam (Fig. 2-9). It became an important hub for obtaining large African logs which often weighed between 10 and 15 tons. Logging equipment was not powerful

Figure 2-8
The logs in the pond are African mahogany and those in the cribbed pile are Cuban mahogany. The ship was built for Ichabod T. Williams & Sons for the purpose of transporting Central American logs to their Staten Island plant.

Figure 2-9
Office and living quarters for company staff of Ichabod T. Williams & Sons in Grand Bassam, Ivory Coast.

enough at the time to efficiently log these huge trees. It was, therefore, necessary to use large numbers of Africans to fell, square up the logs, and to move them to streams where they were carried by flood waters to the coast. African headquarters were moved eventually to Takoradi in the Gold Coast Colony (Ghana). The Belize (British Honduras) operation was similar in many respects. Mahogany logs were moved to stream side to await a flood which would carry the logs to the sea where they were "caught," boomed, and loaded on ocean steamers. Literally, the forest riches of the entire world found their way to Ichabod T. Williams & Sons. Exotic species arrived in New York from India, Australia, Burma, Java, Ceylon, the Philippine Islands, Russia, France, and far-flung quarters of the world.

After nearly a quarter century of operation on Staten Island the City of New York once again condemned their waterfront property for municipal docks and piers (Figs. 2-10, 2-11). Thus, in 1919, and for the sixth time in the company's history, it became necessary for the firm to find a new home. After extensive studies and examinations of potential locations along the Gulf and Atlantic coasts, 70 acres were purchased at Carteret, New Jersey, by the George D. Emery Company, a Williams family subsidiary (Table 2-1). It was perhaps, the finest possible site for their type of business (Fig. 2-12). The property had rail connections with

Figure 2-10
These "sunning racks" on Staten Island contain recently cut mahogany lumber. Exposure to the sun insured that the lumber would be "well colored."

Figure 2-11
Stacked mahogany lumber at the Staten Island yards of Ichabod T. Williams.

Table 2-1
The several corporate faces of Ichabod T. Williams & Sons

- Williams and Dawson Cabinet Makers (1824)
- Williams and Smith (1838)
- Smith and Williams (1851)
- Ichabod T. Williams (1870)
- Ichabod T. Williams & Sons (1882)
  - George D. Emery Company (c. 1911)
  - Otis Astoria Mahogany Company (1932)
    - Astoria Importing & Manufacturing Co., Inc.
    - Compania Astoria Peruana, S.A.
    - Astoria Pan-Americana, Inc.
- Ichabod T. Williams & Sons, Inc. (1957)
- Thomas R. Williams doing business as
  Ichabod T. Williams & Sons (1964)

Figure 2-12
The 70-acre Carteret, New Jersey, complex of veneer and saw mills was very likely the largest integrated mahogany manufacturing establishment ever conceived and constructed.

all the major trunk lines and had an extensive water "frontage on a 30-foot channel." By all standards it was an ideal location for a manufacturing and distributing point. "Construction of the largest Mahogany Saw and Veneer Mill in the world was commenced in 1922 and completed early in 1925" (Williams, 1983).

> The great timber basin, ten acres in area, has an 850-foot frontage, large enough for three ocean going steamers to discharge their cargoes at once, either into the log pond itself or onto a 350-foot concrete wharf, where there is a railroad track directly connected to the Central Railroad of New Jersey; also there is a slip for lighters. Between the dock and sawmill, a large log yard was laid out equipped with standard gauge tracks along side each lane of piled logs. The log yard and basin provided a storage capacity for over 15 million board feet of logs. Two locomotives and four electric derricks eased the handling of the timber (Anon.,1938).

The plant consisted of 16 buildings with a floor area of 250,000 square feet. The latest technology was used throughout for high production and for comfort of the skilled workers. Electric motors were directly connected to most machines. Conveyors and live rollers directed logs, flitches, cants, and lumber from one stage of production to the next. The three band saws were capable of producing 100,000 board feet per day. Veneer flitches passed directly from the sawmill to a storage facility and steaming tanks. Live rolls carried the "steamed" flitches from steel vats to the veneer buildings where two machines sliced as much as 250,000 square feet per day. Veneer saws, capable of sawing flitches up to 24 feet in length, had a capacity of over 40,000 feet per day. These veneers were either dried by automatic driers or racked for air drying. Veneer was also rotary peeled on a Capital lathe from bolts which had been "boiled" in large concrete tanks. As the veneer was cut, it was conveyered to two Coe "roller-driers" located back of the lathe.

> Lumber was inspected in a wonderfully lighted and steam heated (room)...laid out to enable the inspector to devote the most careful attention to grading. All (mahogany) lumber (was) sunned and then piled upon permanent concrete pile bottoms for seasoning. The lumber yard (contained) over two miles of rail road siding, making it possible to ship directly from the lumber stacks to the purchaser. Dry kilns were linked to planers, resaws, and a storage building so that no major rehandling was necessary (Anon.,1930).

This great industrial complex resulted largely from the vision of Thomas Williams and his son, Thomas Resolved Williams. The latter had joined the firm in 1900 at age 19 and became a partner in 1905 (Fig. 2-13). At the time of the move to Carteret, young Thomas was about 40 years of

Figure 2-13
Thomas Resolved Williams, with his father, masterminded the building of the Carteret, New Jersey, complex of mahogany mills. He joined the company in 1905 and was the company's principal shareholder when it closed in 1973.

age and his father in his mid-sixties. The senior Thomas Williams had joined the firm in 1873 and had become a partner in 1888. He remained active in the business until his death in 1935.

Ichabod Thomas Williams, great-great-grandson of the founder of the business, and the only son of Thomas Resolved Williams, became formally associated with the firm in 1930 and was made a partner in 1938 (Fig. 2-14). One of the major problems of every other generation of Ichabod's forebearers had been "to find, to have and to hold a plot of ground large enough and with adequate facilities for handling the steadily expanding business of this great lumber and veneer organization" (Anon.,1938). This was not to be the major problem faced by Ichabod T. Williams in his tenure with the family firm.

His father had become the principal partner in 1935 upon the death of Thomas Williams. The early 1930s and the two decades which followed were traumatic for the company because they brought with them the Depression, World War II, and the post-war changes in markets and life styles. The Depression years were particularly difficult for the firm because of declining prices and disappearing markets. There was also a

Figure 2-14
Ichabod T. Williams became associated with the family firm in 1938. He and brother-in-law, Enos Thompson Throop, failed in their attempt to save the company when their proposals were rejected by Thomas R. Williams.

perceived threat of competition from established European producers of tropical lumber and veneer. Thomas R. Williams wrote to a business colleague in 1931 to this effect:

> I... am quite aware of the possibility of an increased invasion of foreign manufactured veneers into this country. I presume this has been brought about through the activities of Messrs Mallinson of London and their distributors in this country, and it is just one point to show how necessary it is for us to watch our domestic market and guard against the possibility of foreigners monopolizing some of our business here.

Williams suggested solutions included increased tariffs based on domestic values and more equitable adjustments of labor rates throughout the world. The letter also alluded to the dismissal of the "Gillespie Furniture Company case" by the Federal Trade Commission which had been manufacturing "mahogany" furniture out of lauan or "Philippine mahogany." This issue was to haunt the mahogany industry after World War II.

Ichabod T. Williams, in a 1983 interview, recalled that they had:

> ...weathered the Depression because we were better financed than our domestic competitors and had more capital resources available to us. We were also favored by the movement of the furniture industry from its traditional locations in the northern and central parts of the country to the South. We had the advantage of being closer to that market and opened our own sales office in High Point, North Carolina. This office covered all the Carolina areas and the southern part of Virginia in the 1930s. This augmented our salesmen who covered New York state and, particularly, the furniture plants in the Jamestown area and the casket trade in Syracuse, Utica, and Rochester.

The company's financial strength in 1932 made it possible to acquire both the Otis Mahogany Company and Astoria Importing & Manufacturing Company, Inc., (AIMCO) with sawmill and veneer mill located in what was then Long Island City. The J. S. Otis Mahogany Company mill was built in 1920 on the Mississippi waterfront near New Orleans. Known for the quality of its products, the company began manufacturing mahogany lumber in 1866. The completion of the mill coincided with "...the arrival of a cargo of splendid [Honduras] mahogany on the steamer General Currie. J. S. Otis, president of the company, is the authority for the statement that the cargo of mahogany logs received on the General Currie is the largest ever bought to an American port" (Anon., 1920b).

When acquired by the Williams family, the mill consisted of two bandsaws used primarily for the manufacture of mahogany lumber. A new company, the Otis Astoria Corporation, was chartered in the state of Louisiana to accept the stock of both companies. AIMCO continued as a corporate entity primarily for the purpose of managing Peruvian operations acquired in these transactions at Iquitos in the upper reaches of the Amazon basin. This mill had originally been erected by the Aquana Mahogany and Timber Company of Boston in the early 1920s.

Originally known as Astoria Veneer Mills at the turn of the century, the company became the Astoria Veneer Mills and Dock Company when it acquired the Huddleston-Marsh Mahogany Company of Long Island City in 1917. The Huddleston firm had fallen on hard times during World War I because of the "...shortage of bottoms and high freight rates..." although the company had four steamships "engaged in bringing logs from Central American ports..." (Williamson, 1963). Perhaps, the Astoria Veneer Mills company was better managed as it was credited in 1917 as "doing an immense business as storage and custom mills for mahogany, cedar and foreign fancy woods..." (Anon., 1917). The plant had one mile of dock and pier front and yards that occupied 34 acres of land contiguous to the pier. Employing 300 men, the company operated a cedar mill, a battery of bandsaws, and a veneer mill equipped with seven veneer saws, three slicers, a staylog lathe and a rotary lathe. Alexander S. Williams was president of the company and had been preceded by William H. Williams who patented a 32-stroke slicer which was later built by the Coe Manufacturing Company in 1907. By 1920, the Astoria Mahogany Company, Inc., was being identified as the successor to Huddleston-Marsh Mahogany Company, the Astoria Veneer Mills and Dock Company, and F. W. Kirch, Inc. (Anon., 1920a).

A Peruvian corporation, Compania Astoria Peruana, S.A. (CAPSA), was subsequently organized for the purposes of carrying on business activities within Peru. Otis-Astoria through CAPSA held "virgin timber [lands] in Eastern Peru aggregating 6,000 square miles [which] assure[d] uniformity of quality and texture" according to a company advertisement in 1939 (Anon.,1939b). During World War II a veneer mill in New Orleans was added to produce mahogany veneers for the Higgins Boat Company. The waterfront property was leased from the New Orleans Dock Board.

The Iquitos operation was centered around rosewood and mahogany. Veneer logs were brought down the Amazon to the coast on river steamers where they were transferred to a company ship bound for New York. Lesser quality logs were cut into lumber to be forwarded to New York. Third grade rosewood logs were crushed to extract rosewood oil. The company is alleged to have made a fortune just off the oil which was used in the "beauty treatment industry."

Beginning in the early 1940s, the company's attention was largely turned to producing veneers for war-related activities. Mahogany veneers were cut for the CJ4A military glider used in the Normandy invasion. These veneers were fabricated into component plywood parts by the Haskelite Corporation in Grand Rapids, Michigan. Mahogany lumber and veneers also went into the construction of PT boats built by the Elco Boat Company in Bayonne, New Jersey. The hulls of these English designed boats were made out of mahogany strips laid over a form in spiral patterns at right angles to each other with an impregnated fabric placed between the layers of thick mahogany veneer. The sandwich was subsequently copper riveted to produce a strong, but light, hull.

After World War II the company began to feel the effects of a significant change in the public's preference for mahogany furniture. This marketing situation had become an acute economic problem for the mahogany producers by 1957. Life styles in the growing suburbs called for modern furniture designs. The shift away from mahogany was not a new phenomena. It had happened before. "When reproductions of 18th Century furniture became popular, mahogany was favored. The big competitor was generally American walnut. Danish modern and other contemporary furniture styles used teak, oak, and walnut among other species. The lighter woods were said to be preferred by working wives who had less time to dust the furniture or had better things to do with leisure time" (Vollmer, 1982). The company also lost markets to the postwar boat companies which substituted plastics, Fiberglas, and aluminum for mahogany planking and plywood.

Other disquieting business realities were also becoming obvious. Former European colonies were making economic strides in the post World War II period, often assisted by bilateral and multilateral aid entirely, if not heavily, underwritten by the United States government. Facilities for the manufacture of quality hardwood veneers were built in these developing nations where such factories had not existed prior to World War II. The mills were often equipped with the latest technology and were managed by experts provided by foreign entrepreneurs. These countries were also advised to process and export their timber resources as finished and semifinished products in order to capture the "value added through manufacture" and to provide employment for burgeoning populations.

The partnership arrangement of family members and trusted employees which had existed since 1838 terminated on July 1, 1957, when the company was incorporated. The partners on that date received shares in the new corporation in proportion to their former partnership agreements. This was an effort to take advantage of modern corporate business management practices.

Thomas R. Williams by this time was nearly 80 years old. He had in his lifetime survived economic depressions and periodic cycles of mahogany "booms and busts." He was certain that he knew all there was to know about the mahogany business. Williams was also the senior partner, principal stockholder, and a very stubborn person not given to delegating authority to son, Ichabod, or son-in-law, Enos Thompson Throop. Ichabod's partnership share, in particular, was limited despite decades of quality service to his father and the company.

In 1952, the New Orleans Dock Board notified Otis Astoria that it would not renew the lease in order to expand adjoining shipping facilities. Assets were transferred to Ichabod T. Williams & Sons Company and Otis Astoria was liquidated. When the mill machinery was dismantled, some of it was transferred to the Carteret, New Jersey, plant and the rest was sold.

AIMCO operations in Peru were suspended in 1965 and the property sold to the United States Plywood Corporation. In 1966, AIMCO was liquidated. After the dissolution of AIMCO, CAPSA continued operating a by-products division which exported rosewood oil, leche caspi, and leche oje through Astoria Pan-Americana, Inc., a New York Corporation established to act as the sales agency and financing unit for the Peruvian operations. These corporate assets were disposed of in 1971.

Ichabod Williams and Enos Throop were well aware of the daily erosion of the company's business prospects and knew that the company could no longer expect to be a continuing profitable venture. These two men took their concerns to T. R. Williams on several occasions saying in effect: "We think it is time for you to do something about this problem. The foreign mills are undercutting us. We just are unable to compete. We have union labor and high taxes which they don't have. We think that the company either has to diversify or sell out."

The elder Williams refused their counsel and turned down a tentative offer to sell the company to a foreign firm. The die had been cast and the inevitability of liquidating the firm's assets became a reality. Not being able to change Thomas' mind about their vision of the future, the two junior members of the firm asked to be bought out on the proviso that they would continue to work on a salary basis for a limited time period. When interviewed, Ichabod recalled: "I told him I wanted to retire when I got to be roughly 65, which incidentally horrified him. He never retired until the business was liquidated." This occurred when he was 92 years old. He had realized in 1970 that he was running a deficit business and that, for the first time in his business life, he would not be able to ride out the storm to a new day of mahogany prosperity. The plant which had been a technological wonder in 1925 was now an obsolete and costly albatross.

The company's death throes were witnessed by Ichabod T. Williams, who said:

> Once the decision had been made, we immediately stopped buying logs and we manufactured into lumber and veneer those logs we had. The next thing we did was to make discounts on large sales. There were 64 acres of open lumber yards plus seven dry storage buildings filled with kiln dried lumber and crated veneers. Gradually, we whittled down and closed out the inventory. The sawmill and veneer machinery were auctioned off. Some of the sawmill equipment went to Mexico, some veneer equipment to Canada and our two ancient slicers to Cuiaba, Brazil. When everything was sold off, we shut the door at the end of 1973 which was also the closing of a fascinating era encompassing 138 years of America's wood processing history.

Chapter 3

COASTAL MAHOGANY MILLS: PENNSYLVANIA TO LOUISIANA

THOMPSON MAHOGANY COMPANY

Another early pioneer in the mahogany trade was Louis Thompson of Philadelphia. His hardwood lumber yard opened for business at 11th and Ridge in 1858. It was equipped with milling equipment to saw both mahogany and marble which were sold to local cabinet manufacturers. The firm began business as Louis Thompson and Company. This name was later changed to Louis Thompson Mahogany Company which persisted until 1926 when the company was incorporated and the name changed to Thompson Mahogany Company. Louis also had a timber operation near Laguana, Mexico. Mahogany logs were brought down to the harbor at Laguana by rail and trans-shipped by boat to Philadelphia where they were sawed into lumber. "Laguana" has remained the firm's cable address. Robert P. Thompson, a fifth generation family member in the business, in an extended interview, provided much of the following historical information.

> In those early days the mahogany lumber business was a regional type of business because of transportation difficulties. Someone would have a yard in Boston. We had a yard in Philadelphia and somebody would have one in Baltimore. Even Baltimore and Philadelphia were far enough apart so that a cabinet maker would buy from a local yard. We sold pretty much within a 100 miles then because at a greater distance there was probably a closer concentration yard where the local cabinet trade could buy mahogany. In those early days the cabinet trade really consisted of a craftsman and a half-dozen workers.

George Thompson was the second proprietor of the business. While still "reasonably young," he was killed in a mill-yard accident when an unsecured log came loose and rolled over him. His son, Charles H. Thompson, Sr., succeeded him in running the business. Born in 1864, Charles Thompson witnessed the growth of Philadelphia following the Civil War. As Philadelphia grew, the downtown Ridge Street property became very valuable and was sold sometime after 1892 along with the sawmill. The lumber yard and business office was then moved to 18th and Indiana Avenue. He later was said to have regretted selling the sawmill because of the company's vulnerability to custom cutters when business improved and he couldn't get logs cut into lumber at a reasonable cost and in a timely fashion.

> Much of the custom cutting of mahogany logs was done at the Astoria Lumber and Veneer company, a high-capacity sawmill operation located on the East River in New York City (Fig. 3-1). It was more economical to bring logs to the Astoria mill and have them custom sawn there. Many people did that. We were not the only ones. The Astoria company also maintained a log yard where you could come in and buy mahogany logs and have them sawed for your account. You would have to pay a premium but at least you got the lumber.

Charles H. Thompson, Jr., was born in 1892. He entered the firm in 1910 and assumed ownership upon the death of his father in 1946. Both Robert P. and Charles H. Thompson III, representing the fifth generation, entered the family business when their father became president. Charles III was not destined to play a major role in the management of the company. "He didn't mind buying logs and doing some of those things but didn't like managerial responsibilities." Robert P. (Bob) Thompson was

Figure 3-1
An 8-foot Wheland roller bearing bandmill was used in opening up this large African mahogany log.

elected president in 1950 prior to his father's retirement in about 1953. To provide additional managerial support "outside" people were brought into the company beginning in 1950, including Bob's life-long friend, Jack Stayman.

Charles, Jr., was active in the management of the company much earlier than 1946 and made the decision about 1952 to process the company's logs into veneer in Moorefield, West Virginia. This was a mill built by Minor Davis of Indianapolis about 1927 which had "gone-under" during the Depression. Minor Davis survived as a veneer producer at his Central Veneer Company located in Indianapolis. Bob Thompson described the purpose of the mill:

> It was [to be] a veneer mill for cutting basically domestic veneer. Walnut was a main wood in that era. Everything was walnut. By that time the South was growing and furniture business was starting to become more of an industry. People like the Bassetts, Thomasvilles, and the Drexels were starting to mass produce furniture. We knew these furniture people because we were shipping lumber from [Philadelphia] to them in carload lots. We had started custom cutting walnut and mahogany veneer about 20 years earlier. Astoria did not have veneer facilities so we would take the best logs and ship many of them out to Indiana to Hoosier Veneer, Curry-Miller, and National Veneer. Freight wasn't such a big factor in those days. The southern people in particular wanted more veneer but we couldn't get enough veneer cut because of capacity limitations where we custom cut.

The Moorefield location was chosen because of the quality and amount of black walnut available in the area and the existence of a nearly completed veneer mill. The basic machinery installed by the previous owner was augmented by Thompson with the necessary equipment to provide for a more efficient operation. Production at the Moorefield veneer plant ended in 1959 when the Portsmouth, Virginia, facility was constructed.

In the early 1930s the Thompson Mahogany Company began to experience some difficulty in obtaining veneer and lumber logs from Central American countries. To counter the situation, the company turned to the Brazilian mahogany resource in the upper reaches of the Amazon. At Itacoatiara, 900 miles up the Amazon, a number of tributaries join the main course of the river. As Thompson related:

> It was a great place for a mill. Good mahogany grew in all those river areas. You fell them along the banks and when the rainy season came the floods would float the logs so they could be brought down the river in rafts. We ran the saw mill from 1932 until Vargas became president of the country in 1951. One year we cut five million feet. It was the most gorgeous mahogany you ever saw. Boards 40 inches wide and 16 feet long.

One edict of Vargas required that 51-percent of all Brazilian companies be owned by Brazilians. This was circumvented by the Thompsons by selling the entire mill to the Brazilian manager for $1 in exchange for exclusive purchase rights for the lumber produced. This arrangement continued until about 1956 when less expensive African mahogany lumber began to be marketed in the United States. At this point the company "...just gave up on Brazilian mahogany."

> The best veneer logs came from Africa and not Central or South America and we could get plenty of mahogany logs from Africa right after World War II. The Gold Coast was a British possession and very, very well run. The British were anxious to export logs so we sent a man over there who bought all the logs we really wanted. The logs were sent by ship to Norfolk and then by rail to Moorefield and our veneer mill. Rail rates tripled in probably two to three years. That was when we decided to build a mill in Norfolk which was completed in 1959.

Norfolk Veneer Mills, Inc.

Jay Rayner, Alex Schmidt, and the J. H. Monteath Company, supplied part of the working capital for a 25-percent share of the mill. They were veneer dealers and needed assured custom cutting rights at a time when former custom cutting mills were now producing veneers for commercial sales direct to customers. This arrangement was short lived as the Thompsons bought up the remaining 75-percent interest as those shares became available for purchase. Later, a 25-percent interest was sold to James Stenersen, Sr. This relationship lasted four or five years.

By 1954, the Portsmouth mill was annually "consuming" about 6,000 tons of mahogany logs to produce 50-million square feet of 1/28-inch thick face veneer on a one-shift basis. Flitches were cut with a 6-foot bandmill, and two 17-foot Capital slicers and a Capital staylog lathe cut the veneer. High-speed automatic clippers and three 60-foot dryers, completed the basic equipment (Anon., 1954). By the 1980s, mahogany was only about 10 percent of the veneer sliced at the Norfolk mill. The main species were cherry, red oak, white oak, and ash. "Carolina" pine was also being sliced and exported to Germany. Roger Stenersen, scion of the Williamson veneer family, was production manager of the mill and represented the fifth generation of his family to be actively involved in face veneer manufacturing.

Through efficient management and by running a night shift in Portsmouth for the Mengel Company in Louisville between 1955 and 1960, the fixed cost per unit of veneer was dramatically reduced. This profitable arrangement continued until the Mengel Company was purchased by Container Corporation of America which had little need for an operation that supplied only four percent of the Mengel division's revenues. Mengel's veneer-related assets were quickly purchased by the Thompson Mahogany Company. These consisted of logs, veneer inventory, and log commitments they had made in Africa. In Bob Thompson's words:

> It was a perfect marriage. The veneer was all in our warehouse anyway so we could just substitute it for our own veneer. The Dixie Veneer Mill of the Dean Company with headquarters in Chicago was just a quarter-mile away. We used to compare costs. When we did this, they would say, "God, we really can't compete with you guys. This is ridiculous."

The net result was that the Dean organization became joint 50-percent owners of Norfolk Veneer Mills, Inc., on December 1, 1957. Dixie Veneer Company ceased operations at its Norfolk plant on the same date. Thomas A. Dean, chairman of the Dean Company, and Robert Thompson stated at the time: "We feel that by combining the best of our facilities, personnel and technical 'know-how' high production can be maintained economically and service to our respective customers improved" (Anon., 1957). The agreement lasted until about 1967 when the veneer end of the Dean Company business was sold to four former Dean employees who eventually built a new mill in Princeton, West Virginia. Norfolk Veneer again became wholly owned by the Thompson Mahogany Company.

New Directions for Thompson Mahogany

Charles H. Thompson, Jr., died in 1976, a year after the death of his wife. Their two sons, Robert P. and Charles III, each inherited one-half of the company. It had been profitable and had made money every year since the Depression. Retained earnings had been invested in new equipment at Norfolk, new warehouses and a showroom in High Point, North Carolina. "Our investments had always been in bricks and mortar." With their personal wealth fixed in non-liquid company assets, an investment counselor suggested three options: 1) Sell the company as a going business, 2) Liquidate the assets while they were both living, or 3) Let the family liquidate the company after their demise. After due consideration of the continuing profitability of the company and the loyalty of the employees, they decided to sell the company. As Bob Thompson told it:

> Tommy McPherson, the managing director, [of William Mallinson, Inc. of London] was over here shortly after we made the decision to sell the business and I told him what we planned to do. He said, "Well, by gosh, we would like to have some penetration in the USA. We would be interested in talking to you.

The deal was struck the following morning without benefit of lawyers because: "They only muddy up the water." Two independent, certified appraisers evaluated the real estate in Philadelphia, Norfolk, and High Point; and a CPA certified the inventory of veneer and lumber. After accounting for the firm's cash and accounts receivable, the financial analysis was sent to McPherson. Negotiations were completed a week later, on December 31, 1976. "We were subsequently paid 100 percent in cash over a five-year period. It was a very hard thing to do. My God, it had always been a family company."

William Mallinson, Inc., was a subsidiary of William Mallinson & Denny Mott, Ltd., of London. The Mallinson-Denny group was comprised in 1977 of over 60 companies located in the United Kingdom, Europe, Asia, the Far East, Australia, and South America. The beginning of this enterprise was quite humble. William Mallinson opened the doors of his business at No. 1 Wood Street, off Old Street, London, on February

28, 1877, at the age of 22 with the assistance of two brothers. Supplying top quality lumber to specialized manufacturers formed the backbone of his business and the business flourished. Within 10 years he was manufacturing plywood including the aircraft plywood used in the plane flown by the first Englishman to fly a powered aircraft in the British Isles. William was succeeded by his son, W. J., in 1923 and another son, Stuart, in 1943 who founded the Mallinson Group of Companies. The firm of Denny Mott and Dickson Limited was acquired in 1968 by Sir Paul Mallinson, which created the world-wide Mallinson-Denny Group of Companies (Anon., 1977).

At the time of the sale to Mallinson, the Thompson Mahogany Company was the largest East Coast supplier of difficult-to-find hardwoods used by boat manufacturers, furniture makers, a pistol handle manufacturer, and interestingly enough, the copper and paper industries for specialized uses of exotic woods in some processing equipment.

In 1980, Bob Thompson came to the conclusion that there should be a leadership transition of two to three years before he retired as an active participant in the day-to-day operations of the firm. Jack Stayman, head of the company's veneer division, was the logical candidate to succeed him as president.

> I don't know what my title is. I have yet to find out, but Jack is president and I am an employee. I run the lumber end, although we have two young men here who are very aggressive and do a good job. They really run the lumber division to a great extent. The transition has worked out pretty well. Since the sale in December 1976, there have been no major changes in personnel or management. The only requirement is to continue to make a profit and to send a monthly financial statement to London. We don't hear from them from one year to the next.

Although still reporting to the Mallinson-Denny Group, the Thompson Mahogany Company in 1980 became a minor cog in the industrial empire of Brooke-Bond-Liebig when BBL bought out Mallinson-Denny in what was originally a hostile take-over. Bob Thompson explained this rather strange acquisition this way:

> BBL, a billion-dollar English food conglomerate, had saturated the food business. You can't create a greater demand for food as you can when you come out with a new product. They had something like 60 percent of the tea business in England and the rest of Europe. They were able to diversify and also acquire [Mallinson] land in Africa and South America where they raise cattle.

When BBL was acquired by a still larger conglomerate, the apprehensive Thompson Mahogany Company management team determined in 1986 that it would be best for the company and its customers if they rebought Thompson Mahogany and returned it to the status of a closed corporation. One stated objective was to "...bring continuity and better service for customers. The big advantage to customers is that they are now, once again, dealing with the owners, the ultimate owners. They know they can depend on us" (Anon., 1986).

Donald Thompson (a non-relative) handled teak purchases and sales while Eugene Reitz primarily purchased and sold mahogany and other tropical species. Bob Thompson, who was "officially retired," still provided guidance and inspiration during regular visits to his company office. He assisted in the task of maintaining traditions of quality which extended back more than 140 years. He also provided service to customers, some of which were with the company for more than 70 years. In 1988, Donald Thompson and associates purchased the company. Thompson and Reitz became president and vice president, respectively, and James McDonald became comptroller.

The Norfolk Veneer Mill in Portsmouth, Virginia, had been sold somewhat earlier to Universal Furniture Industries, Ltd., in 1985. The company was based in Hong Kong with plants in Taiwan and elsewhere in Southeast Asia. At the time of purchase, the firm specialized in "knock-down" furniture shipped in containers to its reassembly plant in the United States and other foreign locations.

THE WILLIAMSON FAMILY COMPANIES

The Baltimore Connection with the mahogany trade is best illustrated by the Williamson Veneer Company although not firmly established there until 1900. The hardwood veneer dynasty was begun by Marshall Dodd Williamson in Indianapolis about 1850 (Fig. 3-2). More than 20

Figure 3-2
Marshall Dodd Williamson began a hardwood veneer dynasty about 1850. His descendants have been associated with more than 20 veneer mills.

Figure 3-3. Veneer Firms Associated with the Williamson Family

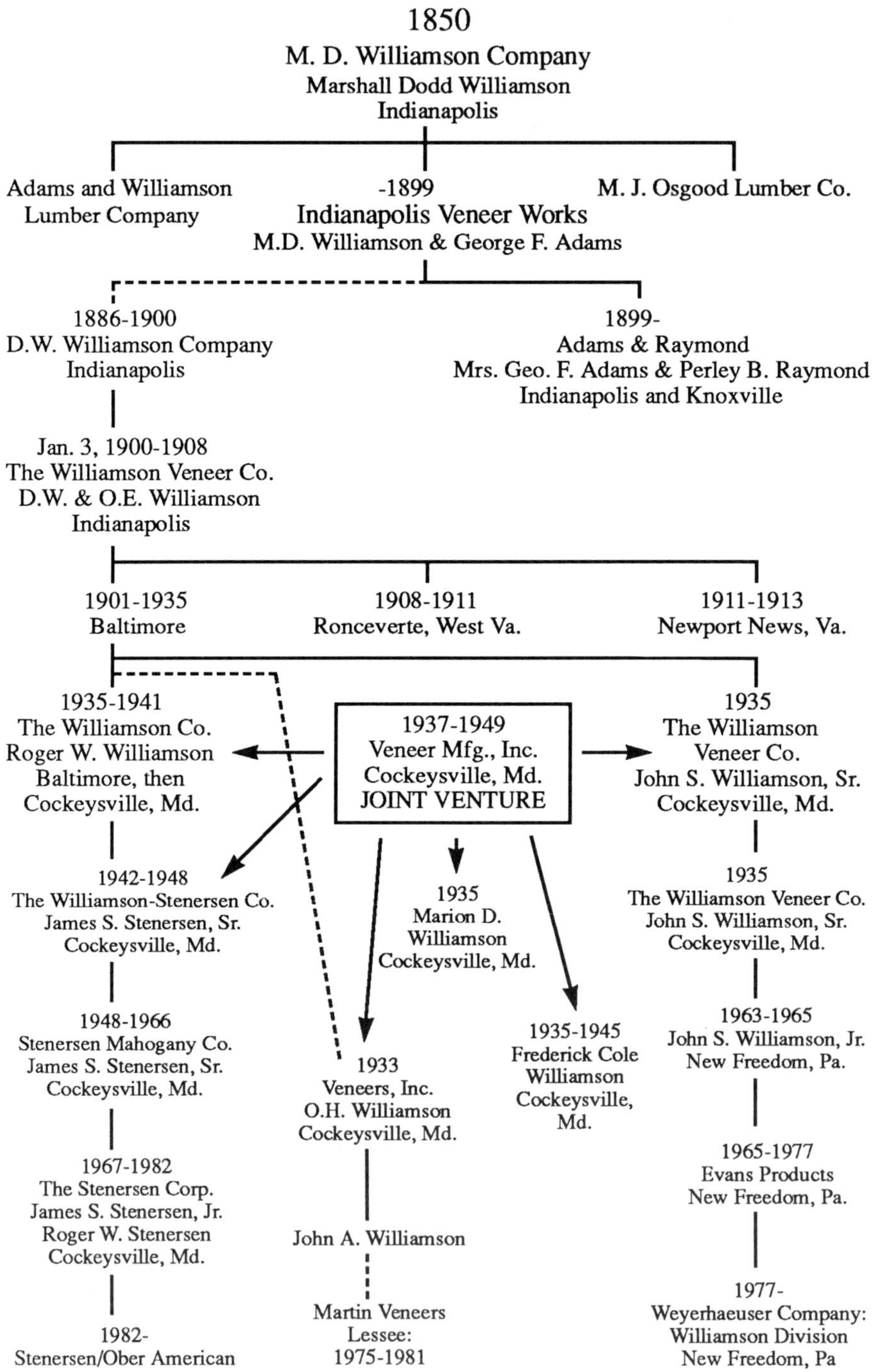

Figure 3-4. Descendants of Marshall Dodd Williamson

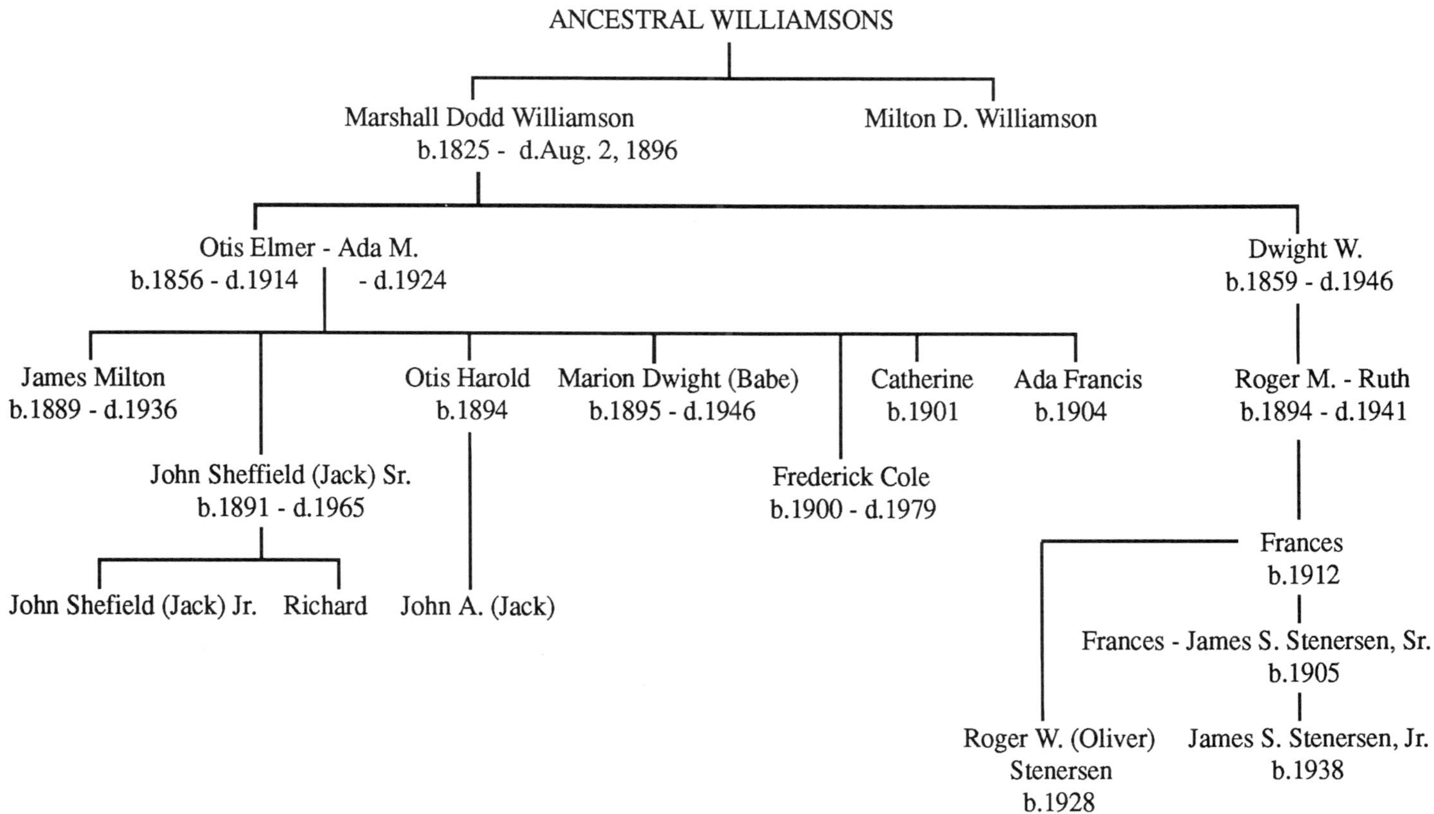

veneer firms identified with the Williamson family (Fig. 3-3) were to evolve in the decades that followed. A genealogy of family members through five generations provides a necessary guide to individual members within the family structure (Fig. 3-4).

Little is known about the early life of Marshall Dodd (M.D.) Williamson other than he was a native of West Virginia and was born in 1826 (Anon., 1883). At the age of 19 he moved to Indiana and resided in the city of Indianapolis beginning in 1845. In Indianapolis he became a skilled cabinet maker. His son, Dwight W. Williamson, with considerable pride recalled seeing some of the furniture his father had made. "The accuracy of detail was a marvel to me. I remember taking hold of the corners of the drawers and slipping them in and out with ease, almost as though they were on rollers—no catching or sticking as in the drawers of most furniture of that day."

M.D.'s entry into the veneer business may have been triggered when he attempted to find mahogany crotch veneer to restore an antique cabinet. Finding none, he travelled to Cincinnati where he purchased a Cuban mahogany plank which "showed" a crotch figure. Returning to Indianapolis, "he put the plank into a vise and with a hand-saw cut off enough 1/20-inch thick veneer to serve his purpose...which was a difficult feat but he was an expert in putting his saws in order and never failed to accomplish what he undertook" (Williamson, 1939).

In 1850, when he was about 25 years old, he began to cut veneers commercially as the M. D. Williamson Company. He used a circular saw driven by a mule walking around in a circle. It was geared on a 10 to 1 ratio in such a fashion that the mule did not have to walk very fast to obtain the necessary rpms for the saw to cut very thin veneers.

Although it is impossible to date precisely all of the entrepreneurial activities of this young man, it is known that he was "extensively interested in a variety of mercantile and industrial enterprises" which included the shipping of walnut logs and lumber to New York (Anon., 1883). Many of the logs were purchased by Kassing's Mill in New York which was "probably the first mill in this country to cut cabinet or figured veneers." He was well known in the trade and ultimately moved to New York City in about 1867 to direct this successful business. However, while watching the loading of a rail car of large logs in Indianapolis, a skid accidentally broke loose and struck M.D. on the left leg. He returned to New York City and was bed fast for two years before the leg was eventually amputated. In the meantime his New York business had been lost.

When he had recovered, M.D. returned to Indianapolis in 1870 to "get his bearings and plan his future." He was aided by M. J. Osgood who had an established lumber business and joined George F. Adams as a company-partner under the name of Adams and Williamson Lumber Company. Somewhat earlier, the Sewing Machine Cabinet Company of Bridgeport, Connecticut, had built a large factory in northeast Indianapolis which was equipped with a rotary lathe and a staylog lathe for cutting veneers. Eventually, M. D. Williamson rented a section of this plant including the staylog machine and began cutting figured walnut, butternut, and ash veneers and continued to do so until December, 1881. These veneers found a ready market in New York and an active business in cabinet veneers resulted. The machine being used was generally unsatisfactory because of its light construction..

Determined to obtain better equipment, he persuaded his partner, George F. Adams, to construct a new mill along side another building Adams owned at the junction of Massachusetts Avenue and the Big Four Railroad in Indianapolis. The complex was comprised of an office building 18 x 36 feet in size, a two-story machine shop 40 x 60 feet, and a two-story building for storage purposes 40 x 216 feet. The veneer equipment was powered by a battery of large boilers and a first-class Corliss engine of large capacity situated in the adjoining building. Two staylog cutting machines "of heavy pattern" were built by Charles E. Titus of New York based on patterns for similar machines he had designed and built for C. H. Burton and Company of Cincinnati. Williamson (1939) wrote in his commentary that "...the machines were of the latest improved construction especially adapted for the delicate work of cutting veneers ...which for perfection of work and thorough adaptability in every way for the purposes intended is not surpassed by any similar mechanism in the Union, and with possibly a single exception not equalled by any in use at the present day." After the machines were installed, Titus was persuaded to remain with the firm to supervise production and did so until January 1, 1883, when he returned to New York.

Indianapolis Veneer Works

When the new mill was completed, Adams and Williamson changed the name of their partnership to Indianapolis Veneer Works. The firm was successful from the beginning and booked large orders of oak and walnut veneers from firms such as the Singer Sewing Machine Company of South Bend, Indiana, and the Pullman Company of Chicago.

On November 11, 1882, Dwight W. Williamson joined the firm after having successfully completed "the mechanical course" at Purdue University. His talents were soon used to devise a better way to grind the veneer knives and to assist in the development of better veneer machines and clippers. Both M. D. and D. W. Williamson patented veneer clippers and M. D. patented an improved staylog lathe machine which was manufactured by the Capital Machine Company beginning in 1887.

Between 1883 and 1886, when a fire destroyed the plant, Marshall D. Williamson, as general manager, was responsible for log buying functions, veneer manufacture and sales and was regarded as the master of the business in all of its details. Sales were represented in New York by I. T. Williams & Sons and in Chicago by Adolph Sturn and other brokers such as the Chicago Veneer Company (Fred C. Beeson and the Lord family). The Indianapolis Veneer Works' sales office in Cincinnati was managed by John De Pinal.

The 1886 fire started in the adjoining furniture plant (D. E. Stone and Company) and destroyed both veneer works and a nearby church organ factory. Dwight W. Williamson, who had left the company in 1885 to start a machine shop business, was called back to reconstruct the veneer mill's machinery and power plant. In the process, the mill was enlarged to accommodate six veneer machines. To do so, the land occupied by the furniture and organ factories was turned over to the Indianapolis Veneer Works business by its owner, George F. Adams. In less than a year the mill was in operation and business flourished once again.

In the early summer of 1896, Marshall D. Williamson became seriously ill while on vacation with his wife at Chautauqua Lake Assembly in New York. He was returned to Indianapolis by train where he died on August 2, 1896. He was remembered by his son as a strong, courageous man who was diligent, industrious, and "...never tired of talking about figured wood or the machinery for cutting veneers" and who was both patient and tolerant yet "...when put on his mettle he was like a storm..."

After Williamson's death, George F. Adams assumed responsibility for operating the business and did so with the help of Perley B. Raymond until Adams died in 1899. After some litigation between Mrs. M. D. Williamson and the heirs of George F. Adams, Mrs. Williamson's widow's share in the business was turned over to Mrs. George F. Adams "at a good price." Perley B. Raymond subsequently joined his sister, Mrs. Adams, in a partnership which continued to operate the "old" Indianapolis Veneer Works establishment under the "new" name of Adams and Raymond Company. The center section of the mill was destroyed by fire in 1925. Prompt action by firemen saved several thousand dollars of veneer stored in the southern section of the two-story building.

D. W. Williamson and Company

Dwight W. Williamson, who had left the Indianapolis Veneer Works in 1885 to start a machine shop, organized a veneer manufacturing company with his brother in 1886 having found "the machinery business too slow and tedious to meet my growing needs." The name, D. W. Williamson & Company, was first adopted to identify the newer mill. D. W.'s brother, Otis Elmer (O. E.) Williamson, had long been associated with the veneer industry but had never been involved directly in veneer manufacturing. As an independent log buyer, he was recognized as a good judge of figured walnut and other timber based on his long experience and close observation of defect signals in growing trees. O. E. had sold logs to the Indianapolis Veneer Works and to veneer manufacturers in the state of New York.

The two brothers built their new veneer mill adjacent to the Indianapolis Veneer Works at 10th and Massachusetts Avenue (Burrell, 1964).

A one-story veneer mill and power plant and a three-story "dry house" were built on this site. A warehouse was constructed across the alley on property purchased from the Indianapolis Veneer Works. Two veneer cutting machines were purchased: a seven-foot Sinker-Davis Company veneer lathe and a ten-foot Capital Machine Works veneer machine. The early days of the business were described by D.W. Williamson (1939) as follows:

> O. E. had a large lot of finely figured Ash logs, which he had [previously] offered to the Indianapolis Veneer Works and they had declined and this wood was our first job. The veneer was very good and we sold it readily at a high price, so that we made a nice profit on the transaction.
>
> About this time Curly Birch came into vogue with furniture people and O. E. was first in the field for the logs. For the new firm, he bought large quantities of these logs without reserve. Carload after carload was shipped from North Carolina and cut into veneers, until our warehouses were filled and then the selling began. In a few months our stock was so low that we had to sell Curly Birch veneer fresh from the knife before it was dried. Still the logs rolled in and we were kept busy with this product for several years until the market began to show signs of weakening.
>
> In the meantime, I had constructed a slicer attachment for the ten-foot veneer machine and we had developed a slice-cut quartered Oak that became very popular with furniture manufacturers and this increased our volume so that we had to put in a third veneer machine to take care of the demand.
>
> Coincidental with the Curly Birch and Quartered Oak veneers on which we had a big run, we had not neglected Figured Walnut. We had cut a considerable quantity of high grade wood and held it in stock not wanting to press the sale of it so long as other business was going strong. We considered Walnut the basis of the veneer business and other woods transient. We had been the first and largest in the manufacture of Curly Birch and sliced quartered Oak so that we really got a large part of the business. Father often came down to see us and at one time he remarked that he believed we were doing more business with our *three machines* than they were doing with *six machines*, which I construed as a compliment to our efficiency.

By 1899, the two brothers were beginning to become concerned about the Indiana timber supply. Large walnut trees were becoming "very scarce in Indiana, Ohio, and Illinois" and some log dealers were buying white oak logs in Kentucky and reselling them in Indiana as Indiana white oak.

> Greater quantities of walnut were being shipped out of West Virginia, Virginia, North Carolina and Tennessee. North Carolina had all the curly red birch and most of the figured ash. West Virginia had a fine quality of white oak of large size. We had found it very severe service for a man to travel out of Indianapolis to the southeastern states in order to buy wood to ship to Indiana. O. E. rebelled against this so we decided to look around for a better mill location.

After seriously considering sites in Charlestown, West Virginia, and Norfolk, Virginia, a Baltimore location was decided upon by Otis Elmer Williamson.

Baltimore was easily accessible to the figured wood production of West Virginia, Virginia, North Carolina, and Tennessee and likewise these states were easily reached from Baltimore. Being on the seaboard, Baltimore was well located [as it turned out] for the importation of Mahogany and foreign woods. The surrounding territory had a considerable amount of walnut of large size, much of it figured, and a fair supply of poplar. But the best timber in this section for veneers was the red and black oak, then which none was ever better. The white oak, however, was not good, because it was flinty and hard. It was better suited for wagon tongues and fellows, barrel staves, ship timbers, railroad ties and bridges.

Four acres were purchased in the Highlandtown section at Baltimore and Haven streets. The Union Railroad tracks were on the east side of the property and the Baltimore and Ohio railroad switching spur was located on the south side. Baltimore, also, had one of the finest deep-water ports in the country. In the spring of 1901 construction started on a 70 x 170 foot veneer mill to house three "heavy pattern" veneer lathes. A sawmill, soaking vat, and a three-story brick warehouse completed the manufacturing layout. Unfortunately, the entire veneer plant was destroyed by fire on July 4, 1904. A fireproof set of buildings was subsequently rebuilt on the same site between 1904 and 1905. Veneers cut from figured walnut butts and long flitches and figured ash were specialities of this operation.

Although D. W. Williamson (1908) was to become wealthy from his veneer operations, he wrote: "That the great reason for the poverty of the veneer business... is that [it] is restricted to such small and feeble operations, such petty and exacting orders,... and such unorganized trade conditions, that it is comparable to no other business at all." To correct price and other problems he proposed that an association "...collect the forces of the infant industry... and coalesce them into a tangible organization. Its objective is to raise the standard of its operation; persuade them to a rigid investigation of their methods and cost; get them to put prices on their products that are commensurate with the cost of production; and then to induce them to stiffen backbones and stand for what is right and fair, as a compensation for honest work." He was to live long enough to see such an organization, The Veneer Association, created under the auspices of the New Deal's National Industrial Recovery Act of 1933.

The mill was well planned and fully equipped. A double sawmill was separated "entirely" from the veneer mill which was equipped with "two 10-ft and four 7-ft veneer cutting machines. The mill building [was] separated from all drying and storage buildings" to avoid the circumstances of the 1904 fire (Anon., 1908).

The Williamson Veneer Company

Meanwhile, on January 3, 1900, the Williamson's partnership in Indiana, was incorporated under Indiana law and the name changed to the Williamson Veneer Company. It was also decided about this time that the Indianapolis site should be sold and the machinery moved to an existing building in Ronceverte, West Virginia.

It was known that the oak trees of the Ronceverte locality were large and ideal for manufacturing quartered white oak veneer. Indianapolis employees W. H. Snyder, Charles Kellogg, and Ralph Young were brought along to manage and be responsible for the day-to-day management of the Ronceverte mill until the mill was dismantled in 1909. From 1909 until the partnership was dissolved in 1914, O. E. Williamson devoted himself primarily to buying Circassian walnut logs for the Baltimore mill from brokers such as Joseph Gardner & Sons of London and Edinburgh, Manuellides Brothers of Russia, Henry Hoffman of Germany and from several other walnut shippers from the Russian forests. Mahogany logs were purchased in London and Liverpool. He also "sold" figured ash and burls to European buyers such as Adolph Frank of Marburg, Germany, Mueller & Sons of Hamburg, Germany, and Miguet & Cie of Paris, France.

In 1909, the Williamson brothers sent W. G. Munyan, a veneer salesman, to build a veneer mill at Newport News, Virginia. The primary purpose was to supply veneer door stock for the Paine Lumber Company of Oshkosh, Wisconsin, reputed to be one of the largest veneer door manufacturers in the country. Paine Lumber provided financial support for this venture and contracted to purchase two rail carloads of 1/20-inch red oak veneer each week. One veneer machine and a veneer dryer were shipped to Newport News from the company's mill in Ronceverte, West Virginia, to get the operation under way.

Two years later in 1911 the business was incorporated under the laws of Virginia as The Williamson Veneer Company of Newport News, Virginia. The incorporators were D. W. Williamson. C. J. Cummins, O. D. Bachelor, J. A. Willitt and Roger W. Williamson.

Roger Williamson, D. W.'s son and only child, was appointed assistant to C. J. Cummins, general manager for the firm. The success of the Newport News business fluctuated with the fortunes of the Paine Lumber Company (which went into receivership prior to 1911) and the more local demands for poplar crossbanding, core stock, and plain oak veneers. The firm suspended operations in 1913. Roger relocated to Baltimore to be with his father. He was a personable and convivial young man and soon gravitated into sales. Among the new clients he brought to the Baltimore firm were the Aeolian Company, the F. G. Smith Piano Company, the Everett Piano Company, Mason & Hamlin, Ivers and Pond, A. M. McPhail, Vose & Sons, and the J. H. Lockey Piano Case Company. D.W Williamson (1939) later wrote: "Naturally, I thought of Roger as my successor in the veneer business."

Early in 1914, the two brothers concluded that it would be best to split up after nearly 30 years of working together as partners. Otis E. Williamson admitted that he was losing interest in the veneer business, was not in the best of health, and suspected that the familial frictions and his independent ways contributed to the decision. O. E. died shortly after the split while on a trip to Chicago. After consultation with O. E.'s heirs, it was concluded that the business should be incorporated under the laws of Maryland with an authorized Capital Stock of $100,000.

The mill was refurbished with a new boiler system and new veneer drying machines to replace the "finger racks" used previously in air drying the veneer. The improvements and extensions made the Williamson plant one of the largest and most impressive in the country. It occupied no less than eight city blocks and consisted of six large buildings in addition to a large yard for storage of logs and lumber. The business was in excellent condition to make money and did. Customers and other associates were treated royally on the "Wilven," Dwight Williamson's 44-foot power cruiser which sailed Chesapeake Bay and other waters of the East Coast in grand style.

In 1916, the company began to manufacture walnut gun stock blanks under a subcontract from Frank Purcell of Kansas City, Missouri. Between 500 and 1,000 gun stocks were cut each day. Williamson in his 1939 memoirs recalled that the federal inspection was very severe and that the slightest defect would cause rejection. "I was burning walnut gun stocks in my fireplace for more than a year after the armistice."

In 1920, the Williamson Veneer Company began to advertise its products. As a trade publication put it: "This advertisement marks a change in business thought and a recognition of publicity value by one of the oldest, best known and most conservative veneer houses in the country. The new plan has brought about abandonment of a long established jobbing connection in favor of a comprehensive, direct-to-the-trade policy" (Anon., 1920a). As a result of this new focus, company sales offices were opened in Chicago, New York City, Jamestown, New York, and High Point, North Carolina.

On March 28, 1922, Dwight W. Williamson acquired the remaining assets (one-half interest in the mill and real estate) still controlled by the heirs of O. E. Williamson. This was accompanied by a great deal of family discord and bad feelings between Williamson and the widow and sons of Otis E. Williamson: Otis Harold (O. H., or Harold), Marion, John, Frederick, and James. Dwight's eagerness to purchase the remaining assets were based on his optimistic anticipations of the next decade based on recent sales experience. However, these expectations ran head-on into the Great Depression Years which followed the Crash of 1929. Recorded sales volumes provide evidence for his anticipation and the stark business realities in the years immediately following 1929. The company had sales of $1,046,392 in 1923. Five years later sales were $100,730.

Williamson's business outlook prior to the Crash and his vision of an expanded business had also led to large scale new investments in additional land, British Honduras timber, new derricks, additional shipping and storage space, a railroad spur, and two new Merritt veneer dryers. Overextended and caught, perhaps, by the duplicity of the bankers who had extended credit, D. W. Williamson was forced out of the business.

He bitterly summed up this experience in 1939, seven years before his death at the age of 87.

> In conclusion we [Dwight W. and Roger W.] decided to let the bank take the property without contest and free ourselves from an incubus that had become intolerable. Thus ended my connection with The Williamson Veneer Company. Under its different names and in its several ventures, I had given it forty-five years of my life and my full capacity of care and attention. I had seen the business grow from a valuation of $40,000 to a net value of $650,000 and then I had seen the assets wasted by the Union Trust Company and Clifford B. Winchester [the executive vice president of the bank].

Winchester, of The Union Trust Company, which now controlled the Williamson Veneer Company, sought out Otis Harold Williamson and asked him to run the business. Harold had a reputation as an extremely smart business man and agreed to do his best to try to pull the company out of the proposed bankruptcy. He had been an aviator during World War I and after the Armistice in 1918 had been employed by The Williamson Veneer Company as mill manager. Although his work was judged to be "efficient and very satisfactory" by his uncle, Dwight W. Williamson, Harold was dissatisfied with his position in the firm and the way the business was being operated and left the company in 1933 to start his own business, Veneers, Inc., the following year near Cockeysville in Texas, Maryland. Two of Harold's brothers (James M. and Marion D.) had also been employed by The Williamson Veneer Company at various times mostly as log buyers, and James M. Williamson had been one of the incorporators of the Baltimore mill in 1914.

His other brother, John S., an engineering graduate of Lehigh University, sold veneer for the company in Grand Rapids from 1924 until the Crash of 1929. His son, John S. (Jack) Jr., when interviewed, said that his father left Grand Rapids for New York City looking for employment "in any field he could find." Needing a man with proven talents, Harold located John in New York and "...found him in 1935 ready to come back to Baltimore and to pick up the pieces of The Williamson Veneer Company." By 1937, John Williamson had pulled the company up "...by its bootstraps" to a point where it was not being pressured by the bank although there were still a number of outstanding financial obligations. To clear the ledger the Union Trust Company suggested that the Williamson Veneer Company property in Baltimore be sold. After the property had been sold, John S. Williamson, Sr., was able to take the Williamson Veneer Company "out from under the bank." In the meantime, Roger W. Williamson had moved across the street and started a veneer sales business, The Williamson Company.

In 1935, four sons of Otis Elmer Williamson were actively involved in cutting and/or selling hardwood face veneers. James, the eldest son, had already left for a more Bohemian life style in New York City and died there in 1936. In 1937, Harold, who had earlier moved to the Cockeysville area from Baltimore, proposed to his surviving brothers and to his cousin, Roger W., that they combine together in Cockeysville with a new production facility for the five owners to be known as the Veneer Manufacturing Company. When they agreed, Harold sold his plant in Texas, Maryland, to a local gravel company. Roger was glad to move out to the "country" at Cockeysville in 1932 and sold his real estate in Baltimore. On money borrowed from Harold he started a fourth sales company. He opened up an office and sample room in a building on York road, and with very little capital conducted business from there throughout the middle and late 1930s.

Veneer Manufacturing Company

Under the proposed arrangement, each owner would have a 20 percent interest in the firm and each "partner" would have the right to cut his own veneers for a two-week period in every 10-week cycle. These veneers would then be sold through the five separate and individually owned sales companies. Operations began with two slicers and one stay-log lathe. Each owner maintained a separate office in Cockeysville as well as individual warehousing facilities, sample rooms, and his own sales company. The principal advantage to this rather bizarre arrangement was scale efficiencies in the production process.

By 1938, Frederick Cole Williamson had opted out of this arrangement and, in a spirit of cooperation, the four remaining competitors united sales efforts in Louisville where all sales were consolidated and represented by the D. W. Lewis Company. Fifty-six veneered species were offered for sale in addition to Cuban, Honduran and African mahoganies (Anon., 1938).

On January 1, 1948, three of the companies merged into a single company at the urging of John S. (Jack) Williamson, Jr.; however, M. D. Williamson elected not to join the merger. Jack, who had joined the consortium after World War II, was considered by his peers and co-workers to be bright and creative. As he later admitted, "It was an impossible situation from everybody's point-of-view. It lasted 15 months. I acknowledged that I had been wrong to try to do it and we busted up again and we went right back doing the same old things."

The family situation had changed, however. When Roger W. Williamson died in 1941, the management of The Williamson Company was soon taken over by his son-in-law, James S. Stenersen, Sr. Marion D. Williamson had also died. His widow, Virginia, continued the operation of his sales company until the early 1960s by having veneer custom cut at local veneer mills when the company ceased to operate.

In 1952, Harold Williamson decided to go it alone. He sold his share of the Veneer Manufacturing Company to Jack Williamson and built a brand new mill. He called it the Beaver Dam Veneer Mill after the stream that ran behind it. Within two years, Harold realized he had made a mistake. He was a machinery man, not a merchandiser. He couldn't sell enough to keep the mill busy. Being nearly 70 years old, he decided to retire and sold the mill to James S. Stenersen (Vollmer, 1982).

The Cockeysville operations of the Williamson family all became heavily involved in cutting mahogany veneers. However, in the middle-1950s, the Williamson Veneer Company essentially left this trade when the market for veneered television cabinets was captured to a large degree by cases made of plastic and metal. As this market disappeared, the company experimented with cutting domestic species, primarily red oak which was becoming popular on the West Coast. Believing that the future of the company depended on the domestic species, a 95-acre tract of land was purchased in 1962 at New Freedom, Pennsylvania, 30 miles north of Cockeysville.

The New Freedom plant began production there on December 15, 1963, with two staylog veneer lathes. At the time of construction the New Freedom plant was considered a model mill in design and efficiency employing the latest manufacturing methods known in the veneer industry. The steadily increasing demand for Appalachian red oak veneer motivated the company to concentrate on that species (Burrell,1964). Upon the death of John S. Williamson, Sr., on November 14, 1963, the company was reorganized. Under the guidance of his son, Jack Williamson, the business soon became the largest producer of red oak veneer and panel faces in the country.

To feed the expanding face splicing demands, another veneer plant was built (1968) near Jefferson, Pennsylvania (one staylog lathe). The Mayes veneer mill at Fleischmanns, New York, was also purchased and converted into a high-speed, full-rotary production plant. As Williamson explained it: "The desire was to have veneer available for faces that were full rotary, half-round, and sliced. A slicer was unnecessary because by careful sorting of the finished veneer enough 'sliced look' veneer was available to satisfy the market." Assisting Jack Williamson in the management of the company were T. H. Gogoll, Jr., David R. Fauntleroy, J. E. Southern, and Robert H. Bommer.

A public relations mimeograph dated September 15, 1966, reported a work force of 110 employees and a payroll of $680,000. More than 3.5 million board feet were converted annually into 40,000,000 feet of final product with an estimated annual market value of $3.5-4 million. In 1965, The Williamson Veneer Company was merged with the Evans Products Company. Jack Williamson stayed on as manager until 1973.

The Evans Products Company was a firm unfamiliar with the raw material business and primary manufacturing. Perhaps, as a result, their New Freedom operation did not live up to their profit expectations. When an opportunity came along in 1977 to sell what had been The Williamson Veneer Company to the Weyerhaeuser Company, Evans Products accepted the Weyerhaeuser offer with some relief. The Williamson name was retained to the extent that the three veneer mills and the face plant operated as the Williamson Division of the Weyerhaeuser Company.

In the early 1980s, the Williamson Division output was entirely red oak although in "normal" times, hickory (primarily), ash, and walnut comprised about 20 percent of the cut. Much of the product was sold on the open market. About 40 percent was utilized within the company at Weyerhaeuser panel or door plants in Hancock, Vermont; West Memphis, Arkansas; and Marshfield, Wisconsin.

Stenersen Mahogany Corporation

When Roger Williamson died in 1941, he left the Williamson Company with "a tremendous amount of indebtedness and no assets." Recognizing the bankrupt condition of the business, lawyers for the estate asked James S. Stenersen, Roger's son-in-law, how he thought the matter should be settled (Fig. 3-5). Stenersen said in an interview that he had promptly "...made them an offer in order to just gain the name. I knew nothing about the veneer business. I had never seen a piece of veneer before." The offer was accepted.

Stenersen was a creative and ambitious entrepreneur. He had been an automobile dealer and was a merchandiser for the Hecht Company in Baltimore before this baptism in the veneer business in 1942. The company name was soon changed to The Williamson-Stenersen Company. Harold Williamson encouraged Stenersen by providing substantial assistance. "He warehoused our inventory and taught me about the manufacturing end of the business." Dick Cavanaugh who had been a Williamson employee for 35 years provided experienced answers to Stenersen's probing questions in other areas of the business.

Almost all manufacturing effort in 1942 was directed toward the war effort. One of Stenersen's first jobs was a contract shared with Harold Williamson to furnish mahogany veneers for PT boats. Later, he cut hickory veneers for toboggans and skis used by Russia in its winter campaigns against the Germans. After the war, Stenersen spent much of his time on the road merchandising the product.

> By being able to turn the veneer over very rapidly, I was able to operate without any money. I turned product into cash almost immediately strictly through efficient merchandising. By doing so, I was able to expand the business to about 14 times what it had been. I established myself throughout the trade and made a lot of friends who were customers. Having established myself, I changed the company's name to the Stenersen Mahogany Corporation. This was about 1948.

Stenersen also learned to fly which expedited his contacts with customers and his ability to operate the mill in Cockeysville. He also maintained managerial control of a partial interest in a veneer mill in Norfolk, Virginia, and operated a sales office in High Point, North Carolina. It was not unusual for him to keep a morning appointment 100 miles from the

Figure 3-5
James S. Stenersen, Sr., a creative and ambitious entrepreneur, successfully rejuvenated the defunct Williamson Company. He later changed the name of the company to the Stenersen Corporation.

office and meet a valued customer at the office and display room shortly after lunch.

After James S. Stenersen, Sr., bought the Harold Williamson Beaver Dam Veneer Mill "lock, stock and barrel," he sold his 25 percent interest in the Norfolk mill to the Thompson Mahogany Company of Philadelphia, Pennsylvania, and his 20 percent interest in Veneer Manufacturing Company to John S. Williamson, Sr. Later, in 1963, he bought the site and buildings from John S. Williamson, Jr., and converted the space into a warehouse.

Stenersen took advantage of his sales ability and exploited opportunities. He is remembered by John S. Williamson, Jr., as "a *tremendous* salesman and well-liked in the trade. We jobbed a lot of Honduras mahogany lumber and he was almost solely responsible for the sales." Stenersen was also heavily involved in providing veneers for TV and radio cabinet businesses when they were booming. "I had done very well when I stepped out of the veneer business in 1966. I had specialized more or less in a car load business and in the true mahogany crotch and swirl veneers...and produced 50 percent of these types of veneer cut in the United States before mahogany went out of style in 1957."

James S. Stenersen, Jr., and Roger W. Stenersen bought the company in 1967 after their father had "retired" to pursue other business interests. Roger had managed the factory for his father and had the reputation for having an uncanny knack of developing the best veneer possible from every log that came in the mill. Jim, Jr., had learned the sales end of the business beginning at age 22 in the High Point, North Carolina, sales office. On his own initiative he expanded the company's sales effort into European and South American markets. Since the company name did not accurately represent the character of the business anymore, it was changed in 1968 to simply The Stenersen Corporation. The sales business in High Point became the Stenersen Sales Corporation. In 1978 Roger sold his veneer business interests to his brother. He later became the highly respected manager of the Norfolk Veneer Mill.

The Stenersen mill operated with a 17-foot Johnson City (Tennessee) slicer, a 14-foot Capital Machine Company slicer and a 10-foot variable staylog Capital lathe. Between 1957 and 1974, most of the veneers had been cut for domestic markets utilizing a diversity of North American species. Beginning in 1975, the business was oriented to the export market with as much as 85 percent of the cut volume destined for overseas markets. The emphasis was on white oak veneers cut from logs procured from Maryland, Virginia, and Pennsylvania. The downturn in European markets in 1981 had caused the firm to produce more domestic veneers and to seek new markets for them in the United States. When these efforts failed, the mill was closed on December 10, 1981 and the machinery put up for sale.

Veneers, Inc.

Otis Harold Williamson organized Veneers, Inc., in Cockeysville, Maryland, about 1936, after leaving the family veneer business in Baltimore. Harold was the second oldest brother of the five sons of Otis Elmer Williamson. He is remembered by James S. Stenersen, Sr., as the "main brother" and by his nephew, John S. Williamson, Jr., "...as an extremely smart businessman [and] a very penurious businessman ... not kindly in most ways." A former production manager remembers him more explicitly as "a slave driver."

After encouraging his younger brothers and his cousin, Roger W. Williamson, to join him in Cockeysville in a joint production venture, Harold found that he was not able to obtain enough cut veneer during his allotted time of three weeks out of ten to satisfy his growing sales potential. In 1948, the newly constructed Beaver Dam Veneer Mill was organized under the Veneers, Inc., corporate structure. The plant used one slicer and one rotary in cutting veneer. In 1955, this mill was sold to the Stenersen Mahogany Company. Harold later (about 1950) built another plant in Cockeysville and equipped it with three slicers and one staylog lathe. This facility was located near the Stenersen Corporation mill. Harold remained active in the business until he died in the early 1960s and the business was taken over by his son, John A. Williamson. John had earlier owned a veneer mill (CHAPAS, S.A.) in Guatemala where he operated a 1927 Capital slicer and cut Primavera, Bianquita, and Rio Bianco for the domestic market in the United States.

Hurricane Agnes in 1972 devastated the Veneers, Inc. mill at Cockeysville. Flood water reached into second floor office space, rushed through the crated veneer warehouse, and carried away stacked veneer flitches together with yarded logs (Ross, 1972). Three years later in November, 1975, the mill was leased to an Austrian who operated the mill as Martin Veneers, Inc., as a custom cutter for European clients until May, 1981, when it ceased to operate because of poor export market conditions. At that time Peter Daup was president of the corporation and Josephine Loidl was export manager. Prior to closing, 70 percent of the cut had been white oak with red oak, ash, walnut, and poplar rounding out the total production estimated to be 30,000 square meters per day. John A. Williamson continued a veneer sales business under the firm's original name, Veneers, Inc. The veneer mill in 1988 was operating as the Veneer and Lumber Corporation.

PENROD, JURDEN and CLARK COMPANY

Another of the giants of the hardwood industry was John Noble Penrod (Fig. 3-6). Widely recognized at his death as the "Walnut King," he built a walnut empire based at his plant in Sheffield, a suburb of Kansas City, Missouri. It was not until eight years after his death in 1925 that the company's activities began to shift to Norfolk, Virginia, and to the exotic timber of Africa, South America, and the Philippine Islands. By this time the firm had become Penrod, Jurden & Clark Company.

Penrod was born on August 13, 1855. The family farm was located near Silver Lake, Indiana, where he received a grade school education. At the age of 23 he was employed by the John H. Lesh [lumber] Company

Figure 3-6
John Noble Penrod began his timber products career in 1878 in Indiana. His greatest success came after he moved to Missouri in 1895 where he processed walnut and oak lumber and veneer.

of Goshen, Indiana and in 1885 became a partner in this firm. Three years later he organized Penrod, Wood & Company at Wabash, Indiana. By 1895 he had prospered sufficiently to acquire the controlling interest in the Des Moines Sawmill Company in Kansas City, Missouri, and moved there to take charge of the business and changed the name to the Penrod Walnut Corporation. When a small rotary machine was installed in 1906, the company name was modified and became the Penrod Walnut and Veneer Company (Anon., 1925).

By 1912, his company had grown to include a 7-foot band mill, two Capital veneer lathes and a Philadelphia veneer dryer. The annual production of black walnut veneer was between nine and 10 million square feet. Oak veneer production totalled about six million square feet per year. Owing to the rapid growth of the business and the high cost of rail transport, a second veneer mill was built at Helena, Arkansas, with similar processing equipment. Although oak veneers formed the major share of the veneer produced, ash, elm, pine, cottonwood, and red gum veneers were also manufactured at the mill.

The Helena operation was incorporated in 1912 as Penrod, Jurden & McGowen, Inc., which later became the Penrod-Jurden Lumber Company. The timber holdings were located on 6,000 acres of the very best Arkansas bottom lands from which 12 to 15 million board feet were cut annually. The company also purchased land with high-grade stumpage in the St. Francis basin about 25 miles from Memphis and 40 miles from Helena. Eighty to 100 loggers were employed to cut and load the logs on railcars for the trip to Helena. The company was active in the export market and maintained an office in London (Anon., 1912b).

By 1918, the Helena plant was equipped with three large Capital rotary lathes and three dryers. After operations began at Helena, the Kansas City plant was devoted almost exclusively to the production of walnut lumber and veneer. Penrod is known to have traveled to the Caucasus to purchase Circassian walnut and to London and Liverpool in 1912 to attend the mahogany auctions (Anon., 1912a).

During World War I, Penrod Lumber and Veneer Company was entirely engaged with war-related work producing gunstocks and airplane propeller lumber utilizing 60,000 board feet of logs every 24 hours. About 2,000,000 gunstocks had been furnished allied governments by the end of the war. John Penrod continued as president during this period and was assisted by Ralph L. Jurden, vice president, and J. C. Rodahaffer, secretary-treasurer. Jurden, his son-in-law, remained president of Penrod-Jurden Company, Inc., in Memphis and was considered to be one of the most progressive figures in the hardwood industry (Fig. 3-7). He had joined Penrod in 1904 and was thoroughly trained in all company departments. He succeeded Alexander Lendrum as secretary and general manager in 1913 and became vice president of Penrod Walnut & Veneer Company in 1917 (Anon., 1919).

Figure 3-7
Penrod's son-in-law, Ralph L. Jurden, was considered to be one of the most progressive individuals in the hardwood industry. When Penrod died in 1925, Jurden became president of the Penrod Lumber and Veneer Company.

Prior to World War I, Penrod made frequent trips to Hamburg, Germany where his company maintained a sales office. He was an active participant in community affairs and business activities. Penrod served as first president of the Walnut Export Sales Company (London) through which the production of most walnut manufacturers in the United States was distributed to European markets. He also was the first president of the American Walnut Manufacturers Association in 1912. When Penrod died on September 24, 1925, at the age of 70 from a malaria attack, his firm was credited by the Kansas City Star as being the largest producer of walnut articles and veneers in the United States (Anon., 1925a).

On October 7th, Ralph L. Jurden was elected president. Since 1916, he had resided in Memphis as president of the Penrod-Jurden Company. During this time he had continued as vice president of the Penrod Walnut & Veneer Company. John C. Rodahaffer became vice president and general manager of the company and in change of operations at Kansas City. Prior to the death of Penrod, Jurden had sold his Memphis business to the Chicago Mill and Lumber Company and later moved back to Kansas City where he owned an extensive farm and raised fine horses. He was reputed to have a "...very dynamic personality and a distinct and unusual social charm in addition to a notably astute business mind" (Anon., 1925b).

Already the world's largest manufacturer of walnut lumber, walnut dimension stock and walnut veneers, Jurden in 1928 arranged a merger with two other "world-class" walnut producing companies: The Des Moines Sawmill Company and Kosse, Shoe, and Schleyer Company, founded in 1905 in Cincinnati. The new corporate structure included Jurden as chairman of the board; Rodahaffer as president and in charge of the Kansas City operation; and vice presidents Vernon L. Clark, Albert Heidt, and Alex Schmidt. Heidt and Schmidt had come to the United States from Germany in 1903 where they had been employed by Theodore Francke, the largest lumber importer in Germany. They joined the Kosse, Shoe, and Schleyer Company in 1913 and quickly rose to the top management positions. Vernon Clark, former president of the Des Moines Sawmill Company, remained in charge of the Des Moines facilities. The size of the new enterprise can be evaluated by its inventory of 30,000,000 feet of walnut veneer and 12,000,000 feet of walnut lumber "on sticks." The total annual plant capacity was 110,000,000 feet of veneer, 20,000,000 feet of walnut lumber, and 10,000,000 feet of imported cabinet woods. In announcing the merger, Jurden was quoted as saying:

> The consolidation has been effected primarily in order that we might better serve our trade. Our business is increasing steadily and we are enjoying a gratifying amount of export trade. Fifteen percent of our lumber is now being shipped to foreign countries...Merging the concerns into one will effect many economies and will enable us to give prompt shipments of any size orders and enable us to more quickly serve our trade in any part of the United States (Anon., 1928).

Vernon L. Clark was a young employee of the Des Moines Lumber Company when Penrod bought an interest in the company in 1895. Clark was an ambitious, articulate individual and an excellent organizer (Fig. 3-8). He was active in Republican politics in Iowa and the "politics" of the walnut industry. It was his initiative, for example, that brought together

Figure 3-8
Vernon L. Clark was an excellent organizer and was largely responsible for the first organizational meeting of the American Walnut Manufacturers Association in 1913.

industry leaders for the first organizational meeting of the American Walnut Manufacturers Association. Clark in 1963 recalled that there "...were but four of us at the original meeting, John Penrod of Kansas City, Ray Pickrel of St. Louis, and Louis Maschel of Pekin, Illinois. Frank Purcell...joined shortly afterwards." The importance of this association of entrepreneurs can be recognized in the following account by Pickrel (1946) about other organizational meetings which occurred in 1913. The dangers of excessive price competition and the necessity of broadening the markets for walnut in the United States are the key elements in this interesting and revealing commentary.

> On June 1, 1913, at a meeting in the Jefferson Hotel in St. Louis, Mr. H. A. McGowen, John N. Penrod, and Ray E. Pickrel first discussed the possibilities of broadening the American consumption, as we had been fighting each other for the American business and each year the selling price was getting lower.
>
> We decided to call in other mills that made a business of manufacturing walnut and try to get them to meet in Chicago, at the Congress Hotel, on June 15, 1913. At this meeting five mills were represented, namely:
>
> | | |
> |---|---|
> | Geo. W. Hartzell | Piqua, Ohio |
> | H. A. McGowen Company | Louisville, Kentucky |
> | Penrod Walnut Company | Kansas City, Missouri |
> | Pickrel Walnut Company | St. Louis, Missouri |
> | Sanders & Egbert | Goshen, Indiana |
>
> Everybody was cautious but it was decided to call another meeting in early July, as we had decided to try to work on the furniture production and we had learned that they had a big meeting or sales period in Grand Rapids, Michigan, the first two weeks in July.
>
> [Our] July meeting was held in Chicago, and that night thirteen men went across the lake on the steamboat to Holland, then by electric line to Grand Rapids. Our agreement before leaving Chicago, was not to talk what our individual mills could do, but to talk walnut, which could be had in any quantity. The call in a body to nearly all the plants of importance in Grand Rapids, was a success and we were well received and most all plant managers were surprised that walnut was available as they thought it was all gone twenty years before and many had discontinued to try to use it on this account.
>
> Although mahogany and oak held the big use at that period and veneers of each could be sawn or knife cut, yet we walnut fellows were up against it, as only one mill had a veneer knife, this being Penrod Walnut Company, a small rotary which had been put in to cut an occasional order for Singer Sewing Machine Company. But we met the challenge and the veneers and lumber were furnished free for the sample line any factory would produce and we all agreed to guarantee to furnish any quantity needed in the future. The first user of walnut veneer was the old Phoenix Furniture Company for dining room tables and this was rotary cut.
>
> The trip was a success beyond expectation. No new stock or samples of walnut furniture were shown in Grand Rapids, in July 1913, except one straight chair and one rocker from Wisconsin Chair Company, Sheboygan,

> Wisconsin. But, in six months' time there were forty-eight manufacturers showing one or more suites in their January 1914 show in the furniture market.
>
> Our visit to [the] Grand Rapids market gave us the key to our advertising which was stocks, inventory and availability of large supplies of walnut. The good qualities of workability, beauty and features of the "best all purpose wood in the world" were established by using it. The game had been won... By the end of the first World War...our market was ready to step into. It was here in America and not in Europe.

Two decades later, the focus and objectives of the Penrod Company were beginning to shift. Albert J. Heidt (Fig. 3-9) was a good friend of Ralph Jurden and was an admirer of the exotic woods of South America, Africa, and the Far East (mahogany, satinwood, and rosewood). He interested Penrod in these woods to the extent that he located and bought the bankrupt Atlantic Veneer Company mill in 1933 in Norfolk, Virginia, after he became vice president of the Penrod organization. The mill had gone bankrupt in 1933 and had been a wholly-owned subsidiary of Hoosier Veneer Company of Indianapolis. Heidt supervised production at both Norfolk and Cincinnati. Mahogany was the source of about 50 percent of the Norfolk output from 1940 until the middle to late 1950s. Heidt was credited with going "...to the four corners of the earth and bringing home unusual and beautiful types of exotic cabinet woods" (Anon., 1936). The balance of the cut consisted of domestic species.

Albert Heidt served the industry well as president of the American Walnut Manufacturers Association and the Fine Hardwoods Association.

Figure 3-9
Albert J. Heidt was instrumental in moving the Penrod, Jurden and Clark operations to Norfolk, Virginia, in 1933.

Heidt was president of Penrod, Jurden & Clark when he died in 1945. Alex Schmidt, a vice president, died in 1948. J. C. Rodaffer, who had succeeded Jurden as president, was chairman of the board when he died in 1961.

Penrod company personnel, with some pride, referred to their business enterprise as: "A little company with a big, long shadow." This reflects on their historical leadership in introducing to customers in the United States a number of foreign woods such as Paldao (Dao) and Golden Narra from the Philippines. The company was active in tropical countries and loaded the first hardwood logs out of the Philippine Islands after WWII. The company also provided the Narra, African mahogany, and walnut veneers for the interior of the 1965 Cadillacs and Buick Rivieras as well as the veneers for Wurlitzer organs and Zenith TV cabinets. At one time during the 1960s, the company had an inventory of more than 35,000,000 square feet of exotic veneers, the largest in the United States. Penrod, Jurden & Clark maintained buyers in Mexico, Guatemala, and the Philippines and worked with agents in Germany and Africa. The company also acquired 20,000 acres of tropical forest in Nicaragua from which it produced logs for export until appropriated by a revolutionary government. The firm had also worked with the Ford Motor Company by veneering logs shipped to Norfolk from Ford's ill-fated Fordlandia rubber-growing venture in the Amazon. The logs were cut from a variety of species in the land clearing and road building process. Some time after the logs had been custom cut and warehoused in Norfolk, agents of Henry Ford unexpectedly arrived at the Penrod office and, acting upon Ford's instructions, demanded that all of the several million feet of sliced veneer be burned. Although the Penrod Company offered to buy these veneers, they were ultimately consigned to the company boilers with Ford picking up the costs of moving the veneers from warehouse to power plant.

The firm's location in Norfolk was seen as being geographically advantageous for eastern veneer markets and became company headquarters by 1950. Overnight delivery could be promised for most East Coast locations, including High Point, North Carolina. The company considered itself primarily a producer of veneers for the domestic market. It purchased mahogany veneer from OBER, a French marketing firm, and sold OBER domestic veneers for its European trade.

In 1981, Penrod was essentially a producer of domestic species and operated a "face plant" in Norfolk, where oak, cherry, and pecan veneers were spliced to form prefabricated sheets purchased for laminating to core stock by furniture manufacturers. Company officers were also beginning to think about future production and marketing strategies for the firm. Their nine-foot band mill had been destroyed by a fire the previous year and the entire site was continuing to be squeezed by the Norfolk International Terminals.

The officers and managers involved in the business in 1981 were:

Edward A. Heidt, Sr.	Chairman of the Board
Edward A. Heidt, Jr.	Chairman
V. L. Clark	Vice President-Retired
John T. Ricks	Vice President, Domestic Veneer Sales
G. Karl Bordner	President
Walter M. King, Jr.	Secretary and Counsel

An end of a long manufacturing era came in 1987 when Penrod, Jurden & Clark ceased to produce veneer. Virtually all the machinery was sold to a firm in New Brunswick, Canada. While domestic manufacturing came to an end, Penrod continued to enlarge its import/export division. The name was officially changed to "The Penrod Company" in 1987 and, by 1989, the company became the largest importer of plywood substraights and components in the United States. In 1989, Penrod purchased the World Trading Division of Masonite Corporation to augment its expansion in this arena. Penrod has regional offices in Hong Kong, the Philippines, South America, and various cities in the United States.

THE FREIBERG MAHOGANY COMPANY

The Freiberg Mahogany Company was located on the Mississippi river in a small community adjacent to New Orleans, Louisiana. The company was somewhat unique among the great coastal mahogany enterprises in its concentration on mahogany resources in British Honduras (Belize). Like the Williamsons and Penrod, Jurden & Clark, the business started in the Midwest and migrated to a coastal port for transportation and other cost advantages. The genius behind this strategy was Harry A. Freiberg, Sr., who was born July 13, 1878, to an affluent Cincinnati family. His father was a whiskey distiller of some repute. Although encouraged to go into the family business after young Freiberg was graduated in 1899 from Harvard University with a liberal arts degree, he declined to do so. Eventually, after a period of indecision, Freiberg was advised to go into the lumber business with a partner, Jim Rauh.

Freiberg's son, Harry Freiberg, Jr., was somewhat more specific: "My father put up some money and built a little mill near the [Crosley Field] ball park at the corner of Findlay and McLean Avenues" (Green,1968). The sawmill produced lumber from the locally available hardwood timber. These rough-sawn products were marketed in the Cincinnati area to a variety of businesses and secondary manufacturers. In 1903, his brother, Albert H. Freiberg, an orthopedic surgeon, suggested that Freiberg and Rauh specialize their production as he had done in the field of medicine. After making market inquiries and contacts in Chicago, Grand Rapids, and Philadelphia, Freiberg chose mahogany as the wood to specialize in. He admitted his "innocence" about buying logs in Mexico and Central America initially and freely admitted later that his first lessons were costly experiences. His first visit to Mexico involved buying mahogany logs owned by the U.S. Steel Corporation, acquired by it in a failed negotiation with the Mexican government. The logs were a compensation for the steel company's development costs.

Freiberg's son, in an interview, remembered how his father often told about his eventful trip which was to change the nature of the company for the next half-century. According to the story, Freiberg was dressed suitably upon his arrival in Mexico with a khaki "uniform" and high-top boots. He also carried a pistol as these were troubled times in Mexico. This was decidedly out-of-character for a man who was a patron of the arts. He was ultimately guided safely to the salt water lagoon where logs were stored. Satisfied with the size and apparent quality of the logs, arrangements were made to transport the logs to Mobile, Alabama, by ship and then by rail to Cincinnati. The volume of logs was such that a whole train of cars was required to transport the logs to Cincinnati. At the time of his Mexican visit, Freiberg was unaware that destructive toredo marine worms will quickly infest logs stored in salt water and he had not asked that any of the logs be rotated in the water so he could inspect the undersides of the logs which were riding low in the water. His error became apparent as the logs journeyed northward from Mobile. The large rotting worms left an almost unbearable stench along the tracks as the train moved Northward. When the logs were opened up at the mill in Cincinnati, the extent of the damage was readily apparent. Most of the lumber and veneer flitches resembled Swiss cheese. Some portions were salvageable, however, and were manufactured into materials sold to the Dornbrecker Furniture Company in Portland, Oregon. A small profit was made on the total transaction in spite of all the difficulties associated with this rather bizarre and involved business deal. Suitably impressed by the potential of the mahogany business and by encouragement from the Pullman Company in Chicago and furniture executives in Grand Rapids, mahogany logs began to flow regularly from Central America to the Cincinnati mill. Freiberg spent a considerable amount of time in the early years of this operation in British Honduras learning about log quality, the measurement of the volume and value of ponded logs, and the logistics of efficiently loading and transporting logs by rail and ship. In 1904 he formally opened an office in Belize, British Honduras, to better organize the company's tropical log procurement activities in the region. To service customers in the United States with both lumber and veneer, veneering machines were added to the Cincinnati plant's manufacturing capabilities.

By 1910, the mills and yard of the Freiberg Lumber Company covered an entire city block. The office was located on Poplar Street near McLean Avenue. An article in the Hardwood Record magazine described the physical plant and products this way:

> This mill contains a band saw, with edgers, trimmers, etc., and is well equipped with labor-saving appliances. In the veneer mill adjoining are

> veneer saws of the latest and most efficient type, and preparations are being made for installing slicing machinery. The company has the distinction of having brought to Cincinnati the largest cargoes of Mexican mahogany in the city's history. The yards also carry a selected stock of oak and Spanish cedar, with a line of general hardwoods (Anon., 1910).

A photograph taken in 1914 shows that the firm was a large industrial complex. The company name, Freiberg Lumber and Veneer Company, was conspicuously painted on both the water tank and buildings. Sensitive to the increasing cost of rail transportation between Gulf ports and Cincinnati, Freiberg began in 1915 to seek a site in the South accessible to ocean-going ships which could carry logs from Central America directly to the point of manufacture.

Equipment was found in Catahoula, Louisiana, a small lumber town east of Lafayette, where a large sawmill and planing mill had operated until all the accessible old-growth cypress trees had been harvested. The mill and its machinery were purchased at bargain prices. Freiberg shrewdly sold off unneeded machinery at a profit and moved the rest of the equipment to Harahan, Louisiana. The site was "...conveniently located alongside the levee of the Mississippi river..." near New Orleans on land leased from the Illinois Central Railroad (1918b).

The site had been used by Illinois Central when it ferried railroad cars across the Mississippi prior to the construction of a bridge across the river. A dike had been extended by the railroad into the river parallel to the bank to facilitate the ferrying process. Upon visiting the property, Freiberg recognized the potentials. A log pond was constructed by closing off the open-end of the "bay" formed by the dike. Incoming shiploads of logs were unloaded directly onto a tramway which lifted the logs over the dike and into the freshwater pond for storage. Eventually, the company had several ships which were constantly engaged in bringing mahogany logs from the tropics. No American mahogany mill was located as close to the source of supply. Freiberg had recognized the peril of dealing in countries "...where government is unstable, labor undependable, and distance is great between the source of supply and final destination" (Anon., 1920). The mahogany lumber and veneer cut at the New Orleans mill were shipped by carload lots to Cincinnati for distribution, completing the entire journey from the stump to the consumer, under the uninterrupted control of the company.

Lumber and veneer production in Cincinnati were phased out in 1916 soon after the decision was made to move to Louisiana. The executive business and sales offices, however, remained in Cincinnati until the company was sold in 1955 because of Freiberg's unwillingness to move from his beloved Cincinnati.

The day-to-day supervision of the New Orleans manufacturing facility was left in the hands of Jim Rauh, Freiberg's trusted partner and minority stockholder in the company. The plant superintendent, Frank

Czerney, was an experienced employee who provided the skills necessary to maintain the daily production schedule of 25,000 to 28,000 board feet for the sawmill. Highly regarded by Freiberg for his technical knowledge, Czerney also was responsible for the veneer mill's equipment which included a veneer slicer and 124-inch Capital staylog lathe which had been brought down from Cincinnati.

During World War I, The Freiberg Mahogany Company supplied mahogany used in ship construction as well as dimension lumber which was converted by other firms into propellers for aircraft built in the United States and by overseas allies. Earlier, in 1915, Freiberg had served as a member of the War Industries Board. Following the war, he became more directly involved in civic affairs and the organization of The Mahogany Association. In 1922, he became president of the association and two years later was one of the founders of the City Charter Committee in Cincinnati which was instrumental in breaking a corrupt political machine. Freiberg remained a director of the Committee until 1933 when he successfully ran for Hamilton County treasurer, an office he held for one term (Anon., 1949).

In the spring of 1937 Harry Freiberg, Jr., left Harvard University where he had been majoring in Modern European History. His father promptly sent him to New Orleans to complete his education. Business was not good in the post-depression and pre-World War II years of 1937-1940 for the company. In 1940, Harry Freiberg, Jr., was assigned to sell the company's products on the West Coast which he recalls as being another important "post-graduate" learning experience. In March 1941, he returned to New Orleans at age 24 to become manager of the lumber and veneer operation. This abrupt transition from salesman to manager had been precipitated by the failing health of Jim Rauh and the death of Frank Czerney.

One of the items the company continued to produce after World War I was mahogany airplane propeller stock. It had been purchased in large quantities by both the British and Japanese governments. Sensitive to the impending military situation and the fact that his younger brother, John, was serving in the Navy, Harry Freiberg, Jr., became determined that the company would not do any more business with the Japanese. Although pressured by the U.S. State Department to change this decision so as to maintain "normalized" business and diplomatic relations with the Japanese, the Freibergs remained adamantly steadfast. Harry, Jr., remembers with some relish a final meeting with a Japanese trading delegation in a New Orleans hotel in the late fall of 1941. The Japanese officials became enraged that the Freibergs would not furnish the propeller stock they had ordered. Threats were made that Japan would take their business elsewhere if the Freiberg company would not supply their needs. Harry calmly responded that in a democracy such as the United States they were free to seek the propeller stock wherever they so desired. And with that, he bade them "Good night." It was a small incident in the prelude before World War II but may account somewhat for the diligence and dedication to the company's war efforts which were to follow shortly thereafter.

During World War II, Harry Freiberg, Sr., served on the Industry Advisory Committee of the War Production Board. The company more than doubled the capacity of its mill to enable it to handle war-related demands. Freiberg supplied 1/16-inch and 1/8-inch Honduras mahogany veneers for the Higgins Company in New Orleans for PT boats and manufactured very thin veneers (1/110 inch) for the Haskelite Company in Grand Rapids which manufactured components for military gliders.

Orders were expeditiously handled, but not without difficulties due to the age of the equipment. A new slicer was purchased and the sawmill rebuilt. Production increased until the smokestack fell and destroyed the steam boilers in the powerhouse. Faced with a problem of maintaining shipping schedules and no short-term solutions to the power problem, Freiberg called his father in Cincinnati to report the catastrophe. Within 24 hours Harry Freiberg, Sr., had worked out a temporary solution by using his close personal ties with officials of the Illinois Central Railroad. A steam engine was immediately dispatched to generate the needed power. Steam was directly piped to a Corliss steam engine that powered the sawmill. Enough energy was supplied to cut the necessary number of flitches required to keep the veneer mill running at capacity. For this and other production heroics, the firm was awarded three Army-Navy "E" Awards for their production efficiencies in the war effort (Anon., 1949).

After World War II the sawmill and veneer mills were rebuilt again. The plant was considered to be one of the most modern mills in the country. It was equipped with two nine-foot band saws, an 18-foot slicer (one of the first Capital Machine Company slicers to be equipped with hydraulics), a 10-foot staylog lathe, and two Proctor-Swartz dryers. A warehousing innovation was also instituted whereby the crated veneer was stored vertically. Instead of boxing the veneers, a crate consisting of a top and bottom lid strapped together with metal bands was used to package the sliced flitch veneer (Fig. 3-10).

In this postwar period, the Freiberg Company cut 85 percent of its total production from Honduras mahogany. Their extensive timber interests in British Honduras led the Freibergs to become financially involved in the development of several other commercial ventures in that country. In 1952, Harry Freiberg, Sr., at the instigation of his son, John, established the British Colonial Air Lines. Harry, Sr., was then 74. This was the first scheduled passenger, freight, and mail service to all points in British Honduras, with some flights continuing to Chetumal, Mexico. John Freiberg managed log procurement and the other tropical business ventures.

During 1948 and 1949, the veneer mill ran day and night supplying mahogany veneers for the evolving TV industry. Veneers were also supplied to the Steinway piano company in large quantities. Beginning in

Figure 3-10
Veneer sold in domestic markets usually remains unclipped on the sides until it reaches the secondary manufacturer. After sample sheets are removed, veneer from individual flitches is crated and warehoused vertically prior to shipment.

1950, substitute materials such as printed hardboard began to be used in TV cabinets. The Freibergs decided to sell the plant and airline in 1955 to Texas Industries, Inc., a cement manufacturing firm based in Dallas, Texas (Anon., 1969). Harry, Jr., and John Freiberg left the Freiberg Mahogany Division of the Texas-based company on February 28, 1958. The mill continued to operate until 1960 when it closed and Hurricane Betsy destroyed the buildings in 1965. The salvageable equipment was sold shortly thereafter. One of the dryers was sold to Central States Veneer Company of Paducah, Kentucky, and the staylog lathe was sold to the veneer mill located at Pleasant Hill, Missouri.

Harry Freiberg, Sr., died in Cincinnati on May 16, 1969, at age 90. Until his death, he continued visiting his office on a daily basis to manage a portfolio of securities acquired after sale of the Freiberg Mahogany Company in 1955.

Chapter **4**

INDIANA'S EARLY VENEER MILLS: 1850 - 1920

The location of hardwood face veneer mills was and is predicated almost entirely on the regional availability of veneer quality timber, cheap transportation, or some combination of resource availability and transportation. Prime market areas for the product also influence the location decision, as does the nature of the local labor pool and business climate. The early veneer mills in the Midwest were particularly attuned to the resource base because of the weight and bulk of the old-growth logs which had to be moved from forest to mill using animal traction power along primitive roads.

Indiana and its neighboring states were particularly blessed with the highest quality and generally accessible walnut and oak timber in the United States. As Midwestern cities grew in population and wealth, the industrial use of hardwood lumber and veneers increased. The manufacture of veneer often was an adjunct to a lumber mill operation when the demand for fancy furniture veneers was recognized. In the late 1800s and early 1900s sawn veneers using a special segmented saw blade were commonly produced by lumber companies. Indianapolis in 1910 was touted as "...the largest veneer producing city in the United States" in a full-page ad in Wood Craft magazine. Nine Indianapolis mills were listed. All produced sawn veneers while three were also manufacturing sliced veneer. The listed Indianapolis mills of 1910 were:

National Veneer & Lumber Company
Long-Knight Lumber Co.
McDonald Veneer Co.
Advance Veneer & Lumber Co.
Talge Mahogany Co.
Central Veneer Co.
Capitol Veneer Co.
Indiana Veneer & Lumber Co.
Gemmer Veneer & Lumber Mfg. Co.

The 1910 list is, probably, only a partial list of the Indianapolis veneer firms existing at the time. Only one of these, Indiana Veneer Company, has remained or survived to the present time. National Veneer & Lumber Company moved the last of its Indianapolis equipment to its Seymour operation in 1952. The Williamson Veneer Company predated these Indianapolis firms by about 60 years and is discussed in detail as an early East Coast mahogany mill having moved to the Baltimore area about 1900. Many other Indianapolis mills were lost to city development planning and business "panics."

The veneer mills of Edinburgh formed another important core of hardwood veneering activity in Indiana. Promotional signs at the city limits in 1989 still proclaimed Edinburgh as the "Veneer Capital of the United States." The veneer mills along the Ohio river at and near New Albany formed the other major concentration of Indiana veneer mills. The New Albany mills matched up with a Kentucky set of mills across the river in Louisville. In 1989, 15 hardwood face veneer mills were operating in Indiana, the largest number of mills in any state.

The oldest mills still operating in Indiana are the Indiana Veneer Company (1892) located in Indianapolis and the David R. Webb Company (1894) in Edinburgh. Both are now foreign-owned and are discussed in Chapter 8 with other hardwood veneer mills having foreign investors or owners. The remaining mills located in Indiana are individually examined chronologically on the basis of establishment date.

ROBERTS AND STRACK VENEER CO., INC.

John Nye Roberts was associated with a number of veneer manufacturing businesses which bore his name over the long course of his life. His son, John R.(Jack) Roberts, during two interviews provided much of the information which follows.

John N. Roberts was born in Richmond, Indiana, on March 28, 1863, where his father was a respected businessman and farmer. As a young man, John Roberts briefly attended Earlham College before moving to Indianapolis to become a skilled cabinet maker at the Indianapolis Desk Company. It was probably not coincidental that John's grandfather was president of the desk company where he was employed. Using his skills and understanding of the wood products business, John Roberts received financial backing from John Lilly, a brother of Eli Lilly, to form the Lilly and Roberts Veneer Company located on Roosevelt Avenue in Indianapolis. The partnership produced lumber and sawn veneers. The principal customer for the white oak veneer they produced was the Singer Sewing Machine Company cabinet plant at South Bend, Indiana. Roberts, in 1883, began cutting veneer with a rotary machine leased from the Sewing Machine Cabinet Company located in northeast Indianapolis and adjacent to the Lilly and Roberts mill. Roberts at this time was well acquainted with Marshall Dodd Williamson who had also rented a section of the sewing machine cabinet plant to cut veneer with the cabinet company's staylog machine.

Roberts recognized the wasteful and time consuming efforts required to manufacture sawn veneer and the difficulty of producing quality veneer with rotary machines. Rotary veneers tended to be inferior because it was difficult to "hold the grain" and to get pressure on the flitch as the veneer was cut. The Singer Company also preferred to purchase quartered (rift) oak veneers for its cabinets. Determined to produce quality veneer efficiently, Roberts designed a revolutionary machine in 1898 to slice rift oak veneer. The machine was constructed for him with the help of Louis Koss, proprietor of the Capital Machine Company in Indianapolis. It was delivered to the Lilly and Roberts firm on December 17, 1900. A patent application was subsequently filed on May 13, 1901, and granted on July 8, 1902, as U.S. Patent No. 704,213. As the application states: "Be it known that I, John N. Roberts,...have invented a new and useful Veneer-Slicer..." Even Roberts probably didn't fully recognize the significance of his invention at the time. The mechanical principals of this slicer can be found in every veneer slicer manufactured in the United States since then. The proto-type machine weighed 85,000 pounds and sliced the veneer as the flitch passed horizontally and below a mounted, adjustable knife.

After extended trial runs, necessary to perfect the operation of the machine, the slicer met all expectations and began producing exceptionally fine veneer. Samples drawn from the warehoused inventory were taken to the Singer Company in South Bend by John Roberts. After the samples had been tested and made into cabinet components, an agreement was reached calling for Lilly and Roberts to supply the Singer Company with sliced quartered oak veneers on a continuing basis. Greatly elated, Roberts returned to Indianapolis by train to discover that the mill and its veneer inventories had been destroyed by fire in his absence. Overextended financially and unable to raise additional capital, Lilly and Roberts were forced to sell their remaining assets and go out of business in 1903. The veneer slicer was sold to a southern veneer manufacturer.

Undeterred, John N. Roberts in 1904 found a new partner, James Connor, a prosperous New Albany book store owner. The new veneer firm was located at lower 9th and River Streets in New Albany on land purchased from Mr. I. F. Force who had acquired the old ship-yard property after the Civil War. It had been called the "Hickory Mill" by the nearby residents presumably because Force was an ax-handle manufacturer. Roberts is said to have been attracted to the New Albany area because of the available timber, the local furniture markets for veneer, and because of the reputation of the local craftsmen. A Capital Machine Company bookkeeping entry in fine Spencerian handwriting noted that "One horizontal veneer slicing machine" was sold to John N. Roberts on August 16th, 1904. This machine was similar to the earlier model although a number of design changes had been made. This plant remained in operation as the Roberts and Connor Veneer Company until 1919 when it was sold to the [Ray] Pickrel Veneer Company of St. Louis, Missouri (Adams, 1963).

The year, 1914, was significant because Roberts' 18-year-old son, Jack, entered Purdue University. He graduated in 1918 with a degree in Mechanical Engineering. After serving in the Navy as an Ensign during the waning days of World War I, Jack returned to New Albany to find that his father was anxious to reenter the hardwood business. Ten acres were acquired in Clarksville with work beginning on the band mill in 1920. The resulting business was known as the John N. Roberts and Son Company. This company also engaged in the veneer business by having some of the company's select logs custom cut at area veneer mills.

The original horizontal slicer, which was delivered to Lilly and Roberts in 1900 and had been sold to a firm in the South after the fire of 1903, was repurchased by Roberts in 1926. Norman Harding, a future owner of the Evansville Veneer Company, had learned that the machine had been through another fire and relayed the information to John N. Roberts during a visit to New Albany. Harding indicated that it could probably be purchased at a reasonable price. Roberts left immediately and purchased the fire-damaged machine. Major portions of the machine were shipped to the Clarksville site for reassembly except for the "slides" which were sent to the Capital Machine Company for repair. A flitch table of newer design was added while the machine was being repaired. An eight-foot Capital staylog lathe was purchased in 1928 to round out the company's veneering equipment.

In 1947, Frank M. Adams, then employed by Roberts and Strack, had an inscribed plate made which was fastened to the horizontal slicer. The inscription read: "Veneer slicer, designed by John N. Roberts, Built by Capital Machine Co., 1902." The 1902 date on the plaque was in error by two years and likely referred to the patent date of the Roberts-designed machine. The machine produced sliced veneer until 1962 when it was sold for scrap metal after more than six decades of service.

The 1929 business crisis and devastated domestic economy weakened the financial position of the company. As a result, a controlling interest (54 percent) in the company was sold to Samuel E. Stout who owned several hardwood plywood firms in New Albany and Louisville including Hoosier Panel and Veneer, Indiana Veneer and the Crescent Panel Company. Stout then "gave" this business interest to his son-in-law, Edwin O. Dulaney, who incorporated the business as the Roberts-Dulaney Veneer Company, Inc. This arrangement apparently lasted only a short time.

Clarence E. Strack, who had managed Crescent Panel Company for Stout in Louisville, was appointed in 1931 by Stout to "run the business." On December 31, 1931, the "firm" was reincorporated when Strack bought into the company after the death of Samuel Stout. The principals who were involved with this new company included: John N. Roberts, John R. Roberts, Clarence E. Strack, and E. O. Dulaney. Strack became president of the company in 1932, a position he held until his death on January 10, 1970 at age of 76. A memorial resolution of the company in the possession of the Roberts family noted that Strack had attended business college and had studied law. He was cited as being eminently fair although he had held a tight rein over the operation of the company and was firm with the officers as well as other company employees.

After this reorganization, Jack Roberts began buying timber for the mill and travelled extensively in the Midwest as well as the rest of the country working with log producers and buying some standing timber. Mahogany was purchased mostly out of East Coast ports such as Norfolk, Boston and New York City although some was purchased through brokers in London and from brokers in Pensacola, Florida. After the demand for mahogany veneers subsided, black walnut became the major species sliced. During World War II, Sitka spruce timber was purchased in Oregon for shipment to Clarksville where it was sliced into aircraft and glider veneers. Oak, tulip poplar, sweet gum and mahogany were also cut.

John N. Roberts died on August 27, 1954, after a distinguished and productive career in the veneer business spanning more than 50 years. A laudatory eulogy in the New Albany Tribune credited Roberts with "...the development of the veneer industry which has brought world fame to the New Albany area...He knew and practiced the philosophy of hard work but with him the work was tempered with a real love for his profession. His keen enjoyment of life continued throughout his ninety-one years" (Anon., 1954).

Jack Roberts maintained an active role in the firm until 1978 when he retired. John N. Roberts' grandson, Thomas Hale, also participated in the business after graduating from Purdue University as a Mechanical Engineer in 1942. He left Roberts and Strack in 1957 to pursue other business interests in the veneer industry. He is best known for the Hale veneer dryer and a 1951 patented staylog veneer machine designed for cutting rift-oak veneers. This machine became a standard for the industry and was manufactured by the Capital Machine Company of Indianapolis.

Maurice C. Strack, Clarence's brother, was a 1922 graduate of Purdue's School of Mechanical Engineering. He served the company until his death in 1967 at the age of 68. Edward K. Lovelace, his successor, became associated with the firm after his marriage to Clarence E. Strack's daughter, Mary K. Strack, in 1955.

In 1970 there was a "changing of the guard" at Roberts and Strack after Clarence Strack died. E. K. Lovelace was elected president and chairman of the board of directors; Gilbert E. Seger, vice president, treasurer, and general manager; and Richard L. Mattox, secretary and company attorney. John R. Roberts continued to serve the company as a member of the board of directors until 1983. E. K. (Ken) Lovelace, Jr., entered the firm somewhat later and was the firm's representative in the company's sales office in High Point, North Carolina before returning to a management position with the firm. The name of the firm was changed to Roberts and Strack Veneer Company on June 1, 1976.

Although Roberts and Strack had in the 1960s and 1970s cut nearly all of the domestic hardwood species and had specialized in walnut, the principal species cut for veneer in the early 1980s was pecan, followed by lesser quantities of white oak and red oak. Pecan logs were purchased from loggers and brokers in the mid-South states of Mississippi, Arkansas, Louisiana, and Texas. Some of the best pecan was purchased in southwestern Indiana in the Wabash river valley.

The firm produced about 70,000,000 square feet of veneer annually (none for the export market) and 600,000 board feet of lumber in its more productive years. The veneer cutting equipment consisted of three Capital Machine Company slicers and one Capital staylog lathe.

HOOSIER VENEER COMPANY, INC.

Harry E. Daugherty was born on November 3, 1870, in Ladoga, Indiana, where his father, Josiah, was in the flour milling business (Fig. 4-1). After Harry completed his education, he was employed by the Ladoga bank for several years until he and partners purchased the local veneer mill. The mill at that time was producing sawn oak veneers for sewing machine cabinets. At age 25, Harry married Letha Scott of Ladoga and their first child, James Carl, was born in Ladoga on July 8, 1900 (Fig. 4-2). The other children born to this family were Anne, Matilda, Andrew Jackson (Jack), and Letha (Anon., 1929a).

Mrs. James Carl Daugherty described the Ladoga mill as being located adjacent to the railroad tracks in order to expedite the handling of incoming logs and the shipping of veneers to furniture manufacturing centers. The sparks from coal burning railroad engines, however, provided a continuing threat to businesses located along the tracks. In 1911, engine sparks fanned by a high wind and fueled by sawdust and other

Figure 4-1
Harry E. Daugherty formed the Hoosier Veneer Company in 1911 after his Ladoga veneer mill burned.

Figure 4-2
James Carl Daugherty provided leadership to the Hoosier Veneer Company when his father died in 1938 and remained president of the firm for 33 years.

waste materials resulted in a fire which totally destroyed the mill buildings. The heavier machinery was spared from extensive damage by the rush of the fire through the shed-type structures.

As the fire cooled, Daugherty and his partner and boyhood friend, Thompson (Tom) VanMeter Ashby, were already planning how to cope with the situation. They knew that the Advance Veneer Company mill located at 3321 Massachusetts Avenue in Indianapolis was standing idle and, after a trip to Indianapolis, the decision was made to lease the property, move the salvaged equipment to Indianapolis, and form the Hoosier Veneer Company. The equipment moved included a slicer which had been recently acquired. Karl Bordner, one of the former presidents of Penrod, Jurden and Clark, related in an interview that his father told him on several occasions of the arduous nature and the difficulties encountered in moving the equipment 40 miles to Indianapolis entirely by horse and wagon. The "...greatest single ordeal [was] the successful move of the huge, heavy slicer." Bordner said that his father should have known all about moving the slicer "because my Dad was the man who drove the wagon." The Daugherty-Ashby partnership lasted until 1912 when Tom Ashby sold out his interest to Daugherty and moved to Jackson, Tennessee, to form the Ashby Veneer and Lumber Company.

Hoosier's first Capital veneer slicer was bought in 1913 and a staylog rotary lathe was added in 1914. Dungan Slat Conveyor Dryers were purchased in 1913. These were used until 1927 when they were replaced by a Proctor and Swartz dryer. Daugherty remained president of his firm until his death in 1938 when his four surviving children inherited the business. He is remembered by Percy Simmons as "a spectacular person, a little flamboyant for those days" and by Mary Daugherty as "...an autocratic man who called the shots. He never was a joiner and never went to a social party connected with veneer industry association meetings."

Nevertheless, Harry made many good friends in the veneer business in his extensive travels in the Midwest, East Coast, Europe and Central America. He often traveled with Frank Montgomery who also made trips to the Far East for the company. Harry never drove a car but always had a car and driver standing by. He was particularly proud of his office building in which each office was paneled in a different veneered species with matching desks. Since he was a proud man, he was particularly incensed that his business was thrown into receivership in the 1930s by the American Fletcher Bank of Indianapolis. The company's financial difficulties had been exacerbated by Harry's unswerving faith in the veneer business which had prompted him to continue importing logs from Australia and Europe after the market crash of 1929. At some sacrifice the business was ultimately saved and restored to financial health.

Jack Daugherty, who was born in 1908 was probably the more adventuresome of the two brothers. Mary Daugherty recalls that there never was "...a finer man than Jack Daugherty." At the age of 21 in 1928, his father sent him to Australia on the first of several far flung adventures searching for exotic timber. A later trip to Mexico in 1930 was made just after he graduated from Wabash College and at a time of political unrest and revolution in rural areas. It was made in the hope that Jack could track down a stand of "gigantic" walnut and oak trees rumored to exist somewhere on the West Coast of Mexico. He later wrote about the trials and tribulations of his successful quest which supplemented a series of letters he had written in 1930. Jim Burrell edited the lengthy manuscript into an interesting and revealing series of articles which appeared in five issues of Plywood and Panel Magazine. The series was published "...as a memorial to the Daugherty's and all the other industry entrepreneurs of their day who, against heavy odds, went 'that extra' mile seeking the best timber for better veneers" (Burrell, 1979-1980).

When Jack reported back for work in Indianapolis, one of his first jobs was to close down a leased operation at the Drinkerd Veneer Mill just a few blocks down the street from Hoosier Veneer and across the street from Central Veneer, Inc. In 1932, Jack was sent by Harry Daugherty to build a new mill (Atlantic Veneer Company) at Norfolk, Virginia. This was not a propitious time to start up a new business of any kind. However, it was perceived correctly that mahogany and other African and South American species were beginning to vigorously compete with domestic veneers and that the furniture industry was moving to the South to take advantage of non-union labor, low taxes, and plentiful timber resources. Lowell F. (Jake) Ross was transferred to Norfolk from

Hoosier's Indianapolis mill as its veneer cutter. During the short time that Atlantic Veneer operated, the mill cut a large volume of Oriental walnut which Jack Daugherty had located in Australia. Some European woods were veneered, as well, in the Norfolk mill. Frank Montgomery was also transferred to manage this subsidiary company. However, the Atlantic Veneer operation could not withstand the deepening depression for veneers and went into receivership. The company was ultimately sold by Harry Daugherty to Penrod, Jurden and Clark Company, a Midwest firm also looking for an East Coast location.

Jake Ross was hired by the new owners and in the next 28 years advanced from veneer cutter to manager of production by establishing an enviable reputation for producing high-quality veneers. In retirment, he served as a veneer manufacturing consultant and trouble-shooter for the Capital Machine Company. Frank Montgomery also remained and operated the Penrod mill until he retired (Anon., 1973c).

Jack had never gotten along with his father in business matters and this may account for his frequent assignments away from Indianapolis. He also "...didn't see eye to eye" with his brother, James Carl, and sold his inherited interest in the Hoosier Veneer Company to Carl soon after their father died in November 1938. Carl assumed the leadership role for the company at this point and continued as president of the firm for the next 33 years. Jack remained in Norfolk and formed the Virginia Wood Products Company. This company imported exotic logs, bought domestic logs, and sold both types to veneer manufacturers in the Midwest and Southeast.

In 1963, Bill Schanbacher joined the company. Schanbacher had close ties with the hardwood industry after World War II and had been employed by the Monteath Veneer Corporation, the National Veneer Company of Indianapolis, Willard Hawes and Company in Norfolk, and the H. G. Kolbe Company of Virginia. The latter two companies were in the business of procuring African timber for American markets. Virginia Wood Products, Inc., proved to be a successful merging of Daugherty's domestic business ties with Schanbacher's foreign procurement capabilities. When Jack died in 1970, Schanbacher purchased the company and managed it until his death in 1982.

William C. Linn, a graduate of William and Mary College, was employed by the telephone company in Norfolk when he met Matilda Daugherty, Harry's daughter (Fig. 4-3). His interests shifted to the veneer industry after their marriage. Linn first worked for the Monteath veneer sales organization for two years. Monteath custom cut veneers at the Penrod mill and Linn learned about the production end of the veneer business at the Norfolk mill before moving into a sales position with the company in New York City. After the death of Harry Daugherty in November 1938, Linn joined the Hoosier Veneer Company in a sales capacity. He was admirably suited for this position and presented a polished, suave,

Figure 4-3
William C. Linn, a son-in-law of Harry E. Daugherty and a suave and polished gentleman, sold Hoosier Veneer Company's veneer in the furniture manufacturing belt of the Southeast before becoming a company vice president in 1942.

and southern-gentleman image for the company in the Virginia and North Carolina furniture territory. He covered these markets until 1942 when he assumed a management role with the company in Indianapolis becoming vice president for sales in 1944.

During World War II the Hoosier Veneer Company contributed to the war effort by producing spruce, gum, and birch aircraft veneers. Spruce veneers cut by Hoosier were used in the construction of Howard Hughes' "Spruce Goose" seaplane which may have been the largest airplane ever constructed. Other veneers were sold to the Roddis and Haskelite companies and used by them in constructing troop-carrying gliders.

Linn remembered that the mill was equipped in 1942 with two 12-foot Capital slicers and an 8-foot lathe. A 10-foot staylog lathe was added later. Two mechanical dryers were used including one which the company had fabricated when it had phased out the slower slat conveyor method of drying veneer.

Robert Bolles was the other Harry Daugherty son-in-law in the business. Bolles attended Wabash College and met Letha Daugherty through her brother, Jack. Bolles became a very adept log buyer and eventually was appointed as vice president for log procurement and remained in this position until his death in 1958. Elwood Douglas completed the roster of company officials and was the long time secretary-treasurer. Don Long assumed this role when Douglas died. Eugene J. Kelley was the company's well-respected salesman for many years.

The Hoosier Veneer Company was a strong supporter of trade associations and contributed to both the American Walnut Manufacturers

Association and the Veneer Association for several decades. Both Carl Daugherty and Bill Linn were officers in these associations.

In 1967, Carl Daugherty bought out his two sisters. Bill Linn stayed with the company until 1969 when he retired from the business. In 1971, the Hoosier Veneer Company was liquidated. Carl had become increasingly debilitated by diabetes, the mill's machinery was antiquated and the mill site stood in the way of Interstate I-70. Carl Daugherty died in 1977 and is fondly remembered as a gentleman, historian, and a producer of quality veneer. Don Smith was the last chief executive and, as a lawyer, arranged for the liquidation of the company assets.

CENTRAL VENEER, INC.

Central Veneer, Inc., was founded about 1908 by H. J. (Jap) Barnard. The mill was sited at Winter Avenue and the Beltway Railroad in Indianapolis. Veneers were sliced with a used, electrically driven, 1906 model Capital Machine Company slicer. The plant was considered to be "...about as modern and convenient as it is possible to make a plant these days." The company made a specialty of sliced quartered oak veneers and reported "...a very nice business" (Anon., 1910). By 1921, the principal line of products included sliced mahogany, walnut, oak, and figured gum and rotary cut gum and yellow poplar.

The output of the plant in 1919 was 16 million square feet and 20 million in 1920. Sawed oak veneers were custom cut "at a nearby mill" (probably the Billy Drinkerd mill across the street) and not included in the production figures cited (Rohr, 1921). Barnard's emphasis on efficiency continued through 1925 when he completed "...his latest warehouse which is the acme in convenience and efficiency...and may well serve as a pattern to other manufacturers" (Anon., 1925a). In 1926 Charles Lykins bought the firm and three years later Lykins' three sons (Cecil, Leighton and Lawrence) bought the company from him just prior to the general business collapse of 1929 (Fig. 4-4). One result of this economic disaster was the closing of Central Veneer from 1929 until the economy recovered prior to World War II. As Harold C. Davis remarked in his office: "The veneer business hasn't always been a good stable business." Davis joined the firm in 1945 and began buying stock in the company. Harold's father and grandfather had also been in the veneer business during the industry's early days in Indiana.

Josephus Davis, Harold's grandfather, was one of the original investors in Indiana Veneers, Inc., in Indianapolis. Minor Davis, Josephus' son, was born about 1883 and was employed by Indiana Veneers, Inc., in 1904 when he was 21 years of age. He later became part owner of H. J. Sanders and Son which operated a small veneer mill at Whittaker, Indiana, between Gosport and Martinsville. When the Sanders' mill burned down, Davis had veneer custom cut at the Billy Drinkerd Veneer mill. The cut veneer was dried at Central Veneer. Minor Davis, at some point

Figure 4-4
Cecil Lykins poses with African mahogany logs in the yard of Central Veneer, Inc.

in his early career, had worked for David R. Webb in Edinburgh or possibly the Muloda Veneer Company which was the predecessor company. It is known that Cash Davis and two others organized the Muloda Veneer Company in 1895 (Otto, 1972).

Minor Davis operated his own sawn veneer business at Gosport, Indiana in 1920 or 1921. The building was owned by his father, Josephus, which is an indication that the father may have been in the veneer business there before investing in Indiana Veneers, Inc. After the business was liquidated in 1926 or 1927, Minor went to Morefield, West Virginia, to assist in the organization of a veneer mill there. The mill fell victim to the Depression and was closed prior to Minor Davis' death in West Virginia on March 30, 1932. The mill was later purchased by the Thompson Mahogany Company of Philadelphia, Pennsylvania.

Harold C. Davis, Minor's son, was born in 1909 and at a very early age began to gain veneer-business experience which was to serve him well when he joined Central Veneer, Inc. Davis, a man with a strong personality, treasured his privacy. Therefore, the details of his interesting formative years in the veneer business have been omitted at his request. In January 1960, the firm was reincorporated by Davis and Cecil Lykins. Davis became sole owner in 1964.

The veneer mill during the early 1980s was operating the 1906 slicer on a two-shift basis producing white oak veneers for export. A 1954 slicer was inactive because it wasn't wide enough to handle all random-length flitches. A staylog lathe was also inactive and had been for the previous 15-18 years. Veneer production was constrained by inadequate drying capacity.

Davis, who became president of Central Veneer, Inc., in 1960, was eventually assisted in the management of the business by his two sons, John C. and Richard. John, vice president and secretary of the firm, was responsible for log buying and managed the yard and sawmill. Richard, vice president and treasurer, handled sales and the export market. An overseas salesman provided the contact with foreign veneer purchasers (Anon., 1960).

Logs were purchased regionally although Davis expressed the desire to buy just Indiana logs because they are "much better logs." Davis also reflected on the age-old problem of buying "Indiana" logs which may have come originally from Kentucky, Tennessee, West Virginia, and Illinois. The firm employed about 60 people and had an estimated annual production volume of about 20,000,000 square feet of veneer. The mill, which had essentially been frozen in time since 1926, ceased operations in 1984.

NATIONAL VENEER AND LUMBER COMPANY

At the Fourth Annual Meeting of the Indiana Hardwood Lumbermens Association (IHLA) in 1903 it was decided to nominate Burton F. Swain as its candidate for the presidency of the National Hardwood Association (Fig. 4-5). In a letter to the Hardwood Record in Chicago the secretary

Figure 4-5
Burton F. Swain was a highly respected hardwood lumber executive prior to purchasing the National Veneer and Lumber Company in about 1910.

wrote: "The candidate selected by the IHLA is Mr. B. F. Swain of D'Heur & Swain Lumber Company, Seymour, Indiana. Although Mr. Swain is a comparatively young man, being 38 years of age, his firm is well known throughout the United States and Europe as a high-grade business institution. He is an eminently successful business man, the result of 'pluck and push' and those applications to the manufacturing and marketing of hardwood lumber." This was the early flowering of an important hardwood manufacturing dynasty in Indiana (Anon., 1903).

Swain had become associated with Clyde J. Roach in 1892 and became a partner with him when the Seymour lumber mill was reorganized to become the Swain-Roach Lumber Company sometime after 1903. The large band mill was operated until 1935 when it was sold and eventually moved to West Virginia. Swain was also involved with the Talbert Lumber Company, probably located in Shelbyville, Indiana.

Another of Swain's ventures was the purchase of the National Veneer and Lumber Company, organized in Indianapolis in 1906 and located at 1635 West Michigan Street and Beltway Railroad. Its 1908 letterhead indicates that it "...sawed, sliced, and rotary [manufactured] oak, poplar, basswood, and mahogany." "Quartered Oak Veneer" was prominently printed in large type in the center of the letterhead. Therefore, it can be assumed that the firm's first slicer, a 201-inch Capital, was bought in 1907 (Lee, 1981). The activities of this Indianapolis mill have largely been lost over time.

B. F. (Burt) Swain, Jr., emerges from the past in 1925 as treasurer of National Veneer and Lumber Company, a post he held until 1933 when he succeeded his father as company president. Other company officers included Clyde J. Roach, vice president; J. B. Marsh, secretary and general manager; and Burrell Wright, treasurer. All had extensive experience with the company ranging from 14 to 44 years. Roach, who had served the company longest, was credited with being one of the best informed veneer men in the industry and controlled about 25 percent of the stock in the reorganized company.

When the Roach family members sold their stock to the B. F. Swains in 1939, Elizabeth S. Patterson was elected company vice president. Two years later, another reorganization took place when Burt bought the business from family members. Mary, his wife, was elected vice president. L. J. Heyne replaced J. B. Marsh as secretary. These organizational changes were part of a business strategy to recapitalize the firm and build the new veneer plant in Seymour. Funds were secured on September 22, 1941, to begin work on the project which was soon halted by the onset of World War II. Because of changed economic conditions after the war, it was necessary for the Swains to secure additional funding in 1945 to complete and equip the new veneer mill. On November 12, 1946, the board of directors authorized the purchase of a 16-foot Capital slicer, a Capital staylog lathe, and a Proctor-Swartz dryer. A particle board operation was started in 1947, one of the first in the country. Veneer was produced in Indianapolis until 1951 when Earl S. MacDonald died. He had been a company vice president and manager of the Indianapolis operation. Veneer operations were later consolidated at the Seymour location the following year. Earlier about one-half of the veneer machinery in the Indianapolis plant was dismantled and moved to the company's location in Seymour which had been the site of the Swain-Roach Lumber Company. The 1907 slicer was sold and moved to Campeche, Mexico. In 1953, the Indianapolis plant was sold.

Burton F. (Ted) Swain III was elected to the board of directors in 1948 and his younger brother, D. Billings Swain, joined the firm as a director in 1953. These two men, both Purdue University graduates, were instrumental in maintaining the competitive edge through mill expansion and modernization coupled with innovative production techniques. Larry Richardson, a former sales manager, has said: "The edge that National Veneer has over other veneer companies is that [they are small enough] to control what [they] do but still large enough to...sell to all [domestic] markets" (Lee, 1981).

Mill operations between 1976 and 1981 were improved by adding new equipment and expanding in areas such as debarker, dryer, computer, shipping room, and sampling area, and by adding 20,000 square feet of new warehouse. A unique staylog lathe, with flitches mounted back to back, is capable of slicing up to 162 sheets of veneer per minute which contributes to the firm's potential of producing 425,000 square feet of veneer per day. The off-bearers stack the veneer sheets from the two flitches together because of the speed of production. During the drying process the veneer sheets are reassembled as cut from the two individual flitches.

The normal complement of workers is about 90 when running one shift on the two upgraded Capital slicing machines purchased in 1946 and two shifts on the two Hale dryers. Electric eyes at the output ends of the Hale dryers measure the widths of dried sheets as they exit. This information is sent directly to a calculator programmed with the length of the flitch from which the veneer was cut. Total square footage of veneer produced from each flitch is then automatically computed and recorded. The company's business computer not only keeps track of production but also supplies routine accounting functions for the firm.

The principal species cut in a recent year were white and red oak, pecan, hickory, and cherry. Some walnut and maple were also being cut. Logs are purchased from timber producers and sawmills. This is supervised by one full-time log buyer. The processed veneers reach markets through the Seymour sales room and sales representatives in High Point, North Carolina, and Memphis, Tennessee.

E. Rex Early, vice president in charge of production, emphasized the company's no waste attitude and the use of veneer residues in the Swain's particle board plant and the conversion of wood waste into heat energy for heating the plant, log vats, and the Hale dryers. J. David Swain was the fifth generation to join a Swain family wood products business. Another son, David M. Swain, followed in 1984. In 1982 National Veneer and Lumber Company was formally split off from Swain Industries, Inc., with D. Billings Swain remaining as president of National Veneer while B. F. Swain III assumed control of Swain Industries and the particle board operation.

EVANSVILLE VENEER AND LUMBER COMPANY

The names of those who founded the Evansville Veneer Company at 100 South Kentucky Avenue in 1904 remain lost in the distant past. However, Charles W. Talge and George O. Worland are credited prior to 1915 with the skills that achieved for the company its "...fame through [their] untiring and continued efforts." Talge, the president, was of Dutch ancestry and eliminated the "von" before his cognomen "...because he was always too much rushed with business to have it pronounced that way." Worland, secretary-treasurer, was said to be "...a raging battleground for shop efficiency, the production of high-grade veneers and the capture of large orders on a quality basis." Together, they succeeded in establishing a reputation for making sawed, sliced, and rotary-cut veneers utilizing a finely-equipped plant, efficiently operated (Anon., 1915).

On February 5, 1925, a $100,000 fire which had started in the drying room destroyed manufactured veneer and a "...good deal of fine mahogany lumber" stored in the building. Two other large buildings were saved through efforts of firemen and Robert R. Williams who had earlier become owner and president of the company (Anon., 1925). Between 1925 and 1932 the company went through some difficult times ending in bankruptcy. Assets were put up for sale early in 1932 and all bids were rejected by the court. In June, 1932, the assets were readvertised. Included were machinery and equipment, a modern brick veneer plant, four-million square feet of veneer and two-million board feet of lumber (Anon., 1932). They were purchased by Norman G. Harding at the auction held on September 15, 1932 (Fig. 4-6). His daughter, Patricia Harding Snyder, believes that at the time of the bankruptcy in 1931 or 1932, Robert R. Williams had changed the name of the company to the Indiana Tie Company. George O. Worland was mentioned as a previous owner. Harding immediately changed the name to Evansville Veneer and Lumber Company. His wife, Blanche, was recorded as being secretary-treasurer. At that time the veneer cutting equipment consisted of a slicer, a rotary lathe, and a large battery of veneer saws.

By 1937, the Evansville Veneer and Lumber Company had restored luster to the company and was "...one of the country's largest veneer producing plants." Harding was assisted in the enterprise by his son, Norman, Jr. American walnut, oak, and cross-banding poplar veneers were the production mainstays (Anon., 1937).

Figure 4-6
The assets of the Evansville Veneer and Lumber Company were purchased by Norman G. Harding in 1932. He managed the company until his death in 1948.

Prior to the Great Depression, Evansville had had a thriving furniture business with 37 plants operating. Most of these companies disappeared in the 1930s and by 1981 only two were still in business. One of those that survived the initial shake-out was the Karges Company which was a major market for Evansville Veneer. In 1938 the Evansville Veneer and Lumber Company added a "face" department which was managed by Ray Tepool between 1938 and 1950 when he became plant superintendent, a job he held until 1958. He later held the critical job of pricing the company's veneer and training new company management staff until his death in 1983. When Norman Harding, Sr., died in 1948, the management of the firm was taken over by Frank Snyder, Harding's son-in-law. (Anon., 1981b).

The Jasper Corporation which began business in 1950 under the leadership of Arnold F. Habig in Jasper, Indiana, soon became Evansville Veneer's largest customer. When Evansville Veneer encountered serious financial difficulties in 1955, the Jasper Corporation acquired the firm, paying off its debts, and modernizing the plant facilities. A former employee, Gus Stetter, credits Arnold Habig with being the driving force in getting the business back on track. "When he moved, he got things done. I had a lot of respect for him."

Hired in 1960, Stetter was sales manager for seven years before becoming general manager. He remained in this capacity until 1975 when he left for a similar position with Swords Lumber and Veneer Company. In reminiscing about his experiences with the Evansville Veneer Company, Stetter said that the company:

> ...began exporting small quantities of veneer to Europe in the early 1960s. The Jasper Corporation at the time had a man traveling worldwide looking for markets to sell their products. Of course, Germany was not too hard to do business with. Those people have always looked for outside suppliers. They were wide open for the purchase of logs and good veneer. The company exported logs to them until 1974 or 1975.
>
> We were never totally set up to cut veneer for the export market. That's why we never did get into large quantities. At Evansville we would go into the sawmill with these German customers and we had a way of keeping back the extra good logs and selling them what we wanted to sell. This is a matter of pricing. When we saw a really super good walnut log we priced it out of reach. That way, we kept it for the domestic production and we would price other logs so they would buy them instead. They would leave and we would process the veneer and ship it to them as we finished it. We finally got to the point where we were shipping full containers of veneer. For a domestic mill not set up any better than we were, this was a rather good accomplishment. We had our domestic business to fall back on as well.

Evansville Veneer and Lumber Company now forms a part of the wood products division of Kimball International, Inc., which includes sawmills at Chandler, Indiana; Cloverport, Kentucky; and Gardensville, Tennessee. John C. Ackerman, a Purdue graduate, was the Kimball executive responsible for production and quality of the veneer operations. These sawmills in addition to the one at Evansville serve as aggregation points for veneer quality logs which are fed into the Evansville veneer mill operation. The parent firm produces Kimball pianos and organs, home and office furniture, cabinets, and other wood-based products.

The Evansville plant was initially operated under the direction of its general manager, Francis M. Habig. "Fancy-face veneers are sliced or cut on the staylog machine from cherry, white oak, walnut and mahogany as well as exotic imported exotic cabinet woods. Rotary cut veneers for cores, crossbanding and backs is peeled from such species as gum, poplar and sycamore." Veneer cutting equipment consisted of two Capital slicers, a staylog lathe, and two Coe rotary lathes. The rotary lathes proved to be money losers and were sold in 1959 (Graham, 1956).

Evansville Veneer and Lumber Company was an early manufacturer of sliced pecan and persimmon veneers. Pecan became the principal species cut. Walnut and the oak species were also important. Pecan logs were obtained in West Texas, Arkansas, Louisiana, and Mississippi. Other logs came from Central America, the West Coast, and the Midwest. Evansville Veneer, along with the Dean Company, may have been the only producers of sliced Sitka spruce veneer in the early 1980s. The veneer plant operates with three Capital slicers (including one obtained from Wood-Mosaic) and a staylog lathe. A metal detector eliminated the costs associated with knife changes due to nicked knives. Before the metal detector was installed, there was an average of 1.5 knife changes per shift due to imbedded metal in the flitches. Another example of the use of modern technology is the automatic and computerized recording of the surface area of veneer as it passes through each of the three Proctor and Swartz dryers. The amount of veneer produced from each flitch is identified back to the log from which the flitch was cut. These data become part of a permanent accounting and quality control record which is not completed until the veneer is priced and sold. New technology is introduced whenever it will cost-effectively minimize labor inputs or simplify materials handling and increase production flexibility (Burrell, 1974).

The mill produces 80,000,000 square feet of veneer annually on a one-shift basis. Veneers are sold on the open market in the United States and world markets as well as being routed to Kimball International, Inc., by way of the corporation's new "face" plant in Chandler, Indiana. Open market sales were managed by Francis Habig through a High Point, North Carolina, sales office.

AMOS-HILL ASSOCIATES, INC.

Edinburgh, Indiana, prides itself on being the hardwood face-veneer capital of the country. Signs at the entry points to the city announce this pride to all who travel to Edinburgh, undistinguished otherwise and not dissimilar to other south-central towns in rural Indiana. The wood products industry has been important to the community for decades. The local veneering of hardwoods can be traced back for more than 75 years. At least four prominent families are responsible for the large-scale veneer establishments that have been located in Edinburgh. Those most closely associated with Amos-Hill Associates, Inc., and its predecessor companies were the Thompson, Amos, Wertz, and Hill families. Partnerships and mergers over the years have led to the corporate structure existing today.

Alexander Thompson was one of the proprietors of Edinburgh when the town was platted in 1822. He had been born in Virginia of Scotch-Irish ancestry. Over the next three generations the family was associated with several commercial ventures in the town including a flour mill, a general merchandising store, and a private bank. The old Thompson home still sits at the top of a hill looking down on the town of Edinburgh "...like a castle over a village [and] is a grand though faded mansion. It has all the trimmings of the wealth its first owner amassed more than a century ago through the mills of Edinburgh" (Balika, 1984).

W. T. (Will) Thompson was the son of John A. and Mary Cutsinger Thompson and the great-grandson of Alexander Thompson. Sometime between 1908 and 1910, Will Thompson organized the W. T. Thompson Veneer Company between Perry and Ward Streets. It was equipped with the latest veneering equipment and was one of the show plants in the industry. Featured were four veneer saws, a concrete floor and power connections located in a tunnel under the floor. The veneer saw room was

45 x 80 feet and efficiently arranged to accommodate a large number of flitches and to have easy access to the railroad siding (Anon., 1913). Rosco Cutsinger and James Edward Thompson, Will's brother, were among the employees.

After W. T. Thompson's death in 1921, the veneer mill was sold to Grafton Johnson of Greenwood who changed the company name to Thompson Veneer Company. Members of the Hill family (Harry, Howard, Tom, and Wilford) operated the plant for Johnson for several years. In 1927, the company was sold again, this time to the Amos Lumber and Veneer Company. The mill was still reputed to be one of the largest and best equipped mills in the country. The veneer saws of earlier days had by that time been replaced by four staylog lathes and two Capital slicers (Visitor, 1927).

The acquisition of the Thompson Veneer Company by the Amos Lumber and Veneer Company more than doubled the Amos company's veneer production capabilities to about 100,000,000 square feet. The Amos Lumber and Veneer Company, itself, can be traced back to December 1912, when Roy Amos and his father-in-law, Henry Wertz, went into business as the Wertz and Amos Lumber Company. These two partners were joined in January 1913, by Roy's father, J. Ora (Jack) Amos (Fig. 4-7). Together, they bought the Henry Maley bandsaw mill on South Kyle Street in Edinburgh. This organization lasted only until 1914 when the business was dissolved and reorganized the following year as the AM and O Lumber and Handle Company, the acronym representing the names of

Figure 4-7
J. Ora Amos became the first president of the Amos-Thompson Corporation in 1929 when the Amos Lumber and Veneer Company and the Thompson Veneer Company were consolidated.

the four partners: Jack and Roy Amos, R. C. Mayhill, and Walter O'Neal. This short-lived firm (1915-1917) became the Amos Lumber Company when Jack and Roy Amos bought out the other partners. By 1922, the two men had formed another partnership with Jesse May and Chris Shepard to sell veneer which they custom cut at the David R. Webb plant in Edinburgh. The Amos Lumber Company began to produce rotary veneer in 1923 and in 1925 their first slicer was installed in a new plant located on the Kyle Street site. The company had by then become the Amos Lumber and Veneer Company.

The Thompson Veneer Company and the Amos Lumber and Veneer Company were operated independently until 1929 when they were consolidated to form the Amos-Thompson Corporation. Jack Amos became president; Daniel Wertz, vice president; and Roy Amos, secretary-treasurer, following the merger (Amos, 1929).

Daniel Wertz had become a stockholder of the Thompson company just prior to the merger. He was the son of Henry Wertz and brother-in-law of Roy Amos and is, probably, better remembered as a lumberman and owner of the Maley and Wertz Lumber Company in Evansville, Indiana. He had followed his father into the lumber business in that company in 1901 and, 30 years later, was described as the Dean of hardwood manufacturing in Indiana and as one who treasured the "good old days" when there were not "too many unnecessary and unreasonable changes in methods" (Anon., 1931). In 1939, Amos-Thompson entered the plastics business and formed the Amos Molded Plastics Division of the company. This was a highly successful venture and an early pioneer in the field of molded plastic products.

Upon the deaths of Daniel Wertz and Jack Amos in 1944, new officers were elected. They were: Roy Amos, president; Bice Roth, vice president; and Dale Amos, secretary-treasurer. Dale Amos had joined the family firm in 1936. Bice Roth, who was to have a long and distinguished career in the veneer industry, began working for the Thompson Veneer Company in 1923 while still in high school (Fig. 4-8). After a year at Purdue University as a mechanical engineering student, he exhausted his funds and returned home for a full-time job with the Thompson Veneer Company. In both the Thompson and Amos-Thompson companies, he held many positions including veneer grader, veneer sales manager, vice president and general manager of the wood division, and, ultimately, president of the Amos-Thompson Corporation between 1968 and 1973. He was widely known and respected for his leadership abilities and served the hardwood industry in numerous capacities. Roth was a two-term president of both the American Walnut Manufacturers Association and Fine Hardwoods Association and was the first president of the American Walnut/Fine Hardwoods Association when the two organizations merged. He also served as president of the Indiana Hardwood Lumbermens Association in 1965 (Anon., 1973a).

The management team elected in 1944 operated Amos-Thompson until 1956 when Dale Amos succeeded his father as president and general manager. Roy Amos continued to oversee the business as chairman of the board of directors. Arthur A. Pfaff and Henry L. Conover were elected secretary and treasurer of the company at this time.

In November, 1959, the Frank Purcell Walnut Lumber Company, Inc., of Kansas City, Kansas, was acquired to expand Amos-Thompson's walnut procurement and lumber manufacturing capabilities. This move to expand production was in addition to an earlier effort in 1948 when a new 8-foot band mill was constructed in Edinburgh which featured a number of labor saving devices. B. Z. (Zack) Millett (Fig. 4-8) managed timber procurement for the company while Clyde Wertz was in charge of the company's woodworking plants. Millet was transferred to the lumber department in 1954 where he was responsible for lumber sales of both Amos-Thompson and Cherry Timber and Lumber Company, a wholly-owned subsidiary of Amos-Thompson. The following year he was promoted again and made responsible for both log procurement and lumber sales (Millett, 1963).

Clyde Wertz, a grandson of Henry Wertz, began his long career at Amos-Thompson on February 15, 1929, in the veneer sampling room. He worked there before being shifted to the veneer plant where he learned to operate the slicers and rotary lathes. With this experience under his belt, he was made foreman of the veneer plant in 1934. He was promoted to superintendent of the veneer plant in 1940 and two years later was made superintendent of all woodworking operations.

Figure 4-8
Zack Millett (left) was the Amos-Thompson executive responsible for timber procurement in 1963 when this picture was taken. Bice Roth (right) was considered by his peers to be a "genius" veneer salesman and was president of the company between 1968 and 1973.

Veneer sales were transacted from a number of offices. The largest was in High Point, North Carolina, where sales were handled for many years by Edward L. Grimsley, a Duke University graduate in business management. Grimsley had acquired 10 years of experience with Penrod, Jurden and Clark beginning in 1932 when the Norfolk mill was purchased from the Hoosier Veneer Company. The High Point sales office was modernized in 1958 when a new showroom was opened. Rudy Barron, later a sales manager for Wood Mosaic and Pierson Hollowell Company, was a young employee working out of the High Point office at this time (Anon., 1950-1960).

The year 1958 was also noteworthy as the year in which the Air Force Academy cadets in Colorado Springs chose Amos-Thompson walnut veneers and lumber to furnish their new quarters. All the desks, wardrobes, cabinets, doors, and trim were made from materials taken from Amos-Thompson's extensive inventory of quality walnut veneer and lumber. Bice Roth has used this incident on several occasions as an example of the veneer business being essentially "an eye appeal business" and as a "shelf business." More specifically, he said:

> You must have cut veneer in stock even it it means higher inventory costs and running the mill in poor economic times so you will have it ready for sale when times improve. It takes money, takes patience, and takes salesmanship to market what you have in inventory. In the case of the Air Force Academy dorms and chapel, we had three to five million square feet of walnut veneer in stock that met the specifications and no other firm had that volume available.

Roy Amos died in 1967 and, two years later, the Amos-Thompson Corporation was sold to N. L. Industries otherwise known as National Lead and for its Dutch Boy house paints. The conglomerate's interest in Amos-Thompson was its plastics division which the Amos family would not sell unless the wood products and the plastics division were sold as a package. With the Amos family no longer a part of the business after 40 years of stewardship, National Lead turned to Bice Roth for leadership. Roth was promoted to company president and manager of the woodworking division. He remained in these capacities until May 1, 1973, when he retired after 50 years of service to the company and the hardwood industry.

Peter W. Cross, a 1961 graduate of the University of Maine with a degree in wood technology, was hired to replace Roth. Cross had had a wide range of experience in spite of being less than 40 years of age when he took this job in Edinburgh. Following several assignments with Weyerhaeuser in Wisconsin and Vermont, Cross had spent seven years with the Dean Company in Princeton, West Virginia, where he advanced from production manager to vice president in charge of manufacturing. Amos-Thompson continued to manufacture a broad line of veneer and lumber products for world-wide furniture and panel markets until 1975 when the company was sold to the Hammermill Paper Company for $2.5 million (Anon., 1973b).

Albert B. (Dude) Cox, Jr., was named general manager of this Hammermill division. Cox was well known in the veneer industry and, particularly so, in the Edinburgh area as a result of his 29-year association with Hill Brothers Veneer Company. In 1976, this division produced 90 million square feet of veneer. In the following year, Amos-Thompson merged with Hill Brothers Veneer Company which Hammermill had purchased in 1967 and operated as a separate division until the merger. The merged divisions became the Amos-Hill Veneer and Lumber Company and gave the consolidated company greater opportunities to pursue both domestic and foreign markets. In addition to the combined manufacturing and storage facilities, Hammermill provided capital for a major expansion and renovation of the plants. Veneer production potentials were increased to over 4 million square feet per week. About 50 percent of the cut veneer was walnut and the remainder white oak, red oak, pecan, cherry, and elm (Anon., 1978).

Harold (Pat) Patterson, a 40-year Amos-Thompson veteran, was in charge of domestic veneer sales and had been since 1954. Dieter Schaefer managed export veneer sales. Patterson had joined Amos-Thompson in 1934 as a worker in the lumber department under the guidance of Walt Myers. Eventually, he was put in charge of sawmill operations and in 1940 was brought into the office to assist the manager of the lumber production department. He became manager of the lumber department after World War II before shifting to the veneer sales department.

Export markets became increasingly important to the firm and, by 1980, Amos-Hill was exporting one-third of its veneer output to overseas destinations. White oak and walnut were the dominant species being cut although red oak was another significant species. On January 1, 1981, the Amos-Hill staff was reorganized when the former sales director, Walter Hoefener, was appointed manager of the Amos-Hill operation. On a two-shift basis the company had the annual capacity of producing 200 million square feet of veneer using its five slicers and one staylog lathe. To operate the veneer and lumber mills, there were 375 hourly and salaried employees. Fifteen log buyers located in the Midwest and South were required to provide the timber necessary to operate these mills. The redesigned and rebuilt Edinburgh sawmill and Purcell Lumber Company division in Kansas City produced about 2.5 million board feet of lumber annually although constrained somewhat by availability of walnut logs of suitable quality.

The optimism which had been generated in past years began to fade as the economic recession of the early 1980s impacted on the users of hardwood veneers and lumber. In August of 1982, Hammermill Paper Company executives in Erie, Pennsylvania, made the decision to close permanently the Amos-Hill plants in Edinburgh with the intention of selling these assets. Officials of the company reported that declining profits, strong competition, and reduction in domestic and foreign sales necessitated the closing. This action was a severe blow to the economy of Edinburgh. Three months later, on October 12, 1982, a group of local investors announced that some help was on its way and that they had purchased the small North plant of Amos-Hill. This plant had originally been the Hill Brothers Veneer Company mill prior to the merger with Amos-Thompson.

The new firm, Amos-Hill Associates, Inc., was led by president Richard L. Wertz who stated a desire to continue producing quality veneers and providing employment for many of the 250 workers who had lost their jobs at the time of plant closure. Other investors-managers included Steve A. Battern, Connie D. Burton, Edgar F. Harpring, and William P. Jones, Jr. The Frank Purcell Lumber Company in Kansas City, acquired by Amos-Thompson 20 years earlier, continued as a division of the company (Anon., 1982).

Wertz was not a stranger to the veneer business or Amos-Thompson. He had been born in Edinburgh in 1947, the son of Clyde Wertz, a long-time employee of Amos-Thompson. He was also the great-grandson of Henry Wertz, founder of an Evansville hardwood lumber dynasty. Richard had followed the pattern of working at Amos-Thompson while in high school, but he had no serious intentions of making a career in the veneer business. After graduating from Murray State University, he taught at the high school level. Recognizing the economic limitations of being a school teacher, he accepted a job offer at the Amos-Hill plant to become the domestic sales manager. After the plant closure in 1982, "...people laughed when [I] mentioned buying the plant." Except for Jones, who was a lawyer and accountant, the other associates were experienced employees of the former company. The success of the Amos-Hill purchase prompted Wertz and his associates to acquire Hammermill's other plant in 1983. In 1984, Amos-Hill Associates employed 160 people in the mills who were efficiently complemented by an office staff of 15 (Balika, 1984).

The Hill brothers, Harry and Howard T. (Teal), were the sons of Wilfred Hill who had owned and managed the Ideal Veneer Company in Franklin, Indiana. They were true entrepreneurs and were alert to promising business ventures. In 1925, they organized the Hill Brothers Veneer Company in Edinburgh, Indiana. A few years later, in 1932, they were granted the first permit to sell beer in Johnson County and operated a beer distribution monopoly in 13 Indiana counties during the remaining part of the 1930 decade. During most of this same decade they also operated a sand and gravel business east of Edinburgh, a hotel in Hot Springs, Arkansas, and large farms in Illinois and Indiana.

Coming from a family with a veneer cutting tradition, it was only natural that they should eventually gravitate to the hardwood veneer business. Their on-the-job experience with the W. T. Thompson Veneer Company between 1921 and 1924 had been a prelude to the establishment of their own independent venture in the industry. For several years they did not own a production facility but custom cut their veneers at the David R. Webb Company veneer mill. After economic conditions improved following the Depression, the Hill Brothers built a veneer mill of their own on Shelby Avenue in Edinburgh. The mill was completed in 1937. Harry's son, Jack, joined the business at that time. The two Hill brothers and Jack became active partners in the veneer company operations.

On January 31, 1941, Harry Hill and Roy Amos applied for a patent on a process for cutting comb-grain veneer. Comb-grain veneer was also termed "needle-point" or "rift" veneer. Regardless of name, the basic idea was to cut the flitch to enhance the medullary rays in oak veneers. This was accomplished by attaching the flitch to a staylog rotary lathe "...using a cutting knife as distinguished from a slicing or sawing arrangement." The application for the process patent claimed this method produced more rift cut veneer from a flitch than any other known method. It was introduced to the panel and furniture trade as "streamline-cut" veneer. The patent was granted on March 26, 1946, with 50 percent of the patent rights being assigned to the two companies involved: Amos-Thompson and Hill Brothers. The invention was an operational success with both companies profiting from this improved technology and the fashion trend toward rift-oak furniture which became popular in the late 1940s and early 1950s.

L. E. (Tom) Newton joined the firm as a partner in 1944 after the tragic death of Teal Hill in a train-car accident which also resulted in the deaths of Hill's wife and two daughters. Newton had been an employee of Hill Brothers between 1925 and 1931. Hill Brothers Veneer Company continued to be operated as a partnership until 1947 when the business was incorporated. George Demuth, who had been a forester employed by the Indiana Department of Conservation, bought into the corporation in 1950. He remained with the company for about 20 years, serving the firm as vice president for a number of years. Harry Hill died in 1959 after devoting nearly 35 years to the veneer company.

Jack Hill, Tom Newton, and George Demuth were the principals in the company until November 1, 1967, when the business was acquired by the Hammermill Paper Company. At the time of the acquisition the mill employed about 75 workers and was capable of producing 80 million square feet on a two-shift basis. Jack Hill retired when the business was sold and died three years later in 1970. Newton stayed with the company during the ownership transition for about six months before retiring. George Demuth was appointed Division manager by the Hammermill company and continued with the Hill Brothers division for more than two years before retiring. He was succeeded by Albert B. (Dude) Cox, Jr. Cox supervised the modernization of the plant which included a program of purifying and recycling steam and process water. This was a first for the veneer industry (Anon., 1965).

THIESING VENEER COMPANY

Like so many others of his generation who established successful, multi-generational, family-owned veneer businesses, Louis A. Thiesing believed that hard work would be rewarded with success (Fig. 4-9). He began his career in the veneer business as a laborer with Adams and Raymond Veneer Company which was the successor of M. D. Williamson's Indianapolis Veneer Works at the turn of the century. He advanced in the trade and was a veneer cutter for Adams and Raymond, Talge Mahogany, and the Indiana Veneer and Lumber Company. In 1914, he left Indianapolis to work as a stockholder and cutter at the Buckeye Veneer Company in Dayton, Ohio, and was with Buckeye until the company was sold in 1919. Returning to Indianapolis, he organized his own mill in 1920 with his son, L. Merrill, who was also a veneer slicer operator. Robert Mathers has commented that: "They were a good team and worked together very well." This relationship ended when Merrill died in 1938. His father's death occurred three years later in 1941 at which time Robert (Bob) Mathers moved up from vice president to president of the firm. The mill was located at 1501 West McCarty Street in Indianapolis. Ralph Beckwith, a log producer who had started to work for Talge Mahogany in 1911, became a part of the firm in 1936.

The Thiesing's business was custom cutting in other mills until a Capital slicer and a Proctor and Swartz dryer were purchased in 1924. In 1928, a Capital lathe was added. The machines were used to cut veneers for others until the advent of the Depression years in the 1930s (Anon., 1981).

Figure 4-9
The Thiesing Veneer Company was organized in 1920 by Louis A. Thiesing and his son, L. Merrill Thiesing. Starting as a laborer with the Adams and Raymond Veneer Company, Louis Thiesing firmly believed that hard work would be rewarded with success.

A letter written by L. Merrill Thiesing to the Pierson-Hollowell Veneer Company provides an insight into the cost structure of custom cutting in 1931:

> We will cut your sliced veneer for four dollars ($4) a thousand and your half-round for four dollars and fifty cents ($4.50) a thousand if you do the flitching; butts for $13.50. We will keep the backing boards. When we are cutting for you, the mill will be yours.

By 1936, the Thiesings no longer custom cut veneers. Producing all their own quality veneers, they developed a growing sales force with representatives in Grand Rapids, Michigan; High Point, North Carolina; Rockford, Illinois; and Jamestown, New York.

Louis Thiesing's two daughters, Irma and Elsie, did not escape from the veneer fraternity. Irma married Russell T. (Bud) Miller of Curry-Miller Veneer Company, and Elsie started in the family business working half-days when she was a junior in high school. She continued to be employed by the firm until 1939 when she married Robert Mathers. They had three children: James J., David R., and Roberta. All three ultimately became officers in the company.

Robert Mathers was inexperienced in the veneer business when he came into the company in 1938 although he had been in the hardwood lumber business prior to this move. Mathers owned Walnut Sales, Inc., a sawmill in St. Louis purchased from Ray Pickrel and was a major partner in another sawmill in Springfield, Missouri. The St. Louis company bought and assembled walnut veneer logs which were forwarded to the Thiesing mill in Indianapolis. Walnut Sales, Inc., was ultimately sold to David Holmes of Kendallville who later closed the mill.

The fate of the Indianapolis location for the business was sealed with the planning for I-465 around the west side of Indianapolis. In a document prepared for the Fine Hardwoods Association, Elsie Thiesing Mathers wrote:

> The highway people were giving us such a bad time about this property. We were running a veneer machine at David Holmes' lumber mill in Kendallville and one in Indianapolis. We also custom cut at Adams Veneer Company in New Albany. We didn't have a warehouse anymore so we had to truck all the veneer up there to Kendallville. It was too expensive when the whole industry went down hill and that's when we closed down for two years.

As one veneer industry observer of the two-year hiatus put it: "Off the record, their mill qualified for first prize for mills of an antiquated nature. It was a terribly antiquated factory. In fact, some of us wondered how it held together, let alone made any money."

In 1966, the Thiesing Veneer Company moved to new quarters in Mooresville, Indiana. New Hale dryers, their 1924 and 1954 slicers, and a 1928 staylog lathe were installed in the 20,000-square foot, single-story building. Mathers indicated that the 1954 slicer had been purchased about

1967 from West Coast Hardwood, Inc., a company that sliced several California species. "The veneer showed promise but the color was never quite right."

In the 1980s, the mill was strictly (99 percent) a custom cutter of veneers for foreign mills. The veneer was about 80 percent white oak, 10 percent walnut, and 10 percent red oak and southern pine. As Mathers explained the advantages of being a custom cutter: "It doesn't take the capital to buy and inventory logs and warehouse the cut veneers."

One of Thiesing's clients who has custom cut at Thiesing Veneer in recent years was Gunther Rodatz, president of G. R. Wood, Inc. He described his company as "...the largest custom cutter in the United States. We believe that the Far East is the future market..." and sees a shift from Europe to Asia because of labor cost advantages. "The only thing they don't have is the raw materials, like veneer and logs, which they need to buy in the United States" (Anon., 1987). The G. R. Wood facilities are located adjacent to the Thiesing mill and include 10,000 square feet of warehouse space, a "water yard" to store logs, and a main yard for loading shipments. The business generated by G. R. Wood, Inc., represented about 65 percent of Thiesing's capacity.

Reflecting more broadly on export market conditions and other factors affecting the veneer industry, Mather stated that:

> When the Germans first came over here in the late 1940s and early 1950s, they made it awful rough on American veneer mills because the logs went out of the country. They must have really needed them because they paid more for the logs than our veneer mills could pay or we could expect our customers to pay for after we made the veneer. So we changed species and went to pecan. We just had to get away from walnut logs, for example. We never got completely away from oak, however. Finally, the freight costs to Europe reached the point where it was a little cheaper to cut in this country than over there. So we happened to be on the spot at the right time to custom cut.
>
> Right now (1981), the dollar is strong against the German Mark so it's hurting our business. A year and a half to two years ago, when the dollar was weak, it was much easier to do business with the Europeans. As soon as recovery starts, which may be months away, I look for a very substantial increase in veneer for domestic and export and especially export.
>
> South America has veneer. Asia has veneer. They have many species and they have big forests, but the species that are suitable for face veneer are limited in those countries. The United States is blessed with having real top-notch hardwoods that have character. They have their exotic woods, but exotic woods don't lend themselves to being made into a line of case goods. You can take our oaks, walnuts, and pecans and you can make thousands of units and all of them look about the same and are beautiful. Other countries can't come up with woods of quite that caliber.

In 1982, Bob Mathers became the chairman of the board. James J. Mathers was elected president at the same time. He had been employed by the firm since 1967 and had been mainly involved with plant maintenance and production. His brother, David, was elected vice president in 1976, having been employed by the firm since 1973. Roberta Borlik, a former school teacher, was employed by the family in 1977 and succeeded Margaret Antle as secretary the following year. The firm normally employed 80 people and had the capacity to cut about 100,000,000 square feet annually.

B. L. CURRY AND SONS

The Lapping brothers (George and Robert) and Guy Theis built a veneer mill in 1917 on the former site of the DePauw glass works in New Albany. All three had learned their woodworking skills while working for Wood-Mosaic company. The name they gave their firm was the Floyd County Veneer Company (Adams, 1963). The mill specialized in the manufacture of quarter-sawed oak veneers. The sawed veneers were dried "...in a textile drier conveniently located in a drying room adjoining the cutting department." In 1920, the plant operated night and day to produce 500,000 square feet of veneer per week (Anon., 1920). In 1921, the veneer mill was moved to the abandoned railway roundhouse at the end of the Monon line in New Albany. This is the present site of B. L. Curry and Sons.

When the ownership changed in 1923, the company name was appropriately changed to the Monon Veneer Company. The new owners were Vergil and Fred Theis, the sons of Guy Theis. The entire plant was destroyed by a fire in November, 1930. The plant had been idle since March of that year because of a lack of business. It burned on the day that production had resumed (Anon., 1930). The mill was apparently rebuilt by the Theis brothers. In 1940 or 1941, Vergil Theis became the sole owner of the business and changed the name of the firm to Monon Veneer and Lumber Company. When the business came up for sale in 1952, it was purchased by Bernard (Ben) L. Curry.

B. L. Curry was born in Ridgefarm, Illinois, in 1892. After three years in the army during World War I, he began his lifelong association with the wood products industry at a sawmill in Bainbridge, Indiana. However, two years later, Curry moved on to the thriving veneer industry in Indianapolis. He first worked for Jim Meyers at the Indianapolis Sawed Veneer Company between 1921 and 1925, ultimately becoming his veneer mill foreman. In 1925, Curry was approached by Russell Fortune who was a prominent producer and merchant of "oriental" veneers. Fortune apparently was having difficulty in having his wood cut into veneers and suggested that, if Curry started a custom plant, he would give him an exclusive contract to cut all his veneers. Suitably stimulated by this proposition, Ben Curry leased facilities at Gosport, Indiana, between 1925 and 1927 and began cutting veneer. The equipment consisted of a single staylog lathe. The business appears to have been successful. In 1928, B. L. Curry moved back to Indianapolis and built a new veneer plant (Curry Veneer Mill) which was ready for start-up in the fateful year of 1929. A year later, Russell T. (Bud) Miller joined him as a partner. Miller bought out Curry's interest in the business in 1951. Three months later, Curry bought the Monon Veneer and Lumber Company and changed the name to B. L. Curry and Sons.

Curry's son, Jack B., born in 1924, joined his father in the business in 1951. In 1953, James C., a son born in 1929, followed suit. Jerry S., the youngest son, born in 1943, was employed by the firm in 1962. B. L. Curry died on October 21, 1977. With his passing, the firm was restructured with Jack becoming president; James, executive vice president and treasurer; and Jerry, vice president and secretary. The third generation of the Curry family (Jack's son, Jon, and Jim's son, Bradley) became a part of the company a few years later.

Another restructuring took place in 1983 when the older officers were each promoted up one step at the time the position of board chairman was created. Jon and Bradley, the younger generation, took positions in the company as vice presidents.

The mill produces veneer with a 17-foot slicer and a staylog lathe. Including managers, the company has about 60 employees. The firm has specialized in pecan veneers and was one of the first to cut this species, beginning in 1964. By 1966, pecan veneer was the principal species cut. Pecan logs were bought in south-central United States, from Texas to Louisiana. The oak species were obtained locally and cherry logs came from New York and Pennsylvania.

Jack Curry was optimistic about the future of the veneer industry when interviewed. He expressed the philosophy that greater specialization and creativity will be the hallmark of the future as veneer timber becomes more scarce and demands of customers become more specific. He emphasized the need to "cater to customer's desires."

The 38,000,000 square feet of veneer produced annually by B. L. Curry and Sons have never been exported. Domestic sales are handled by a sales agency representative in High Point, North Carolina (Erath Veneer Sales); Grand Rapids, Michigan; and Louisville, Kentucky.

Chapter 5

INDIANA VENEER FIRMS ESTABLISHED AFTER 1920

PIERSON - HOLLOWELL COMPANY, INC.

An extensive interview with Robert E. Hollowell, Jr., revealed that his father, Robert Ertal Hollowell, Sr., and Paris C. Pierson had been business associates and partners in a number of ventures prior to the incorporation of the Pierson-Hollowell Company in 1935. The senior Hollowell was born in Chicago on May 26, 1892, and had grown to maturity in Winona, Minnesota, where family members owned a general store (Fig. 5-1). Following a year at the University of Illinois and two years with the U. S. Army Corps of Engineers, Hollowell was employed as a purchasing agent in 1915 by the Mitchell Wagon Company of Racine, Wisconsin. It was an opportunity which was to direct him to his life's work. The job offer had come from his father, John D. Hollowell, who had become general manager of the wagon manufacturing company in 1914.

One of the firms Robert Hollowell dealt with was the Eel River Falls Lumber Company of Spencer, Indiana. The driving force in this lumber company was Paris C. (Bill) Pierson. Pierson's father, Joseph L. Pierson, had also been in the lumber business and had operated as the J. L. Pierson Lumber Company until his death in 1907. Two years earlier, Bill Pierson and his brother, Temple Guy Pierson, had started their own wholesale lumber business with yards in Spencer and Terre Haute. A report from the R. G. Dun Company (predecessor company to Dun and Bradstreet) on the Eel River Falls Lumber Company, dated February 13, 1917, contained the following information:

> Paris C. Pierson, age 31, single, has been engaged in this line of business since January, 1914, having started new. He is also in the lumber and coal business as a member of the Pierson Lumber Company. Is well regarded as to reputation and habits and is experienced in this line. Is reported prompt pay. Are still investigating.

Bill Pierson did much of the selling on the road and "called upon" the Mitchell Wagon Company during 1915 and 1916. As a traveling lumber salesman, Pierson observed other wagon manufacturers tooling up to produce motorized wagons. Since the Mitchell Wagon Company had not responded to this competitive challenge, Bill Pierson was able to convince Robert Hollowell in 1917 that he might better himself if he became a general partner in the Eel River Falls Lumber Company. Hollowell agreed and moved to Spencer, Indiana. When World War I was declared a short time later, the lumber sales company was liquidated and both men joined the Army. Both served in France before returning to Spencer in September, 1919. They reentered their partnership and continued to do business as the Eel River Falls Lumber Company at Spencer until the spring of 1921 when the office was moved to Indianapolis and the name of the firm was changed to the Pierson-Hollowell Lumber Company. The firm's financial status may have been questionable because Hollowell's worried father requested R. G. Dun & Company to run a confidential check on his son's business. The July 2, 1920, report states:

> They do not respond to our request for a statement. Hollowell is not known to have outside means of consequence. The business was started in a small way by P. C. Pierson about 1914 and about 1917 R. E. Hollowell became interested. T. G. Pierson is a silent partner. Hollowell was previously employed as a purchasing agent. The firm takes the output of two mills and buys logs and has them sawed by other mills. The net worth of the firm is estimated at $40,000 to $50,000.

On August 1, 1922, Pierson and Hollowell joined William C. DeVore on a one-third equal shares basis in a sawmill venture in Tilton, Illinois. This mill was destroyed by fire soon thereafter. A new mill was built in Danville, Illinois, where it was operated as a partnership until December 31, 1922, when the business was incorporated as Pierson-Hollowell Hardwood Co., Inc. Two years later the name was changed to Pierson-Hollowell Walnut Company, Inc. Satellite sawmills were located at Brazil, Indiana, and Effingham, Illinois. The company supplied walnut, ash, and hickory lumber for the Chicago and other regional markets. Frank H. McConnell was associated with the Illinois operation. Ads in county weekly newspapers and advertising posters were used extensively to solicit the sale of walnut timber from landowners. The posters were tacked on poles along the highways and were placed in stores, banks, and other business places. One Pierson-Hollowell Walnut Company poster dated to about 1926 offered to buy "...Any amount, from truck to trainload. Logs to measure 12 inches in diameter and up and in lengths from 8 to 16 feet. Will pay $50 to $150 per 1000 feet standing in tree. Sell direct to mills and put the dealers and jobbers margin in your own pocket" (Pierson-Hollowell Co., 1946).

A third firm, Pierson-Hollowell Veneer Co., an Illinois Corporation, was formed in 1924 to manufacture and sell face veneer. It was a spin-off from the Pierson-Hollowell Hardwood Company which had been custom cutting veneer earlier. Veneer manufacturing by the new company continued to be done on a custom cut basis at a number of veneer manufacturing plants in the Midwest. This company operated until June 30, 1933, when the assets and liabilities were taken over by the Pierson-Hollowell Walnut Company (Anon., 1972).

Figure 5-1
Robert E. Hollowell, Sr., left the Mitchell Wagon Company in 1917 to go into a lumber sales business with Paris C. Pierson in Spencer, Indiana. This association developed into the Pierson-Hollowell Company in 1935.

On January 1, 1924, the Pierson-Hollowell Lumber Company of Indianapolis was incorporated with an authorized capital of 1,000 shares having a par value of $100 per share. A total of 600 shares was issued and equally divided between the two partners. Subsequently, Frank H. McConnell purchased 40 shares and Oliver L. Fevrier bought 10 shares so that, by January 2, 1934, a total of 650 shares had been issued (Anon., 1944).

The period between 1930 and 1935 was particularly difficult for the company. Pierson had invested heavily in speculative stocks in the stock market prior to 1929, pledging his Pierson-Hollowell stock as security. In 1935, the American National Bank of Indianapolis assumed control of the company. On March 31, 1935, the Pierson-Hollowell companies in Illinois and Indiana were reorganized into a single firm, Pierson-Hollowell Company, Inc., under Indiana's 1929 Corporation Act. The firm's capital stock was increased to $150,000, with Hollowell receiving 57 percent of the distributed shares. Pierson was subsequently bought out by the Hollowells and left the company in 1936.

In 1937, the Pierson-Hollowell company seriously began negotiating for the purchase of the Batesville Veneer and Lumber Company at Lawrenceburg, Indiana, near Cincinnati. The Batesville Company was founded in Lawrenceburg in 1902 and produced lumber and sawed veneers primarily for furniture plants in Batesville, Indiana. Later it produced sliced veneers. A Capital Machine Company invoice of May 12, 1906, records the sale to the Batesville Lumber and Veneer Company of "One 12-foot veneer slicing machine [and] one 162-inch Automatic

Knife Grinder furnished complete as per agreement [for] $6,000.00." The company was located in a three-story building in Lawrenceburg. The mill is known also to have provided custom cutting services. The business had some successes, but by 1920 it was in receivership. Elmer L. Snyder operated the company during its extended receivership status prior to the acquisition by Pierson-Hollowell.

On January 8, 1937, the Dearborn County Circuit Court ordered the sale of the Batesville Lumber and Veneer Company to the Pierson-Hollowell Company for $27,000. This transaction was not completed because of the devastating 1937 Ohio River flood which damaged the buildings and carried flood waters into the third floor of the principal building. On April 7, 1937, the Court again ordered the property sold to Pierson-Hollowell, this time for $13,000 because of the depreciated value "...due to the recent flood that damaged the buildings..." (Anon., 1937).

Pierson-Hollowell immediately began a program of extensive repair and improvement which continued between 1937 and 1942. The company's sawmill operation was moved to Lawrenceburg in 1938 following an arson fire that destroyed the sawmill in Danville. W. C. DeVore was put in charge as Lawrenceburg plant manager and given the responsibility for producing quality face veneers and band sawn lumber for the furniture and paneling markets. The equipment consisted of a 54-inch bandmill, two old slicers, and a 10-foot rotary staylog lathe (Anon., 1965a).

During World War II, the veneer mill produced aircraft veneers. Most of the production was red gum veneer although tulip poplar, Sitka and Engelmann spruce, and Port Oxford cedar were also cut. The sawmill provided walnut lumber blanks for gunstocks. Pierson-Hollowell and a consortium of other hardwood veneer and lumber companies marketed gunstock blanks through a joint venture known as Walnut Gunstocks, Inc., which was headquartered with the Pierson-Hollowell Company at 530 Postal Station Building in downtown Indianapolis. William W. Fobes was the general manager of this marketing organization which at one time or another during World War II provided marketing services for Amos-Thompson; Chester B. Stem; Wood-Mosaic; Penrod, Jurden and Clark; Purcell Lumber Company; and Pierson-Hollowell (Graham, 1955).

Following the war, extensive improvements were made to the Lawrenceburg veneer mill in 1947 and 1948. An 8-foot bandmill was purchased in addition to a 12-foot rotary lathe and a 17-foot Capital slicer. The venerable 1906 Capital slicer was completely reconditioned down to its original cast-iron frame. Walnut, white oak, and mahogany (mainly from Central America) were cut in large volumes. Other species cut over the years included hard maple, cherry, butternut, hickory, elm, teak, and rosewood. Native species were purchased generally in Indiana, Kentucky, and Ohio, with lesser quantities originating in Illinois, Iowa, Michigan, Pennsylvania, West Virginia, and Missouri. Another band sawmill was built in Keokuk, Iowa, in 1946 to process black walnut logs. This mill operated until 1957 when it was sold to the Baxter Lumber Company.

In 1947, R. E. (Bob) Hollowell, Jr., a recent Purdue University forestry graduate, at the age of 21, began his studies at Yale University where he concentrated on the international trade aspects of wood products (Fig. 5-2). Upon graduation from Yale in 1948 with a Masters degree in Wood Utilization, Bob Hollowell joined the family firm and served in a number of capacities before becoming president of the company upon the death of his father in 1962 (Anon., 1974).

Bob Hollowell's academic pursuits fit well with the company's longstanding interest in exporting its products. Bob's father, for example, was involved at an early date with the Walnut Export Sales Company. This organization was formed by walnut lumber producers in 1919 as a gunstock sales association which was permitted under the Webb-Pomerene Act. Ralph Jurden of Kansas City was a prime mover in organizing the association which included the Hartzell Lumber and Veneer Company (Piqua, Ohio), C. H. Barnaby Lumber Company (Greencastle, Indiana), Pierson-Hollowell Company, and a number of others. The association headquarters moved several times during the 1920s and 1930s and was inactive during most of the World War II era when the federal government performed this export function.

The Walnut Export Sales Company, Inc., (WESCO) was a later version of the joint trading venture. The member firms consisted of Amos-Thompson Corporation, Pierson-Hollowell Co., Inc., and Chester B. Stem, Inc. WESCO was organized to facilitate the export of veneer. The

Figure 5-2
Robert E. Hollowell, Jr., joined the Pierson-Hollowell Company in 1948 after receiving degrees from Purdue University and Yale University. An industry leader, he also directs the management of 3,500 acres of company-owned walnut plantations and forests.

three firms in 1944 sought to expand the activities of WESCO by cooperatively importing logs and lumber from tropical areas without violating the Webb-Pomerene Act. They had been cautioned not to attempt this by Allen C. Phelps, Director of the Federal Export Trade Office in a letter dated September 19, 1944.

> With reference to your inquiry concerning your desire to make limited selected purchases of mahogany and other logs in certain foreign countries, to be distributed to your three member firms, it is the opinion of this office that the association would not be permitted, under the terms of the Webb-Pomerene Act, to engage in the business of importing such products into the United States. The Webb-Pomerene Law provides that associations organized and operating under its provisions shall engage solely in export trade. The term "export trade" is defined as being "trade or commerce in goods, wares, or merchandise exported, or in the course of being exported from the United States or any Territory thereof, to any foreign nation." In view of this restriction, it is not believed that your association, as such, can lawfully purchase and import logs into this country.

The three firms subsequently set up a companion corporation in 1947, the WESCO Trading Company, to centrally purchase logs and lumber for their separate businesses. The WESCO Trading Company also operated out of the Pierson-Hollowell office complex in Indianapolis. Lumber buying was centered in Campeche, Mexico. Logging concessions in Guatemala and Honduras produced logs which were shipped to member mills. Merritt Wilson directed WESCO operations between 1947 and 1956. As the tastes and preferences of American consumers for mahogany furniture waned, so did the fortunes of the WESCO Trading Company. WESCO Campeche was sold to U. S. Mahogany of New York City. Log imports continued until about 1960 when it was apparent that consumer interest in walnut furniture and paneling was going to be a dominant factor in these markets. The trading company was finally "retired" by Pierson-Hollowell in 1975 after an active life of about 25 years.

Beginning in 1960, Pierson-Hollowell began to actively consider the expansion of their production capacity. Because of limited space for expansion at their Lawrenceburg mill site, the company bought 64 acres at Shelbyville with the intention of relocating there in a new one million dollar plant to be constructed on this property. These plans were frustrated by the economic recession of 1964. In 1965, the Borg-Warner plant, with 140,000 square feet under roof on a 10-acre site, became available near Lawrenceburg at Greendale. Pierson-Hollowell purchased the property on September 30, 1965, and took possession on April 1, 1966. This provided about a 40 percent increase in building space and also permitted their lumber operation to expand with improved facilities at the old East Center Street location. The sawmill continued to produce flitches for the veneer mill which was now about two miles away (Anon., 1965b).

The added space in the Borg-Warner plant made it possible to expand the veneer manufacturing operation and warehousing capabilities. Veneer equipment installations were completed in January, 1967. The slicing equipment consisted of a 12-foot (1906) slicer, a 16-foot slicer (1944), and a 12-foot rotary lathe. These were augmented in 1973 with another 12-foot Capital slicer. The plant was equipped with two Hale dryers and 16 covered vats equipped with automatic controls and temperature recorders (Burrell, 1967).

The overall development strategy of the firm also included the organization of a subsidiary company, the American Face Veneer Corporation, to be located in Shelbyville, Indiana. To this end, the firm completely renovated an older building in 1965. The building had previously been occupied by a diverse group of manufacturing firms including the Diamond Veneer Company which had constructed the building in 1920. Homer Cutsinger was the entrepreneur who had organized the Diamond Veneer Company with the goal of making his firm the largest sawn veneer plant in the country. Cutsinger, however, was running against the technological current for manufacturing veneer; and the company ceased operation as a result of productivity problems and the economic crises of the late 1920s. American Face Veneer also had its share of problems in producing spliced, cut-to-size veneer faces for the furniture trade and plywood manufacturers. Robert E. (Bob) Hollowell, Jr., took on the added duties of president and was assisted by Walter H. Hoefner, vice president and general manager of the new veneer enterprise. Hoefner came to the company with valuable experience, having been manager of Wood-Mosaic's veneer splicing department before becoming manager of the J. A. Habig Veneer Company in Montgomery, Pennsylvania. In 1968, American Face Veneer broke even. In 1969, the firm made a small profit. However, changing markets and the difficulties of finding and keeping good managers became too much for the financially-strained company. The company "...was closed early in 1971 after severe losses." The equipment was sold the same year to Donald McLean Bell of Louisville who had left Wood-Mosaic to head up his own company (Anon., 1966a).

The Pierson-Hollowell Company continued to produce quality veneers and to maintain a healthy interest in foreign markets. When the European market began to increase its demand for American veneers, Pierson-Hollowell was ready. The company entered into a joint venture with Dieter Johnson who later became manager of veneer sales for Pierson-Hollowell. Donald S. Speer, son-in-law of Robert E. Hollowell, Sr., was responsible for export sales after Johnson left the firm. This trade grew and flourished from the late 1960s to 1980 when international exchange rates became less favorable for the export of American products. In 1983, Rudy Barron became coordinator of domestic and export sales. Barron came with sales experience at Amos-Thompson Corporation in Edinburg, Indiana, and Wood Mosaic, located in Louisville, Kentucky.

In the 1980s, Bob Hollowell functioned as both president and chairman of the board. Donald S. Speer served as vice president. He had been employed by the firm in a sales capacity since joining Pierson-Hollowell in 1958. William F. Beckwith, who retired as vice president in 1983, had been with the company since 1972 after a long career in the forest products industry. Between 1941 and 1969, he was employed by the Thiesing Veneer Company and later by the Holmes Lumber Company in Columbia City, Indiana. Fred G. Black, secretary-treasurer and an accountant, joined the firm in 1958. These officers managed a business with about $8 million of yearly sales utilizing 150 employees in Indianapolis and Lawrenceburg, Indiana.

WESCO Trading Corporation and the Walnut Export Sales Company were still carried within the corporate structure in 1989 as wholly-owned, but inactive, subsidiaries. In January 1984, Pierson-Hollowell consolidated the sales and production functions in Lawrenceburg when the sales office and showrooms were moved there from Indianapolis (Miller, 1984).

Dun and Bradstreet once characterized the Pierson-Hollowell Company as "clear, strong, and steady." Perhaps this was because these characteristics reflected personal traits of both Robert E. Hollowells. Both were also able to provide needed leadership to their industry. Both served as presidents of the Indiana Hardwood Lumbermans Association and were active participants in the American Walnut Manufacturers Association and the Fine Hardwoods Association. Bob Hollowell, Jr., also has had the distinction of serving as president of the National Hardwood Lumber Association and of being an important leadership contributor to the Society of American Foresters. He was also the first president of the Walnut Council of America. This last honor stemmed from his activities in encouraging and supporting modern forest management in the hardwood industry and from his pioneering work in black walnut plantation culture. Unlike most face veneer manufacturers, the Pierson-Hollowell Company owns and manages 3,500 acres of hardwood timber lands and walnut plantations in Indiana and Kentucky. These are managed by Robert D. Burke, a Purdue University forestry graduate and long-time employee of the company.

CURRY - MILLER VENEERS, INC.

B. L. (Ben) Curry founded the Curry Veneer Mill in 1929 as a custom cut veneer mill. He had operated a similar business in Gosport, Indiana, for four years and had acquired earlier experience as an employee of the Indianapolis Sawed Veneer Company. When he decided to return to Indianapolis, he was 37 years of age. Financial aid came from a Gosport banker who provided capital for this venture and came into the business as a full partner. Neither were able to recognize that the booming 1920s decade would end in an economic disaster in 1929. The business was assisted materially by the timely intervention of Russell T. (Bud) Miller who bought the Gosport banker's 50-percent interest in the business on September 23, 1930 (Fig. 5-3). The company was incorporated as Curry-Miller Veneers, Inc., in 1931.

Figure 5-3
Russell T. (Bud) Miller spent more than 70 of his 86 years in the veneer business. His creative mind led to production innovations and skillful marketing strategies.

Bud Miller began his hardwood industry career in 1910 when he was employed by the Cincinnati Veneer Company at age 14. However, his earliest acquaintance with the veneer business was through his step-father who worked for the Talge Mahogany Company in Indianapolis when Miller was a youngster of eight or so. His tenure at the Cincinnati Veneer Company was cut short by a mill fire shortly after he got the job. Bud's career in the veneer industry was not resumed until April 12, 1912 when he secured employment with the Buckeye Veneer Company in Dayton, Ohio. The veneer mill was under construction and Miller helped to mix the concrete foundation for the slicer as well as measuring and inspecting the first piece of veneer that was produced by it.

After three years, Miller left Buckeye and was hired by the H. C. Hossafous Veneer Company, also of Dayton, and remained with that firm until called into military service during World War I. After receiving his discharge, he returned to his old job. The firm in the meantime had been sold to J. W. Fry who renamed it the Dayton Veneer Company. Within two months he had been made mill superintendent. His first decision was to lay off seven men to *increase* productivity. His last decision was to leave Dayton Veneer at the prompting of Louis Koss of the Capital Machine Company. With a smile, Bud Miller related that:

> Everytime Louis Koss came to Dayton, he would come down to the mill and talk to me for a little bit. One time he said, "I want you to buy half interest in a veneer mill." I said I would buy it if it looked like a veneer mill. So, I bought it and after I bought it I went out and sat under an elm tree and said to myself: "What the heck did I get into here?" The

Proctor-Swartz dryer wasn't paid for, the vats leaked, and the mill could only run about three days a week.

What Miller did was a classic case of entrepreneurial genius. He went on the road selling veneers in an almost non-existent market, capitalizing on his reputation and business contacts made during more than a decade with the Dayton Veneer Company. He hired Billy Campbell as salesman and in the first month had one of his Dayton log buyers ship in 72 cars of logs. "I put on a night shift. The other Indianapolis veneer companies were boarded up or weren't doing anything and we were running day and night." Within a year, a new warehouse had been constructed to facilitate the handling of the inventory. This building was also financed by Louis Koss who had been impressed by the company's progress toward solvency and rate of repayment on the $60,000 Capital Machine Company indebtedness. According to Miller, the company was on a strong economic footing within two years with all the bills paid or on a plan of scheduled repayment.

His innovative skills can be seen in this personal account of how he used a simple strategy to market veneer:

> The people in the South all knew me. So, I got a rubber stamp made about six inches long. It said: "Manufactured by Curry-Miller." I had the inspector stamp each side of the veneer sample with it and the outside of each crate containing the sliced veneer from the entire flitch. That went on for about six months until one day when one of our best wholesale customers came in here and said: "Say, are you still stamping the veneer." I said that we were and complied when he said that he didn't want his stamped any more. In less than a month he came back and wanted us to start stamping it again because his customers wouldn't buy the veneer without it.

Bud Miller is also known as the man who figured out how to get an even color through the entire walnut flitch. The contrast between the creamy white sapwood and the dark heartwood was a particular problem. Miller had studied this and other veneer problems such as the buckling of sliced veneer when he had been at the Dayton Veneer Company. His solution to the walnut problem involved the way in which the veneer flitch was treated while in the heated water vat prior to slicing.

> We put the flitches, as they were milled, in the vat and I kept a record of that vat. I got to know how much temperature to bring it up to and then let the temperature drop to let the wood breathe. This drew out the color so it all became the right color. [At the time] my competitors never did find out how to do it. They thought it was something I put in the water. Some even came at night to steal the water out of the vat and have it analyzed.

It provided a competitive edge. It is said that when several veneer salesmen were visiting a potential customer they wouldn't even show their walnut veneer samples if a salesman from Curry-Miller was present.

The original Curry-Miller structure housing the veneer equipment was built during 1917-1918 before B. L. Curry's purchase of the property in 1929. It is still used although obscured by the substantial additions which have been made since then. The 1929, 12-foot veneer slicer is still in operation in addition to a staylog lathe. None of the veneer is exported and no custom cutting is done for other firms or brokers.

Bud Miller was also a mechanical innovator. Prior to 1930, Capital slicers were driven with a flat belt that limited the machines to 21 strokes per minute. By using a slip-ring motor and a rawhide pinion gear, he was able to increase the strokes per minute and still cut the veneer without "fuzzing" the veneer faces (Ross, 1972).

Until 1982, Miller was chairman of the board, with more than 50 years of service to the company. Other family members in the firm were Bud's son, Bill; son-in-law, Bob Glass; and grandsons, Tom and Ben. Salesmen for the firm were located in Grand Rapids, Michigan; High Point, North Carolina; and Chicago, Illinois. Log buyers for the company operated out of the mill with a system of scouts and producers who located potential logs and timber which Curry-Miller might be interested in buying as logs. The log buyers traveled to the location where the logs had been aggregated and made purchase decisions at the site.

At the age of 86, Bud Miller died in Canada in 1982 as a result of an accident while on a fishing trip.

CHESTER B. STEM, INC.

The early veneer manufacturers were often known for their love of wood, strength of character, and lofty personal goals. Chester B. Stem was one of these colorful individuals (Fig. 5-4). He was also a hunter, fisherman, and a charismatic man of good humor. A hunting accident, which resulted in the loss of one arm, never seemed to dampen his spirit or the fires that kindled his energies. Born in Michigan on February 24, 1894, young Stem first worked for his father, Kellar Stem, who owned and operated the Grand Rapids Bookcase and Chair Company in his hometown of Hastings. One of Chester's jobs for this firm involved the purchase of lumber from Ray Pickrel of the Pickrel Walnut Company of St. Louis, Missouri. Periodic business contacts between these two men resulted in Stem being hired by Ray Pickrel in September 1922. Pickrel was already a giant figure in the walnut lumber and veneer industry (Fig. 5-5).

Stem (1964) wrote about Ray Pickrel and his own experiences in what must have been the "good old days" in his memory.

> As I remember it, Ray Pickrel came to St. Louis in 1912. Previous to this time he had been a walnut log producer and operated a circular mill at Chandlerville, Illinois. In 1912, Ray was 28 years old. Ray had with him a partner by the name of Greg Haskell who had been a lumber inspector at the old East St. Louis Walnut Company.
>
> Before coming to St. Louis, Ray had been selling quite a few walnut logs to the East St. Louis Walnut Company and had accumulated a modest amount of money. With this he tried to buy into the East St. Louis Walnut

Company but they told him that they felt the walnut in Eastern Missouri and Central Illinois had been pretty well cleaned out and that they were contemplating going out of business. I have figured out to the best of my ability about the footage of walnut that Ray brought into St. Louis after he moved there—that is from 1912 to the middle or late 1940s when he liquidated the old mill, and to the best of my knowledge he must have brought in and used not less than 90 million feet of walnut timber. This is a good example to show how little individuals know about the amount of walnut on hand at that time.

When Ray built the mill in St. Louis, he brought down a second-hand, seven-foot band mill from an old walnut operation in East Chicago. Not long after the mill got going the first gunstock contracts came along from England and he participated in these contracts. And eventually when the United States got into the war, he then participated in the United States contracts both for gunstocks and aircraft propellers.

Several years before I came to St. Louis to work for Ray, I bought considerable lumber from him and I remember one contract for 500,000 feet of 4/4 clear face walnut shorts, three, four and five feet long. The price was $50 per thousand feet, delivered at Hastings, Michigan. As I remember, the freight was around eight or ten dollars. The lumber business along in the 1918, 1919, and 1920 period was, of course, very brisk and at the first of 1920 the average market price on 4/4 FAS was $300, #1 Common—$200, #2 Common—$100.

We enjoyed very good business throughout the 1920s in walnut lumber and in that period the prices settled down to $210 for the FAS, $150 for the selects, $100 for #1 Common, and $40 for #2 Common. Peculiarly, we paid about $115 delivered for the walnut lumber logs in the 1920s and today our price is $150 but there is a vast difference in the quality as the logs we get today are much poorer but again our yield on grades is much higher than it used to be in those days before the industry had rewritten the grading rules to a great extent.

Ray liquidated the mill at the old site in St. Louis in the late 1940s as I remember it and moved out farther west of the suburbs and erected another circular mill. This mill was eventually sold to a combination of Thiesing Veneer Company and the boys at St. Joe, Missouri.

On December 1, 1919, the Pickrel Veneer Company purchased the John N. Roberts & Co. mill. The officers were Ray E. Pickrel, president; Guy C. Houston, vice president and treasurer; and B. M. Minigus, secretary and manager. This was a mill that had been purchased by John N. Roberts in 1905 on the banks of the Ohio River at West Tenth Street in New Albany (Anon. 1920).

Stem's account about Pickrel continues:

This mill consisted of a bandmill for flitching only and the famous twelve-foot European-type, underslung slicer—affectionately known to us as "Sally Open Bottom." The hourly production on this slicer used to hit as high as 3,500 feet while the modern slicer today reaches 12,000 feet easily. There was also an eight-foot rotary and a six-foot rotary. The rotaries ran steadily five and one-half days a week, cutting rotary walnut for the North

Figure 5-4
One of the most colorful individuals in the hardwood veneer industry was Chester B. Stem. A man of charm and good humor, he was also an excellent manager of his company.

Figure 5-5
Ray E. Pickrel was a prominent St. Louis, Missouri, walnut sawmill owner. In 1919, he bought the John N. Roberts Company mill in New Albany, Indiana. The mill was later purchased by Chester B. Stem in 1937.

Carolina and Virginia trade. This sold at an average price of about $18.50 delivered and the yield ran right around that figure. On Saturdays, the mill cut stump wood but we were never a large producer of stump veneer. This mill was under the supervision of B. M. Minigus, who was a vice president of the Pickrel Walnut Company and a stockholder.

In 1934 Pickrel sent C. B. Stem to manage the mill, and in January 1937 the site was inundated by 25 feet of the fast-flowing flood waters of the Ohio River. The mill was destroyed except for the flood-damaged machinery. Stem bought the site and machinery in May of that year. Slowly, the buildings were "...rebuilt, the old steam engines, live shafts and belts put back in working order. In those days the mill [produced only] walnut veneer and manufactured only a small amount of lumber. Another serious flood occurred in 1945 and because of [this] it became obvious that the plant would have to be moved. This decision was made just prior to World War II" (Williams, 1975).

Stem had already bought "some high ground" at 2710 Grant Line Road in New Albany and construction was started in 1945. Due to material shortages at that time, the new plant was not operational until 1946. It was reputed to be "...the most modern [hardwood] veneer plant in the country" and one of the most respected in this highly competitive industry (Adams, 1963?).

The Stem mill specialized in quality. Its program for a better product was based on competent men, high morale of the workers and efficient equipment. Efficiency in the use of labor combined with innovative techniques and technology were keystones in all phases of the business. Significant increases in production with only nominal increases in effort were reported in trade publications (e.g., Saul, 1973).

Richard K. (Dick) Stem took over the presidency of the firm on October 4, 1958, when his father became chairman of the board. Dick Stem had an early start in the business, beginning as a young teenager in 1939. Following graduation from the University of Illinois, he joined with his father to build the firm from a struggling business to its outstanding position in the industry. His passions extended to cultural activities such as music, art, orchid growing, as well as to the "good life" pursuits of horseback riding and cattle raising. His two sons, Charles C. and Robert K. Stem, added a third generational attribute to the firm as vice presidents in charge of production and log buying. The patriarch of the Stem veneer family, Chester B. Stem, passed away on June 18, 1971. Others who played dominant roles in the company included E. S. Miller, T. B. Orbison, and Dewey Hickman.

Since the early days of the business down on the river, the company diversified the species it cut and warehoused. More than 100 foreign and domestic species were commonly maintained in the veneer inventory (Fig. 5-6). The company was particularly known for its architectural veneers which are found in major buildings across the country. This type of veneer was heavily promoted in architectural trade journals and was unusual for a veneer producer. In addition to exotic logs and veneers which the firm imports, it also stocks teak lumber for the boat trade. To salute the fact that the company was one of the largest importers of teak, Richard Stem took what "...used to be the world's homeliest warehouse"

Figure 5-6
The Stem Company was noted for the care given in the manufacture of veneer and the great variety of exotic and domestic veneers which it produced.

and created a Southeast Asia, pagoda-like facade. Painted on the front of the building was a working elephant to show how teak is logged in Burma and Thailand (Williams, 1975).

The firm employed about 100 employees in the 1980s and had the capacity to manufacture about 100,000,000 square feet of veneer annually with three slicers and one staylog lathe. The company maintained sales offices in High Point, North Carolina, and in New York City. Stem representatives were also located in San Francisco, Buffalo, and London. The firm also owned 14,000 acres of farm and forest land, most of it scattered throughout southern Indiana.

In June, 1989, the mill was sold to James W. Robinson and David E. Wunderlin. The company name was changed to StemWood, Inc.

ADAMS CUSTOM VENEERS, INC.

Frank M. Adams is a distinguished and scholarly member of the veneer manufacturing fraternity. Prior to starting his own business in New Albany in 1968, he had been associated between 1936 and 1967 with Roberts and Strack Veneer Company, Inc., located in nearby Clarksville, as plant superintendent, plant manager, and company director. Adams Custom Veneers, Inc., was first located in a leased building on Indiana Avenue between 1968 and 1980 before moving to 300 Culbertson Avenue. The present building and site are located in a reindustrializing section of "downtown" New Albany. The ample space provided in this older structure facilitates the movement of materials on the main floor and the warehousing of processed veneers on the main floor and in a conveyor-accessible storage area in the second-story portion of the structure.

When interviewed, Adams said that his veneer was cut using a 16-foot 1950 Capital slicer purchased from Wood-Mosaic in 1969 and a Capital staylog machine purchased from Amos-Thompson Corporation in 1967. The sliced veneers are dried with a Proctor and Swartz dryer. The firm's 37 employees produce only custom cut veneers. All of his key employees have worked with Adams for up to 30 years.

Much of the firm's recent production has been for European sales company entrepreneurs who have brought logs to the Adams mill for processing. Adams also custom cut and provided warehouse space for domestic contractors. Quality manufacture was particularly important in this business. "You have to do a better cutting job...than you would be doing if you were running [your own logs through] your own mill. Here, every flitch has to be cut nicely—no one cares how fast you cut, just so every single one is cut *nicely!*" (Burrell, 1972).

Traditionally, custom cutting has been an entry tactic used by loggers and veneer company employees in establishing their own veneer cutting businesses. Experience is obtained in log procurement, veneer slicing technology, and the identification and understanding of marketing methods. When established markets and sales experience have been obtained, the next step is to produce the veneers themselves, thus, completing a vertically-integrated operation starting with logs and ending with the sale of veneers to customers. This route was chosen by two former employees of Chester B. Stem, Inc., who used the services of Adams' mill prior to the building of their own plant (Norstam Veneers, Inc.) near Mauckport, Indiana, in 1979.

Adams has had a number of especially difficult types of assignments requiring non-traditional forms of slicing and the experimental treatment of unusual tropical species which were being tested for potential market acceptability. Other "wild" ideas are subjected to experimental treatment with patience. Whatever the reason, Adams cuts the logs sent to him the way the owners want them cut, samples the veneer, and warehouses it until requested to ship it.

NORSTAM VENEERS, INC.

Norstam began production in January, 1979, at a new plant located near Mauckport, Indiana, on a bluff overlooking the Ohio River. The original owners all had received their beginning veneer industry experience at

Chester B. Stem, Inc. in New Albany. Kenneth J. Staugas, company president and a graduate of Michigan State University, had been vice president and sales manager at Chester B. Stem, Inc., where he was employed between 1955 and 1970.

Norstam vice president, George Northrup, a graduate of the University of Maine, was first employed by the Stem company beginning in 1955. Later he became vice president of Robbins Veneer Company in Highpoint, North Carolina, between 1964 and 1970. Patrick Mitchell, active locally as a CPA, has not been a participant in the day-to-day mill operations.

Prior to their investment in a production facility, the group had custom cut veneer at Adams Custom Veneers, Inc., in New Albany. These veneers were sold in both domestic and export markets. The Norstam group had been incorporated in Kentucky in 1971 and received authorization to do business in Indiana the following year.

Veneer is produced at the Norstam plant with a 12-foot Capital slicer and a staylog lathe. A seven-section Hale dryer purchased new in 1979 dries the veneer. Norstam employed 40 people in 1983. White oak was the principal species cut although some walnut, ash, pine, cherry, and red oak have been cut. Mark Fitzgerald, in 1989, was actively involved in the company's domestic sales program.

They shared quarters for a time with Pacific Rim Veneers, Inc. which did some custom cutting along with their own production of specialty veneers. Pacific Rim later bought Breece Veneers, Inc., in Kenova, West Virginia, and moved its operations to that location. Pacific Rim has since gone out of business. In 1989, Staugas bought Northrup's stock when he retired from Norstam Veneers, Inc., and became the sole owner.

MILLS WITH LIMITED HISTORIES

A number of Indiana Veneer mills existed and passed from the business scene and community without leaving an historical presence or a "paper trail" which can be easily discovered or followed. The defunct Indiana veneer mills which were discovered in the course of this study are revealed below in alphabetical order.

American Veneer and Walnut Company, Inc.

The American Veneer and Walnut Company of Roachdale was incorporated with a capital stock value of $72,000 in late 1924 by Robert L. Conner, Elmer D. Fule, and Walter A. Eckstein (Anon., 1925a). Eckstein is the grandfather of Larry R. Frye, present executive director of the Fine Hardwood Veneer Association. The veneer and lumber company existed prior to incorporation. In June, 1925, the plant burned, sustaining a loss of $75,000. "The Veneer department was overhauled by the company and new machinery was placed in the factory six months before" (Anon., 1925c). At the time it was not known if the company would rebuild the plant. No document was found to indicate that it was.

Grafton Johnson may have owned the mill earlier. A 1921 letterhead indicates that he owned the Roachdale Veneer and Lumber Company as well as band mills at Osgood, Indiana, and Grayville, Illinois. Johnson, headquartered at Greenwood, was a part owner of the Thompson-Cutsinger veneer plant at Shelbyville. In 1920, he purchased "...the W. T. Thompson veneer factory at Edinburg, Indiana, of which the late W. T. Thompson of that place was the founder and president. The price paid is said to be more than $200,000, most of which was in cash. Roscoe Cutsinger was retained as manager" (Anon., 1920).

Ideal Veneer Company

Ideal Veneer Company of Franklin was operated by Wilfred Hill, probably, prior to World War I. Hill was the father of Jack and Howard Hill who founded Hill Brothers Veneer Company in Edinburg in 1922 or soon after.

F. M. Bachman Company

The Bachman company was an Indianapolis firm with its roots extending back to about 1880. "The history of this business, which during forty years has been producing a line of high grade Oak Veneers, Hardwood and Yellow Pine Lumber, is inseparably linked with the best traditions of the famous hardwood operations of Indiana...Its facilities...are a monument to successful and honest business" (Anon., 1921a).

Dearborn Veneer Company

The company was incorporated in 1925 and capitalized at $50,000. The incorporators were Oran A. Province, C. M. Shepherd, and Charles R. Rogers (Anon., 1925b).

Diamond Veneer and Lumber Company

The Branigan History of Johnson County makes reference to Martin Cutsinger moving from a farm into Edinburg in 1890. He was employed by his father, Sam, in the starch business until 1893 when he bought into a grain and veneer business with John A. Thompson and David R. Webb. Cutsinger later bought the Diamond Veneer and Lumber Company in Shelbyville. It is referred to as a new plant in 1920 which was founded by Cutsinger who "...had hopes of making it into the biggest sawn veneer plant in the country at that time" (Anon., 1966b). An earlier mill, Diamond Veneer Company, of the 1910 era existed in Edinburg and may have evolved by 1927 into the H. I. Cutsinger & Company mill. The Diamond Veneer Company in Edinburg produced sawn veneer "...in all thicknesses" (Anon., 1910).

The Shelbyville mill "...incorporated many of the latest ideas in power-house and mill construction. It is...a progressive concern whose object is to manufacture for the trade native hardwood lumber...and sawed oak veneers. The veneer mill at present contains a battery of five

Louis L. Seibert veneer saws. Another veneer saw will soon be added." It is of no small interest that company officials recognized that sliced veneer was a competitive system which should be recognized. "In the very near future another building will be added to house a slicer and a rotary machine." Homer I. Cutsinger is credited with being the general manager and C. A. Walker as the sales manager (R. H. W., 1920).

Fangman Veneer Company

The mill was operated between 1950 and 1964 by Ben Fangman in Osgood, Indiana. According to Harold C. Davis of Central Veneer in Indianapolis, "That old man down there didn't even wear shoes part of the time. He came up here and bought a rotary, put it on a truck, and trucked it down there. They said that the truck will never get you back home and Ben says: 'I never yet let the sun set on me outside of my home town and it ain't goin to tonight.' He made it and started cutting veneer. He didn't save much. He didn't quite know the business and in the second place put the veneer in the hands of a broker in Chicago. I don't know what happened but it sure wasn't the right way to go." The equipment was auctioned off on October 20, 1965.

Hoffman Brothers Company

The Hoffman Brothers original plant was located in Fort Wayne and was established in 1867 by A. E. Hoffman, W. H. Hoffman, John W. Sale, and Jacob R. Hoffman. Incorporated in 1904 to manufacture hardwood lumber, the company also produced sawn veneer at an early date. From an historical perspective, Jacob R. Hoffman would be remembered as the man who first perfected a band saw capable of sawing hardwood logs in the United States.

> After the close of the Civil War, Hoffman, who had a lumber business in the vicinity of Charleston, West Virginia, went to Indiana with his two brothers...In 1868 at Fort Wayne the Hoffman Brothers Company began to manufacture "Sawed Chair Stuff, Wagon Fellows and Gearing, and all kinds of Scroll Work" and to deal in black walnut timber. Sometime in 1868...Hoffman...began to experiment with the idea of using a band saw for cutting hardwood logs (Loehr, 1949).

After overcoming bandsaw problems by importing high quality French saws, the company began manufacturing bandsaw mills commercially by 1873. The business' main activity continued to be the processing of logs into lumber. In 1886, the Hoffman Company stopped producing band mills and turned over the manufacture and sale of the machines to a firm specializing in metal work.

In 1919, the company was operating two band mills, one in Fort Wayne and the other in Kendallville. Lumber and veneer production utilized between four and five million board feet per year. Veneer was cut by six veneer saws at the plant in Fort Wayne with a production capacity to manufacture 500,000 square feet per month (Anon., 1915a).

Indianapolis Sawed Veneer Company

Located at Fourteenth and Missouri Streets in Indianapolis, this veneer company was in existence prior to 1908. It was destroyed in that year by a fire, resulting in a net loss of $10,000 to $12,000 "...including the veneer mills, sawmill and engine room, also a large quantity of finely-figured quartered oak veneers and hardwood lumber." Mr. J. D. Mavis was president of the company and indicated that the "...office and warerooms of the place were saved from destruction [making it possible] ...to take care of orders promptly until provision is made for rebuilding" (Anon., 1908).

The plant was again rebuilt in late 1921 and early 1922. "The new building [costing $25,000] will be of brick construction and fireproof throughout. It will be 90 feet wide and 110 feet long" (Anon., 1921b). Apparently, the Indiana Sawed Veneer Company had switched by then to a single 120-inch Capital lathe belted to a 50 horsepower motor (Anon., 1921b). Jim Meyers was reputed to have been later associated with the company which presumably went out of business prior to 1930. Pearce Veneers, Inc., occupied the buildings on a lease basis in March, 1930, when the structure burned entailing a loss of $60,000 (Anon., 1930).

Mitchell Veneer Corporation

Curtis Cochran was a custom cutter of veneer at the Curry-Miller Veneer Company in Indianapolis. William C. Linn said that Cochran "couldn't meet the public, but he could make veneer...and he had a sales organization to peddle the stuff." Percy Simmons contributed that the offices at 3711 East 10th Street were not far from the Curry-Miller plant and that in 1946 Cochran built a "beautiful plant" at Greensburg, Indiana. "It was one of the first modern veneer factories in the state of Indiana. He eventually became very successful and was reputed to be one of the biggest and best in Indiana." Ernest N. Rubush served for several decades as vice president of the company.

Consolidated Veneer Corporation of Greensburg was an affiliated company. "The two plants produced more than 60 million feet of veneer every year for dining room and bedroom suites, tables, chests, radio and television cabinets, doors, and wall panels" (Anon., 1950). The mill closed and was liquidated in about 1964 when Cochran was 70 years old. Financial and family problems are rumored as factors leading to the closing.

Long-Knight Lumber Company

Long-Knight was an Indianapolis business prior to 1910 and operated a single-band Sinker-Davis saw mill. In 1919 it was known that the firm also operated three veneer saws. The veneer saws were individually coupled to 50-horsepower Fairbanks & Morse Company electric motors. Mahogany, walnut, and quartered white oak were the principal species veneered (Anon., 1919b).

Sewing Machine Cabinet Company

This was a company based in Bridgeport, Connecticut, which "...built a large factory in northeast Indianapolis. Along with other equipment they bought two veneer cutting machines, one a rotary and the other a half-round rotary lathe. My recollection is that they were made by the De Quoin Iron Works. At any rate, they were light weight and poorly constructed and would be comparable to the veneer basket machines in use for many years. John N. Roberts [then an employee] used the rotary machine for cutting sewing machine veneers and [M. D. Williamson] rented a section of the plant, including the staylog veneer machine, and began cutting figured walnut, butternut burls, and figured ash veneers. He sent the veneers to New York, where he found a ready market and thus started an active cabinet veneer business." These events took place beginning about 1883 (Williamson, 1939).

Singer Manufacturing Company

The Singer company, when located in South Bend, was one of the largest consumers of hardwood lumber and veneers used in the manufacture of sewing machine cabinets. The company was described as having the world's largest wood-working plant which ultimately encompassed 10 million square feet of floor space and employed 30,000 workers (Pollock, 1931). Ross Kreamer remembered that, although most of the firm's veneer requirements were met by outside purchases, Singer was "...the biggest veneer cutter in the country. I don't remember how many slicers they had there. It was a big operation." Existing historical documents retained by the Singer Company do not mention this activity specifically, undoubtedly because it was a small operation in scope when compared to the company's primary interest in manufacturing sewing machines (Fig. 5-7).

Talge Mahogany Company

One of the most colorful and controversial hardwood veneer manufacturers in Indiana was John H. Talge. He was born in Louisville in 1867, the son of a chair manufacturer. The family moved to Indianapolis when Talge was nine years old. "As a youth he learned the upholstering trade, in which he became a skilled workman, and in 1889, when 22 years of age, he journeyed to the West, finally locating in the City of St. Joseph, Missouri, where he engaged independently in the work of his trade, in which connection he developed a large and successful operation...until 1904 when he returned to Indianapolis." Work on the plant had actually started in 1901 and it became operational at Lewis and East Thirteenth Streets by 1904. Talge was both president and general manager of his company (Dunn, 1910).

"In the Talge Mahogany Company, at its big plant, Mr. Talge has established a business which now has grown to be one of the outstanding

Figure 5-7
Most of the history and records of the Singer Manufacturing Company have been lost. The South Bend plant was reputed at one time to be the largest woodworking plant in the world and the biggest veneer cutter and purchaser of veneer in the United States.

concerns in the city. Timbers and logs, from which veneers are made, are shipped to Indianapolis from tropical countries in such volume that the company has come to be the leader in the industry. Rosewood, satin woods, and other imported woods, as well as mahoganies, are used in the Talge plant for the manufacture of wood products which are nationally used" (Brown, 1923).

These somewhat self-serving biographies also point out that Talge was broad-minded, a member of various civic and industrial organizations, and a church. Talge also had contemporary detractors. Bud Miller was familiar with John Talge's reputation:

> The court had me appraise all of his buildings and machinery. Nobody could trust him. He used to go out and buy a fine flitch and use it for nothing but samples...Then send [the customers] some trash veneer. He did something like that to the water company. He had one of his workers dig down and tap the line. Believe it or not, they said it was 20 years before he was caught after he fired the man and the man reported what had happened to the city. I knew him when I was a little boy, my step-father worked there. Talge was never satisfied unless he had six law suits against him.

The Talge Mahogany Company probably was bankrupted by the Crash of 1929 and the subsequent depression. A letter from the Talge company dated April 29, 1929, is on file with the Pierson-Hollowell

Company and was the last indication of the company's existence uncovered in the study of Indiana veneer mills. The Indiana Industrial Directory supports this supposition by no longer showing the Talge Mahogany Company existing after that date.

Other Indiana Veneer Mills

Very sketchy information was obtained on other hardwood veneer firms that existed in Indiana. These include the following:

Advance Veneer Co., Indianapolis	? — 1911?
Albott and Thomas, Milan	1905 —
American Timber Products Co., Indianapolis	1916 —
J. C. and J. Beer, Milford	1907 —
H. C. Brenson & Co., Indianapolis	1890 —
Capitol Veneer Co., Indianapolis	1916 — 1922
Clover Leaf Manufacturing Co., Marion	1890 —
W. R. Drinkard Veneer Co., Indianapolis	1910 —
Goshen Veneer Co., Goshen	1895 — 1941
Gosport Veneer Co., Gosport	1922 —
Heimberger & Drinkard, New Albany	1907 —
Henderson Veneer Co., Franklin	1929 —
Indianapolis Veneer Manufacturing Co., Indianapolis	1882
Interstate Veneer Co., Indianapolis	1899
Milan Veneer, Milan	?
New Albany Veneering Co., New Albany	1916 — 1929
Nenemacher Veneer Co., New Albany	?
Putnam Veneer & Lumber Co., Roachdale	1916
H. J. Sanders & Son, Whittaker	1916 —
Wabash Veneer Co., Indianapolis	1913 — 1922

There probably were others and some of those listed above may not have been directly involved in producing hardwood face veneers for the furniture and other quality markets. Although obscure, additional research might bring to light mergers and the flow of used equipment into the production units of more successful firms which made a greater impact on the historical record of the industry in Indiana.

Chapter 6

MIDWESTERN VENEER MILLS

KENTUCKY

Midwestern states, coterminus with Indiana, shared in the bountiful hardwood forest resource for which Indiana was noted. Oak species and black walnut were the mainstays of the sawmills and veneer mills attracted to these areas. Louisville was the center of these activities in Kentucky. Two of the early firms flourished to become "giants" through successful vertical and horizontal integrative efforts. Whether large or small, all had succumbed to obsolescence and competition by 1960. The dynamic and colorful individuals who brought world-wide recognition to Wood-Mosaic and C. C. Mengel Company deserve greater attention then they receive here. Unfortunately, for them and many other veneer industry entrepreneurs, we can only have an imperfect glimpse due to scarcity of written accounts and the fading memories of those who knew them. Nevertheless, it is not surprising that these mills, along with the rest of the firms in the face-veneer industry, had unique business histories. Their past might include a roll-of-the-dice origin, strong entrepreneurial personalities, technological contributions to the industry, or seemingly random economic and environmental events. Often these historical and fortuitous events subtly molded the character of these firms.

Wood-Mosaic Company

The Wood-Mosaic Company, located in Louisville, was one of those with a history containing several of these elements; but the roots of the company were in Rochester, New York. In 1883, a physician, Dr. C. E. Rider, decided that his love affair with woodworking had reached the point where it was necessary for him to forego the rewards of a successful medical practice and to devote his energies to that which had been his avocation. For many years prior to this time, he had been fascinated with the infinite variety of color, texture, and grain found in fine domestic and exotic hardwoods and had become a gifted craftsman and woodworker. The name he gave his new business stemmed from his initial endeavor of creating a mosaic-patterned living-room floor in his home by using contrasting colored blocks of white oak and walnut. Teak had been used to border the room. Friends and associates who saw this unusual and attractive floor soon persuaded Dr. Rider in his spare time to design and install similar floors in their homes. As the demands for his skills and flooring increased, he recognized the opportunity to establish a commercial business and subsequently built a small factory. The hyphenated name he gave his company, Wood-Mosaic, was chosen to distinguish his product from other types of mosaic floors which used stone, glass, and tiles as materials (Anon., 1943b).

Dr. Rider bought most of the native wood used in manufacturing his flooring from the Hugh MacLean Lumber Company, also located in Rochester. Hugh MacLean was joined about this time by his younger brother, William Archibald MacLean, who had been born in Thurso, Quebec (Fig. 6-1). William soon became convinced that the flooring company was going to be a successful venture and bought stock in the company. Upon the death of Dr. Rider in 1905, William A. MacLean became president of Wood-Mosaic. Prior to Dr. Rider's death, Wood-Mosaic built a mill at New Albany, Indiana, in 1898, near the corner of 18th and Galt Streets. The New Albany plant manufactured wooden flooring and sawn veneers. It was destroyed by fire in 1914 but was rebuilt almost immediately. A sawmill was added to the growing business in 1907 when the company purchased a sawmill for the Hugh MacLean Lumber Company in Louisville, Kentucky (Adams, 1963?).

In 1913, the MacLean Company bought and rebuilt another sawmill in Louisville. "That mill [was] still standing at the 5000 Crittenden address of Wood-Mosaic [in 1979], probably because it was the first mill in Kentucky made with steel and concrete. It was soon after this...that W. A.

Figure 6-1
William A. MacLean became president of Wood-Mosaic in 1905 and served in that capacity until 1933. It was during this period that the company established world-wide connections and was a leader in the industry.

broke off with Hugh...Prior to the break-up, Hugh MacLean was also operating sawmills at Cloverdale and Bedford, Indiana" (Burrell, 1979).

During World War I, the Wood-Mosaic Company produced walnut gunstocks. Shortly after the Armistice, additional factories were built or acquired at Monticello and Russell Springs, Kentucky; Kenova, West Virginia; Hannibal, Missouri; and Jackson, Tennessee. Veneer mills were eventually located in Louisville, Kentucky; Huntington, West Virginia; and Woodstock, Ontario. The Theodor Franks Erben sawmill in Cincinnati was also acquired at the time of World War I and operated by Alex Schmidt and Albert Heidt. Ed Dunlop, Schmidt's brother-in-law, later managed the mill for Wood-Mosaic.

The Wood-Mosaic Company "...was built along strong business lines with a thorough knowledge of the business as a foundation. A study of the organization is interesting because it fairly bristles with Scottish aggressiveness and pep. [T]he things that stick in the mind in a mental survey of that remarkable organization are its unusual ability and organization for maintaining an unceasing supply of high grade logs and its remarkably thorough equipment for and knowledge of the production of specialties in all lines..." Angus MacLean, son of W. A. MacLean, by this time had joined the company and assisted his father as company vice president (Anon., 1918c).

Wood-Mosaic installed its first veneer slicer and rotary veneer machines at the Louisville site in 1920. W. A. MacLean had worked earlier with John N. Roberts in perfecting some of the emerging veneer-slicing technology and employed this knowledge in a one-million dollar veneer and hardwood sawmill plant completed in 1920. In addition to the slicing machines, two rotary lathes cut yellow poplar cross-banding. The double band mill was able to cut 35,000 board feet of walnut and white oak lumber on a daily basis. A battery of dry kilns had a 500,000-board foot monthly capacity to complement air-drying lumber yards which could handle seven million board feet of quality lumber.

The single band mill in Cincinnati cut "...nothing but the highest grade of quartered white oak and walnut, while the band mill at Jackson, Tennessee, produces a general line of hardwoods. The third single band mill operation at New Albany is devoted to the careful production of plain and quartered oak" (Anon., 1920a).

New Albany was also the location of Wood-Mosaic's parquetry flooring factory. The company's veneer plant at New Albany boasted of six modern veneer saws, all of which produced high grade, quartered, white oak veneers. Wood-Mosaic continued as a New York-chartered company until November 10, 1921, when it went under a Kentucky charter (Anon., 1931). In 1922, the company headquarters were moved to Wood-Mosaic's new plant which had been built on Crittenden Drive in South Louisville. All of the New Albany operations were moved to Louisville by 1930, and the last section to be transferred was the flooring operation. The change was made to reduce overhead expenses and to consolidate production activities (Anon., 1930e).

William A. MacLean died in January 1934. He had served as president of the company from 1905 to 1933. He was succeeded by his son, Angus D. MacLean. Other sons of William A. MacLean were also active in the firm at this time. Charles Stuart MacLean was a vice president, Paul Robert MacLean was treasurer, and Donald Hector MacLean served as a company director. Rounding out the corporate structure were J. J. Egan, vice president and general manager, and W. F. Franket, secretary. It is interesting that W. A. MacLean had decreed that Egan would retain his position for a period of 10 years following "my" death. At the end of the 10 years, Egan was summarily dismissed by MacLean's sons.

The firm continued its specialized interest in high quality hardwoods and created world-wide demands for its lumber, veneers, and flooring products. Its raw material sources for prime logs also shared this world-wide arena (Anon., 1934b). In 1936, the Wood-Mosaic Company pioneered the production of prefinished hardwood parquet flooring. When World War II intervened, the flooring manufacturing facilities were converted to the production of gunstocks for the United States, British, and Canadian governments. Other Wood-Mosaic departments produced ship timbers, aircraft veneers, and other war-oriented materials (Figs. 6-2, 6-3). These efforts culminated in the company being presented, in December 1942, the first Army-Navy "E" to be awarded to a lumber and

Figure 6-2
The deHavilland Mosquito bomber was a fast, maneuverable airplane which was successfully deployed during World War II to attack railroads, bridges, and other specific targets. Except for its powerful engines and mechanical parts, the plane was almost entirely made from veneer bonded to form plywood which met exacting design specifications. Several hardwood face-veneer mills in the United States were involved in producing the veneer.

Figure 6-3
More than 15,000 of these troop-carrying gliders were constructed during World War II. Pulled by C-47 airplanes to their destinations, these plywood and fabric gliders were used extensively in the invasion of Europe.

veneer manufacturer for excellent and efficient contributions to the war effort. It later received four additional government citations for excellence during the war years (Anon., 1958).

After World War II, the company resumed its traditional operations and continued to grow and become once again the largest manufacturer of specialty and custom flooring in the United States. In 1960, the company had consolidated net sales of $17.7 million divided about evenly between flooring, lumber, veneers and plywood.

Evidence of its commanding position in the industry and "...its planned progress and present-day aggressiveness" were noted by Burrell (1962). The company, through its subsidiary, Wood-Mosaic Industries, Inc., operated sawmilling and veneer operations in Colombia, South America, and in the Philippines in addition to "...widespread import and export activities which have grown to sizeable proportions." The subsidiary operated its import/export business out of offices in Louisville; New Orleans; Norfolk, West Virginia; and Houston, Texas. Mahogany, teak, rosewood, ebony, satinwood, and lauan were among the species imported. A fully conveyorized face veneer operation was added in 1963 to the Huntington, West Virginia, corestock and crossbanding plant. Combining the face veneer production from both Louisville and Huntington made Wood-Mosaic one of the largest producers of quality face veneer in the United States. Total output at the time was estimated to be about 10 million square feet per month.

Paul R. MacLean succeeded his brother as president in 1962 and remained with the company in this capacity until 1969. Angus MacLean became chairman of the board of directors when Paul became president. Jon Sig Gudmundsson was vice president and treasurer; Omer E. Hanbrock, vice president; and Eva M. Streepey, secretary. Donald MacLean Bell, grandson of William A. MacLean, formally joined Wood-Mosaic after his graduation from Purdue University in 1953 although he had been a summer employee of the company for several years prior to that time. He remained with Wood-Mosaic until 1969 and was a vice president and member of the board of directors when he left to organize International Hardwoods, Incorporated, also located in Louisville.

Over the long tenure of the MacLean family, Wood-Mosaic established world-wide connections. As a result, many young people gravitated to Louisville from foreign countries and the United States to receive on-the-job training in the Wood-Mosaic "college" as it became known in the industry.

At the beginning of 1969, the Wood-Mosaic Company employed 800 people at plants in Louisville; Huntington, West Virginia; and Tween, Ontario. The Louisville plant employed about 420 of the total. However, substantial profits proved to be elusive. For the fiscal year which ended April 30, 1968, Wood-Mosaic reported net earnings of only $291,681 on sales of $17,167,858. Wood-Mosaic was purchased in 1969 by Olin Mathieson Chemical Corporation.

The following year, Olinkraft, a wholly-owned subsidiary of Olin Mathieson Chemical Corporation assumed operational responsibility for Wood-Mosaic as a part of its Wood Products Division. Paul R. and Angus D. MacLean were retained as consultants. Paul died in 1971 and Angus in 1981. Olinkraft, based in West Monroe, Louisiana, then operated 22 plants in the Midwest, South, and in South America. E. A. Freeman, a former vice president of Olinkraft, became the new president. The sales manager was Charles Roberts (Anon., 1969c). In 1975, Olinkraft became a separate corporation and in the following year Wood-Mosaic became a division of that company. "At the time of acquisition in 1969, Olinkraft saw in Wood-Mosaic an ideal hardwood complement to its pine wood products, paper mill, and converting plants, and a partner in international trade [with Olinkraft's] mill and converting operations in Brazil. The integration of Wood-Mosaic into Olinkraft went smoothly..." until the industry-wide doldrums of 1975 (Anon., 1976). Corporate management bet heavily on new marketing concepts for increased sales and emphasis on foreign markets. Marketing leadership was entrusted to a salesman recruited from the plastics industry. The serious problems of maintaining and increasing the output of veneer with aging equipment apparently were secondary considerations. Howard R. Flock was given the unenviable task of circumventing these problems as vice president in charge of production and the development of export sales.

In an interview, he said that he was well qualified for the job, having literally grown up in the veneer business. He began his career in 1937 with the Pierson-Hollowell Company where his brother, Dutch Flock, was employed. Except for time spent in the military and a few years when the mills were "down," hardwood veneer and lumber filled his working life. Between 1945 and 1950, he was with the E. E. Mitchell Company of Indianapolis which built a mill in Greensburg, Indiana in 1946. During 1951 and 1952, he was plant superintendent at Swords-Morton in Rock Island, Illinois. Subsequently, Flock was with Veneers, Inc., at Cockeysville, Maryland, and in 1955 worked at the Kentucky Veneer Works in Louisville until the mill closed in 1956. He was hired by Wood-Mosaic the same year. Between 1965 and 1975, he was on the board of directors as well as being general manager of Wood-Mosaic in 1974 and 1975. Flock retired in 1976 when he could no longer tolerate the business practices of Olinkraft. Flock represents a number of similar men who dedicated their lives to the production of quality hardwood veneer. The failures of firms such as Wood-Mosaic to fulfill the aspirations of large, multi-faceted companies can most often be traced to a lack of knowledgeable leadership brought into key decision-making positions and poor strategies in matching sales and marketing techniques to conditions in the world economy.

Johns Manville Corporation purchased Olinkraft in 1978 and the following year sold Wood-Mosaic to Katz Werke AG, a West German corporation, headquartered in Baden Baden. Katz Werke AG was reputed to be the oldest corporation in West Germany. The firm produced a diversified line of lumber, ceramics, veneers, paper and paperboard, plywood, and concrete products. Wood Mosaic (unhyphenated), was organized as a Division of Katz of America. The consultant called in by the Katz organization after it had signed a letter of intent to purchase was Michael Boel. As he later recounted:

> I had never seen in my life a more rundown, more obsolete, a more terrible state of affairs in a veneer plant. Here is this old, old plant, rich in tradition, but nothing behind it. It had to be inefficient. [They were] doing a little bit of everything and nothing right.

As part of Boel's restructuring, the labor force was cut to 150 employees. The flooring department which had been at center stage in the company for nearly a century was eliminated along with other low-profit areas of the business. As much as 40 percent of the veneer produced by this plant in the 1979-1981 period was exported to Europe. However, in late 1981, 80 percent of the veneer being produced was sold to domestic markets. Showrooms were located in High Point, North Carolina; Los Angeles, California; and Montreal, Quebec. Sales representatives operated in other market areas such as Grand Rapids, Michigan. Principal species sliced were white oak, red oak, and walnut. Maple, elm, and cherry were also sliced along with smaller quantities of other species. Capital

Machine Company equipment was used in the veneering process and consisted of 1979 and 1981 model-year slicers and two older staylog lathes.

The plant also was equipped with the latest in materials handling, drying, and "crating" technology. A European-style clipping system for export orders utilized a computerized measuring system which prepared the identification tag for clipped bundles as well as recording this information in a computer inventory file.

In the summer of 1983, Larry Richardson and Wayne Bryant formed the Richardson & Bryant Corporation and took over the assets and name of Wood Mosaic. Their objective was to strengthen Wood Mosaic's stature in the domestic and international veneer industry. The organizational goal was to produce 120 million square feet of veneer per year. Richardson, who had sales experience at National Veneer at Seymour, Indiana, had primary administrative and marketing responsibilities as president. Bryant was vice president in charge of operations. Both had been long-time employees of Wood Mosaic (Anon., 1984).

Although expressing optimism about the future of the company in 1984, as Wood Mosaic celebrated its 100th anniversary, Richardson announced the following October that the company had ended veneer production. The company's obituary, as excerpted from the Southern Lumberman, provides some of the specifics and pathos associated with this once proud company.

> The sale included the business, its 16-acre complex, machinery, inventory, and accounts receivable, plus a $16 million indebtedness, $7 million of which was an industrial-revenue bond which the city had issued Katz of America, Inc., when it purchased the whole company in 1980. The bond, one of the largest of its type at that time, was issued on assurance from Katz it would save the 300 jobs at the plant and add more as soon as possible. Unable to meet that goal in three years, the company was sold to Richardson and Bryant.
>
> Richardson and Bryant improved the atmosphere for the workers, employed more women, provided day-care facilities for workers' children, pushed the work force up to 150 and focused on the sales and customer relations. Richardson said he would not change anything he had tried, but "I've learned a few things along the way...like you cannot operate without money" (Anon., 1985).

Mengel Company and Foreign and Domestic Veneer Company

One of the largest midwestern hardwood products companies was the Mengel Company of Louisville. Its early beginnings began in 1875 when C. C. Mengel moved to Louisville at the age of 20. His first business involved the marketing of leaf tobacco. This led to an interest in manufacturing containers for tobacco products and to the company he purchased in July 1877. The small box plant, renamed the Mengel Box

Company, was acquired from a local bank which had foreclosed on the previous owner (Anon., 1931). Mengel was later joined by his brother Clarence H. Mengel to form C. C. Mengel, Jr., & Brother Company, which was incorporated in 1888. Two years later, a younger brother, Herbert W. Mengel, joined the company and performed an important role in the rapid growth of the company. These Mengels were all born in Gloucester, Massachusetts, where the family had been prominent manufacturers of woolen goods for several generations.

In 1899, the business was split with C. C. Mengel retaining control of the box company. His brother, Clarence, obtained as his share the C. C. Mengel Company which was the lumber production division of the former company. In 1919, the Mengel Box Company acquired the controlling interest in the C. C. Mengel Company. C. C. Mengel emerged as the prime mover in the consolidated business.

From this humble beginning, the operations of the Mengel Company grew to be world-wide. The company had "...immense timber holdings in the South and ten large plants in the wood business, including sawmills, veneer and plywood mills, automobile body plants, box plants, a paper mill, etc. It also [had] a huge corrugated shipping container plant. It [owned] vast forest tracts on the West Coast of Africa for production of mahogany logs, brought to this country by its own steamers for manufacture at the mahogany mill in New Orleans, and the mahogany and veneer mills at Louisville" (Anon., 1935b).

The Louisville mill was actually one of a very few large mahogany mills in the United States. Lumber, dimension stock, and high-grade veneer were cut from African and Honduras mahogany. Mexican and Nicaraguan cedar were also cut in lesser quantities. The mechanical equipment in the veneer mill included a 16-foot Capital slicer, a 12-foot Smith machine, two veneer saws, drying equipment, and veneer clippers. A large veneer warehouse, sample room, and factory office were in an adjoining building (Anon., 1920a). The mahogany department became the domain of Herbert W. Mengel between 1900 and the time of his death in 1930. He had come to Louisville in 1890 to join his brothers after stays in Brooklyn, New York, and Bay City, Michigan, where he had gained early experience in the timber business (Anon., 1930).

To provide a continuous flow of tropical logs, the Mengel Company had to commit considerable amounts of human and capital resources to the task. This was illustrated by several of its advertisements in the 1930s. One shows four appropriately-dressed Mengel employees resting in a small African village enroute from the coastal city of Axim (Gold Coast) "...into Mengel logging operations far in the interior." One of the men (F. L. Stork) is quoted as saying: "We think enough of African mahogany to be willing to walk after it—not a mile or several miles, but from 20 to 25 miles per day and for several days in succession." The advertisement also indicates the degree of dedication of these men in seeking the best mahogany for Mengel's customers by describing the "hotels" as mud huts with thatched roofs and the necessity of having food and bedding carried with them using assistance provided by native carriers. The Mengel Company also introduced the use of tractors in transporting logs from interior logging areas to river locations. From these locations logs would be floated to the coast during the rainy season. There ocean steamers would pick up a million or more board feet of them for the voyage to New Orleans. Prior to the use of tractors, teams of native laborers literally dragged the logs to stream banks by hand. The introduction of tractors permitted the logging of mahogany trees at greater distances from water courses and increased the number of economically accessible trees. The strategies used in Africa were also applied by the Mengel Company in Central America and Mexico where extraction conditions were similar (Anon., 1934a).

Although the financial rewards from these tropical investments were no doubt substantial, the economic and human risks were great as well. One of the greatest tragedies was the death of Clarence H. Mengel, Jr., at Grand Bassam, Gold Coast. At the time of his death from "one of the African fevers," Clarence was 26 years old and had been "...regarded as a young businessman of unusual promise and one who would climb high on his own energy and intelligence. He had already demonstrated splendid ability, being quick to grasp ideas and resourceful and energetic in putting them into practice" (Anon., 1920c).

The death of C. C. Mengel on November 8, 1933, may have signaled the beginning of the end for the Mengel Company although the company continued to operate until the late 1940s. Many of the departments were affected by changes in the demand for the company's traditional products and the reluctance or, perhaps, the inability of an aging management staff to understand the significance of market dynamics and to invest in technology and equipment required to remain competitive. One after another, markets slipped away. The Pullman Sleeping Car Company of Burnside, Illinois, no longer required hardwood lumber, and fancy veneers and automobile frames were no longer made from wood. The furniture plant was sold to Kroehler and the paper department to Container Corporation of America. One of the last elements sold was a veneer plant in Thibodaux, Louisiana, which was barely a year old when put up for sale. The company died by degrees. The basic C. C. Mengel plan was to integrate company operations both horizontally and vertically. The plan worked brilliantly for over a half century before being derailed by economic and managerial factors beyond the control of the Mengels and other managers who followed.

Two sons of Clarence R. Mengel refused to follow the urging of their father to consider the Mengel Company as a career opportunity. They were Sam and J. Torbit Mengel, who, after serving their country in World War I, turned down the opportunity for a promising future with the family business. When Sam Mengel, Jr., was interviewed, he said that the two brothers had an intense desire to "make it" on their own. After unfolding a map of the United States and discussing timber prospects, they had zeroed-in on the Knoxville area and agreed in 1922 to build a veneer mill there.

The decision was influenced by the fact that Sam was employed in Knoxville by the Brunswick-Balke-Collender Company which was producing lumber and veneers. Sam bought a 1912, 12-foot Capital slicer from his former employers and hired its operator, William L. Taylor. He was to be the Foreign and Domestic's only veneer cutting room foreman between 1922 and 1969. Taylor had begun working with the 1912 slicer in 1914 at Brunswick-Balke-Collender where he was paid at the rate of $0.075 per hour. His son, William L. Taylor, Jr., followed his father into the business and began his work experience on the same 1912 Capital slicer. By 1972, the younger Taylor had been promoted to plant superintendent, replacing Floyd LaRue, a company vice president, who had held this position for some 50 years.

Although the mill was located in Knoxville, the business and sales offices were located in Louisville where Sam Mengel preferred to be in his capacity of salesman for the business. Torbit (Torb) was more the engineer who could spot production and maintenance problems and was responsible for mill operations. As the business grew, the company added three 18-foot Capital slicers, two staylog lathes and three dryers (two Proctor-Swartz and one Coe). The last equipment purchases were in 1969 when an 18-foot slicer and a Proctor-Swartz dryer were added. One of the older 18-foot slicers was subsequently sold to a firm in Colombia, South America (Anon., 1939).

Torb Mengel was highly regarded for the high quality veneer cut at this mill. Ross Kreamer reiterated that his reputation was well deserved. "Torbit Mengel was one of the best veneer operators in the business. He was an excellent mechanic [and he] cut beautiful veneer. An enviable reputation was built up for turning out fine veneers in both native and imported species."

In the 1950s, the company's principal production was devoted to mahogany, cherry, and a few other domestic species. The cherry logs came from Pennsylvania, the hickory from Tennessee, walnut from Kentucky, and the mahogany from the Gold Coast of Africa. In the 1970s, Foreign and Domestic was best known for its African mahogany veneers which had long been stock items. Foreign and Domestic log procurement specialists were employed in Africa for many years seeking the best logs to be shipped to Knoxville. Mahogany logs entered the country through the ports of Norfolk and New Orleans. As market requirements changed, the mill cut mainly cherry and oak veneers and tulip poplar veneer from logs obtained near Knoxville. Maple, birch, and pine veneers were also manufactured by them. In the past, exotic species were a larger share of their production, including Zebra wood and Tiger wood veneers. Single shift production capabilities were 50 to 60 million square feet per year with a work force of 90 people.

The markets served were mainly in Virginia and North Carolina although Kentucky and Indiana firms such as Hoosier Panel Company, Jasper Wood Products, and Paramount Plywood were good customers. The company maintained sales representatives in Grand Rapids, Michigan, and High Point, North Carolina.

After Sam Mengel, Jr., joined the firm, he worked out of the Knoxville mill between 1945 and 1948 buying logs and standing timber and learning about the production end of the business. The following year was spent in High Point, North Carolina, gaining sales experience. His father reassigned him to the Louisville office in 1950 where he remained to manage the company as president and chairman of the board after the death of his father in 1961. Sam, Jr.'s, son, Sam Mengel III, ultimately became responsible for the Knoxville mill and was appointed vice president of Foreign and Domestic in 1975 after an intensive eight-year internship in company operations.

Sam, Jr., was philosophical in 1982 about timber resource availability in the future. He recalled that: "One time when I lived in Knoxville and was out buying logs, I met an old log buyer who was 85 or 90 years old. He told me that his 'pappy' had told him to get out of the logging business because it wasn't any good. All the timber was gone. Times haven't changed a whole lot." However, the firm was selective in picking the areas where they competed for timber. "We don't cut walnut veneer because, if we did, we would have to pay extra freight from Indiana and Illinois to Knoxville. That extra freight would knock us out of competing with Hartzell and all the other walnut boys. We would have to charge a little bit more because of our extra freight." However, Sam did admit, in the veneer doldrums of 1982-1983, that "...once the furniture market and the panel market get going again, then it will be a scramble for wood."

Foreign and Domestic Veneers enjoyed several boom periods in its history, beginning in the early days when the principal product was veneers for radio cabinets. A later surge came in the late 1920s and early 1930s when outside plumbing was moved indoors, creating a big demand for toilet seats. These were made primarily from 1/4-inch basswood and buckeye veneers. After World War II, booming sales of television cabinets caused another peak period.

"We don't know what or when the next boom will be," Bill Taylor, Jr., was quoted as saying, "but we'll be ready for it" (Moore, 1975). However, by 1985, Sam Mengel, Jr., with an almost impossible task of operating with an outdated plant and a changing market, closed the doors of the Knoxville plant to end a long era of veneer manufacturing by members of the Mengel family.

Kentucky Veneer Works, Inc.

When this Louisville firm was founded in 1899, it began with two veneer saws and two more were added the following year. In 1904, a Capital staylog lathe was installed to augment production. During the next 15 years, many other improvements were made, including changes in drying equipment as this technology dramatically developed about the time of World War I (M. F. J., 1922a). Frederick C. McCracken, owner of the plant, marketed his veneer as Kay-Craft and specialized in oak and yellow-poplar veneers. The plant was located on the Louisville and Nashville Railroad at 1380 South 13th Street, near the downtown area.

In 1920, a modern addition was constructed to accommodate a larger Capital staylog lathe and a clipper. Following a fire the next year, the building was enlarged for a smaller lathe and a veneer dryer. Belt-driven devices were replaced by individual electric motors for each machine. At some point, McCracken added a slicer and retired the 1904 staylog lathe.

John Joseph (Joe) Egan joined the firm in 1945 as manager and was elected president in 1951. Egan had been with Wood-Mosaic for 40 years prior to the move to Kentucky Veneer. He began his career with Wood-Mosaic in 1910 as a clerk when it was located in New Albany, Indiana. Between 1917 and 1919, Egan had worked at a company subsidiary in Huntington, West Virginia. Returning to Wood-Mosaic at the new site in Louisville, he rose rapidly through the ranks to become treasurer in 1923 and vice president and general manager in 1930. He continued this managerial role between 1935 and 1945 when active management shifted back to members of the MacLean family.

By 1951, Kentucky Veneer Works was in its death throes. There had been no capital improvements since the 1930s and increasing competition was aggravating internal problems of inventory control and sales policy. Egan's death in 1955 prompted the sale of the firm to the Midwest Walnut Company of Council Bluffs, Iowa, a company which desired to add veneer production to its already large lumber operation. This was a short-lived marriage and the investment was liquidated in July, 1956, to take advantage of the tax loss. The equipment ended up at the Union Veneer Company in England (Anon., 1956).

Louisville Veneer Mills, Inc.

Daniel E. Kline was born in 1850 in Cincinnati, Ohio, where he was educated and went to work at age 18 for the Harry Albro Company, a pioneering veneer cutting organization in the Midwest. He left in 1889 to establish the Louisville Veneer Mills at 1100 River Road, Louisville. The brick building originally occupied by the company had been erected in 1887 by a company which had gone into bankruptcy (H. W. M., 1927). The plant was "...located, not directly on, but near the Ohio river...at a place locally known as the 'Point' where it is in close company with the three river sawmills of Louisville, and a box factory" (Anon., 1907a).

The small business Kline founded grew steadily and branched out at an early date to manufacture plywood. It was reputed to be the oldest plywood producer west of the Blue Ridge Mountains except for the Grand Rapids Veneer Works in Michigan. The equipment in 1907 consisted "...of two [Coe] staylog lathes, two [Capital] slicers, two veneer sawing mills, and extensive panel equipment for making panels and built-up lumber of all kinds." The buildings to house these machines had grown into a substantial industrial complex. Annual consumption of logs was about 2.5 million board feet of which oak was the principal species. Other domestic species were also processed along with large quantities of mahogany.

Harry E. Kline joined his father's business in 1890 and became an integral part of the management team. Harry was known for his violent temper. Howard Flock noted that "...if you hadn't been cussed out by Harry Kline, you didn't know anything..." He was even rumored to have killed a worker who failed to move a rail car when ordered to do so. His influential father was said to have manipulated the legal system to keep him out of jail. Otherwise, he was known as a "mill" person and an excellent veneer cutter.

By 1916, the Klines were utilizing individual electric drives on their machines and had eliminated line shafts and belting "...reducing the liability insurance rates, and [leaving] unimpeded space for the workmen, with better light and circulation of air. The plant [at that time was] producing nearly $500,000 worth of veneers and panels [and it is expected] the output will be largely increased" (Anon., 1916b).

Apparently, the last major improvements in plant facilities, mechanical devices, and building additions took place in 1920. The sawmill, housed in a four-story factory building, was operated for cutting flitches for the veneer saws and slicers while also producing large quantities of lumber. Cranes, conveyer systems, and a well-planned organizational layout of the several departments completed the industrial picture of this firm. The veneers produced in this modernized plant included mahogany, walnut, quartered oak, and gum. Figured red gum was particularly featured (Rohr, 1921).

Daniel Kline died in 1929, at the age of 79, as a highly respected businessman and civic and industrial leader. He was one of the original directors of Southern Hardwood Traffic Association and served 10 years as a director of the National Veneer and Panel Association. Between 1917 and 1919, he was president of this organization and was instrumental in obtaining a reclassification of freight rates on lumber and lumber products beneficial to both manufacturers and shippers (Anon., 1929b).

Harry Kline succeeded his father in the business. He was widely known in veneer circles for his trade association activities, which included being past president of both the National Veneer and Panel Association and the Louisville Hardwood Club. He also was a director of the National

Hardwood Lumber Association. Harry continued the company's trend toward specialized production of figured red gum veneers. The logs were mainly purchased from Nickey Brothers and Anderson-Tully, headquartered in Memphis, both owning immense acreages of old-growth forest land in Tennessee and southward in the Mississippi River bottoms.

Louisville Veneer Mills ceased operation with scarcely a whimper about 1957, a difficult time for most veneer mills. Inexpensive plywood panels from Japan were flooding the market and competition from domestic veneer mills was excessive. Harry Kline died in 1959, the last symbol of a once powerful company.

Southern Veneer and Manufacturing Company

The Southern Veneer plant was located in Louisville at 2201 Standard Avenue near 22nd Street. George F. Kretschmer was the prime mover in building the mill about 1900. Kretschmer's brother, Louis, was also involved as a company officer and sold the company's veneers in a region centered on Chicago. At some point, John Hunlow became sales manager for the firm.

The original facilities burned in 1910 and, by late 1911, the plant had been rebuilt. "The company advises that although hardly over the effects of the fire, the plant has been running at full capacity for several months. It is equipped throughout with improved machinery and is making a specialty in the manufacture of oak, mahogany, and cabinet [quality] veneers" (Anon., 1911a). Veneer production was carried out with three veneer saws, a Capital slicer, and a Capital staylog lathe. The veneer saws were not sold until 1948.

In 1918, 60 percent of the logs utilized came from the company's own timber land "...where men and mule teams are continually employed in cutting and transporting logs to points of shipping." Two systems were used in drying the veneer: a Proctor drier for the sliced veneer "...and a specially ventilated warming room, where sheets of veneer (principally rotary-cut crossbanding) are suspended from overhead wires till dry." Some of the output was further manufactured in the "face" plant by jointing, matching, and taping face-veneer pieces into dimensional faces. Many exceptionally fine figured veneers were produced from red gum and Kentucky white oak (Anon., 1918b).

By 1920, Southern Veneer had switched from belt-driven to electrically-driven machines in order "...to obtain greater productive capacity and facilities." A new warehouse was constructed as part of the modernization program, along with a new steam boiler and automatic sprinklers in the veneer mill. A new Capital lathe complemented existing veneer cutting machines (Anon., 1920d).

Near the end of the business life of the company in 1955, the veneer machines consisted of two veneer slicers, two lathes, and three Proctor and Swartz automatic dryers. "Employing about 200 workers, the plant produces an average of 30,000,000 square feet of domestic and foreign veneers and lumber each year" (Anon., 1955). The company closed its doors prior to 1960 because of George Kretschmer's age and because of domestic and international business competition. Kretschmer died in 1962.

Freeman Corporation

The early beginnings of the Freeman Corporation in Winchester, Kentucky, can be traced back to George E. Tomlinson who was born in 1868. According to Tomlinson's grandson, Eugene E. (Gene) Freeman, Jr., George had begun his life-long career in the timber industry as a saw filer in Michigan. At the age of 18 he moved to Kentucky and worked as a saw filer in a band mill at Pineville. When this mill went out of business, Tomlinson moved to Ford, Kentucky, and later to Clay City, Kentucky, about 20 miles from Winchester where he worked in another sawmill. Around the turn of the century he moved to the Winchester area and became an entrepreneur by owning a retail lumber business with a partner. Leaving that business, he started a tobacco hogshead business and also manufactured a hardwood banquet table which he had designed and patented. The required lumber was sawn with a 54-inch bandmill.

During World War I, Tomlinson cut walnut lumber with his bandmill for P. V. Shoe Walnut Company, located in Cincinnati, and rough gunstock flitches for Wood-Mosaic in Louisville. By the late 1940s and early 1950s, lumber sales were less than $50,000 per year and the mill ran only sporadically. In 1953, a fire destroyed the primary sawmill building.

When Tomlinson died in 1953, Gene Freeman, then 24 years old, took over what was left of the business. Freeman, an alumnus of Washington and Lee University, had just recently graduated from Syracuse University with a Masters degree in wood utilization. The company had seven employees, a small inventory of lumber, and a few orders for fold-up banquet tables which were still being sold to churches and schools. The company did a "rousing" $73,000 of total business in 1953, which Freeman said was better than 1952 "...because that was the year a lot of the plant's effort was turned to rebuilding Grandfather's boat!" Some of the early decisions were to drop the table business, expand lumber production, and, in 1958, to begin exporting logs. "We ultimately learned that the company lost $7 on every table made and if somebody had ever suggested we try to make it up on volume, I'm sure we could never have moved quite fast enough." In the mid-1960s log exporting became less attractive financially and, in 1967, the firm began producing sliced veneers.

This was the first veneer mill of any type in the area, much less one specializing in walnut. The decision was ill-timed as it coincided with a fairly severe recession and a "shake-out" in the veneer industry. However, in spite of economic adversity and start-up problems, the plant was efficiently producing large quantities of veneer by the end of the year. The problem of inexperienced personal for this type of operation had been solved by training some supervisory employees at firms where veneer logs were custom cut for the company and by hiring experienced people. All the rank and file labor had to be trained at Freeman's mill when veneer operations began. One of the larger problems which had to be overcome, according to Freeman, was "...that of getting key people to understand the expense involved in each minute of downtime which is amazing to those who do not have experience in this business. Saving 30 seconds in changing flitches on the slicer...amounts to a cost reduction of about $160 daily" (Anon., 1968).

Dr. Aubrey Wylie, who served for a time as executive vice president of the company, was largely responsible for the configuration of the veneer mill. Freeman became acquainted with Professor Wylie at Syracuse University. Wylie later transferred to Michigan State University where he and his students designed and made a scale model of a modern veneer plant. It was this model which became an actuality at the Freeman Corporation plant at Winchester. A second Capital slicer was added in 1980 along with a single-deck Hale dryer.

The business has not been without some setbacks which were deftly turned into opportunities. A fire totally destroyed the sawmill in November 1976. Freeman immediately began a rebuilding program which resulted in the construction of a new sawmill designed so that wood waste could be used as a fuel in the new boiler. The mill had the capacity to be one of the largest walnut lumber production facilities in the world. It was equipped with a McDonough bandmill, a Corinth carriage, and a Lasaline guide light system for the saw line. Rebuilding and modernization in other departments soon followed. The double-deck jet dryer was rebuilt and a unique belt tracking device was added (Anon., 1977).

All lumber produced in 1982 was walnut; veneer production was mostly walnut, although red and white oak veneers were also cut. Logs were obtained primarily from sawmills and producers in Kentucky, Indiana, and Ohio. Some logs were also purchased occasionally in Iowa, Illinois, and Tennessee.

Veneer was sold through business representatives in North Carolina, Michigan, Illinois, California, and Canada. About 40 percent of Freeman's veneer was sold in North Carolina and Virginia by the High Point, North Carolina, representative. In recent years, as much as 20 to 25 percent of the company's veneer was exported and sold to European wholesalers. With changes in economic conditions in domestic and foreign markets, it was decided to concentrate sales efforts on a few of the firm's best customers and to specialize in the domestic walnut market while retaining the capabilities to export when a favorable marketing situation existed.

The firm employed 100 to 125 people in the early 1980s, with about 70 of them directly related to veneer operations. Gene Freeman's sons, George T. and R. Reid Freeman, were both active in the firm's management activities in 1989.

From annual production of about 60 million square feet, 12 to 13 million square feet of veneer have been maintained as inventory in the 50,000-square foot Winchester warehouse. Freeman has always taken the "hard line" on plant housekeeping by stressing the importance of a clean mill. Foremen followed established procedures on quality control and maintenance. These procedures have paid off in increasing productivity, output, and sales in the range of about $6,000,000 per year.

Burnside Veneer Company

In 1888, the Pullman Sleeping Car Company organized a subsidiary, the Chicago Veneer Company, to cut veneers required in the construction of its sleeping cars. The Pullman Company was located at Burnside, Illinois, on the south side of Chicago. The cars were ornately paneled, including the upper berth which pulled down at night. The exposed panel for this berth needed to be sufficiently strong to carry the weight of the bed. It was also curved to provide additional headroom for those seated in the cars in the daytime. Therefore, it was necessary to have a material that was light, strong, attractive, and which could be molded to the desired configuration. Hardwood plywood consisting of a tulip poplar veneer core and a walnut, oak, or mahogany "face" met these specifications.

In 1902, the company's veneer operation moved to a site on the Cumberland River which eventually became Burnside, Kentucky, in order to be "...in the heart of a fine yellow poplar timber belt."

> The company claims to have manufactured more rotary-cut poplar during the last fifteen years than any other three veneer mills in the country. The Burnside mill has been running full force, and is turning out large quantities of rotary-cut oak, chestnut, ash, basswood, yellow pine, walnut, etc., in addition to the poplar veneers.
>
> For several years the company maintained offices at Danville, Kentucky, but owing to the tremendous growth of the business it was decided to move the offices to the large warehouse of Fifteenth Place and Western Avenue, Chicago, Illinois. The Chicago Veneer Company of Illinois was organized, which will act as selling agent and handle all orders for the Chicago warehouse, as well as both the Burnside and Clarendon mills. The officers of this company are Benj. W. Lord, president; S. J. Glanton, treasurer, and H. A. Webster, secretary (Anon., 1917a).

Although the mill was primarily used to cut yellow poplar core stock, oak and walnut face veneers were also produced. Additional veneer requirements were purchased from other suppliers such as the Dean Company of Chicago. In 1911, the Burnside cutting department was equipped with two rotary lathes. A third was added by 1916 (Anon., 1911b, 1916a).

In 1920, the Pullman Company, believing that it had cut out all the good timber, sold the veneer mill to Stoddard J. Glanton and H. A. Gable, including an inventory of half, third, and quarter sheets. The Pullman

Company had accepted only full-size, 50-inch x 98-inch, sheets of unblemished veneer, which meant that there was a lot of manufactured veneer that didn't meet this specification. The new owners promptly sought out buyers for this inventoried veneer. It was sold to the Canadian Pacific and Canadian National railroads and used in the building of their sleeping cars. These sales paid for the mill which was reincorporated as the Burnside Veneer Company.

The plant was equipped with three full rotary lathes and a staylog lathe for cutting walnut stumps and burls. On each operating day, the mill could cut a carload of quality veneer. Prior to 1928, veneer was dried by non-mechanical means. In the "hang" house, the veneer was fastened with spring clips, one or two sheets at a time, and hung from the ceiling to air dry. In the "groove" house, the veneer was inserted lengthwise in "finger" racks which held the veneer vertically upright (Fig. 6-4). Air was circulated to assist the drying process.

In 1928, Ross Kreamer went to work for the Burnside Veneer Company. He had been born of pioneer stock at Hotchkiss, Colorado, in 1903. He was well acquainted with the Colorado wilderness and had worked

Figure 6-4
Finger racks for holding the drying veneer were common prior to the advent of commercial dryers. Air was sometimes circulated by fans mounted above the veneer or by convectional currents of air heated by steampipes located beneath the floor.

summers for the U. S. Forest Service in Oregon, Idaho, and Montana and "had timber in [his] blood." At Northwestern University he met and married the daughter of S. J. Glanton. That led to a job at Burnside where he spent two happy years on the river supervising the procurement of logs for the Burnside mill. When interviewed, Kreamer was obviously pleased to recall those years on the river (Fig. 6-5).

> In these early days there were not many serviceable roads back into the hills of Kentucky. Rivers were the principal avenues for commercial traffic. Burnside used both the Kentucky and Cumberland Rivers for transporting logs to the mill. We went as far down the river as Celina, Tennessee, and sometimes to Nashville. Old-growth tulip poplar, white oak, red oak, and walnut logs were moved to the river banks by teams of oxen and mules where the logs would be loaded on barges. The exchange of money for logs was somewhat unique. The logger would leave a jar often filled with local moonshine whiskey along with a note indicating his name. The jar was then stashed in the piled logs. When the tow boat arrived, the barge crew would drink the whiskey and load the logs after which I would pay for the logs by putting the money payment for the scaled volume into the same jar. The logger would return for his money the next day or, maybe, a week later. It would be there. No one ever touched it. Everybody was honest back then.

About 400,000 square feet of veneer were produced daily at the Burnside mill with the four veneering machines and a work force of 140 people. The veneers were marketed for the most part in northern furniture factories in the Chicago area, Grand Rapids, Michigan; Jamestown, New York; and other furniture centers in Wisconsin and Indiana. The

Figure 6-5
The Burnside Veneer Company used both the Kentucky and Cumberland Rivers for transporting logs to the mill on barges. Few roads in the hills of Kentucky were serviceable for log hauling at the time.

numerous piano manufacturers which existed in Chicago and along the East Coast were also customers of the Burnside company.

Kreamer was named president of the firm in 1932 and his wife, Louise Glanton Kreamer, was named treasurer. S. J. Glanton retained a role in the firm as board chairman. The mill stayed financially solvent during the Great Depression of the 1930s because of astute financial planning, cash reserves, and the ability to cut and sell some veneers during the extended depression years. During World War II, the mill supplied aircraft veneers to the Hughes organization as it built the "Spruce Goose" and to the de Havilland Corporation which manufactured the Mosquito bomber. Both types of planes contained yellow poplar and maple veneer in addition to Sitka spruce veneer.

When S. J. Glanton died in 1945, the Kreamers acquired ownership of the company. They continued to operate the mill until 1950 when a U. S. Army Corps of Engineers dam on the Cumberland River began to flood the mill site. The Capital lathes were sold and the other machinery dispersed. Kreamer continued in the veneer business as a custom cutter and as salesman of architectural veneers with his business based in Memphis.

Central States Veneer Company

Ed Dulaney bought the Paducah Box and Basket Company of Paducah, Kentucky, about 1962 and renamed the firm, Dulaney Veneer Company. The company at an earlier time was owned by Max E. Miller who died in 1950. Julien C. Miller and T. C. Ezell were later associated with the mill. The plant was resold in 1964 to Lloyd Cramer. "Ding" Hamann (1975) wrote about starting up the Central States Veneer Company for Cramer and the "...many long hours coaxing an old basket factory into becoming a hardwood veneer mill." Cramer's principal business was the merchandising of veneers through his veneer sales organization, Cramer Veneers, Inc., in High Point, North Carolina. The Paducah mill supplied his High Point operation primarily with pecan veneers although some other domestic and foreign veneers were cut and shipped for the sales organization. The veneer samples were priced in and sold from the High Point sales office. However, shipments to customers were made from the warehouse located on the mill site in Paducah.

The mill manager in 1982, Fred Kirkfield, had been involved with the veneer industry for 36 years. According to Kirkfield, the first 17 years were spent with the Freiberg Mahogany Company in New Orleans until it closed in the 1960s. During the next 17 years, Kirkfield worked for the Amos-Thompson Company in Edinburg, Indiana. During these years at Freiberg and Amos-Thompson, Kirkfield became experienced in log buying and selling, veneer cutting, and in managing various operations of the veneer business. His move to Paducah in 1980 as manager of Central States Veneer Company was, therefore, a natural progression in overall responsibility in the veneer business. Lloyd Cramer maintained control of the business as president of the company. His wife was vice president. Ownership of this veneer mill was not a new experience for the Cramers as they once owned a veneer operation in Guatemala.

Central States purchased pecan logs through producers located in the mid-South. Logs bought in this region were considered to be superior in overall quality to logs cut from timber grown further north in the Mississippi Valley region. Because of this, the Central States people believed they had a competitive edge over "northern" mills specializing in pecan veneers. Flitches were cut into veneer using a 17-foot Capital Machine Company slicer and a Johnson City slicer patterned after the Capital slicers. The veneer was dried with a Proctor-Swartz dryer acquired from the Freiberg Mahogany Company veneer mill (when its equipment was dispersed after a hurricane destroyed the plant in 1965) and a somewhat newer Schilde double-deck dryer of German manufacture. In 1982, the Paducah mill employed 52 people and produced annually about 40 million square feet of veneer. The mill was shut down about 1985 because of plant obsolescence and competitive factors.

THE OHIO VENEER MILLS

Acme Veneer Company

This veneer company is typical of many other mills which had successful, flourishing businesses and then vanished with very little ever being written about them. It is known that about 1910 a new three-story factory and warehouse were constructed at the corner of Eighth and Harriett Streets in Cincinnati. The proprietors were Frank Sudbeck and Harry Hartke. The veneers they produced were targeted for regional furniture and woodwork manufacturers.

Between 1929 and before 1937, the veneer mill was moved to Lawrenceburg, Indiana, with offices and warehouse remaining in Cincinnati. The three-story building occupied in Lawrenceburg was across High Street from the Court House and was known as the "Old Woolen Mill" to local residents.

Harry Hartke died in 1942 or 1943 leaving the fortunes of the firm in the hands of his son Harry (Jack) Hartke, Jr., who had previously been in sales for the company. Interestingly, one of his better customers was the Ohio penitentiary at Columbus. The prison used the veneer in manufacturing furniture for state institutions and commercial sales. Jack Hartke's business associate was James Brown who had formerly been a salesman for the Hartzell organization in Piqua, Ohio.

The Ohio River flood of 1937 devastated the firm. However, the Acme Veneer and Lumber Company, as it was later known, remained a viable corporation until it ceased operations, ultimately a victim of obsolescence and the flood of inexpensive veneers shipped in from overseas during the

1950s. The Pierson-Hollowell Company eventually bought the site and building when they moved their manufacturing operations from Indianapolis to Lawrenceburg.

Albro Veneer Company

Established in 1838 by Henry Albro, this company became one of the earliest and most important veneer concerns in the Midwest before disappearing from the industrial environment of Cincinnati, Ohio. Located at 1932-36 Eighth Street, Albro continued to operate as the Henry Albro Company until his death at about the beginning of the Civil War. His son, Edwin, succeeded him and proceeded to reorganize it under the name of the E. D. Albro Company. This company was among the pioneers in the production of veneers and, later, as panel manufacturers. Just after the reorganization, about 1878, the company was operating a number of veneer saws, a 7-foot rotary machine and a staylog lathe. Later on, two 10-foot slicers were installed. The rotaries were used to cut walnut veneer from plain logs used primarily for the side rails of bed steads. At this time there was practically no call for thin veneers in cheaper woods. "The cutting machines used in the decade ending in 1878 were all of such light construction that it was impossible to cut anything heavier than 3/16-inch poplar, and it was necessary to have well steamed logs to do this." (Bardin, 1922).

Another writer, in 1895, reported that "The E. D. Albro Company, Cincinnati, has a world-wide reputation for its fine, high class veneers, of which it makes a specialty. It is one of the very largest concerns of the kind in the world, and the largest west of New York City, and as far as the matter of manufacturing is concerned, is probably the largest in America" (Anon., 1895). By then, the company manufactured quality mahogany and other foreign and domestic lumber and veneers in addition to cigar box lumber. Its ultimate fate after 1908 could not be determined.

Cincinnati Veneer Company

Frank Unnewehr was a Cincinnati entrepreneur who not only organized the Cincinnati Veneer Company in 1898 but also the Frank Unnewehr Company which produced cigar and packing boxes in Cincinnati. He also organized the Bay Poplar Lumber Company in Mobile, Alabama, which manufactured cigar box lumber. The company was located at 1361 West Sixth Street and was reputed in 1910 to be the largest manufacturer of veneers and thin lumber in Cincinnati. This was prior to a fire which destroyed the plant that same year.

The company relocated to 4411 Estes Avenue in 1929 or 1930. This new plant was equipped with two rotary machines, a slicer, and a veneer saw. Specializing in oak and poplar veneer, the company also cut walnut, gum, maple, and other fancy veneers. Later it may have gone into the production of plywood and veneer faces which it advertised as late as 1948 (Anon., 1930a).

Manley, Thompson & Moffett Company

Another of the Cincinnati veneer mills which was a significant enterprise was the Manley, Thompson & Moffett Company. The company was located at the west end "...of the Eighth Street viaduct...[and was] one of the sights of the west side of the city, covering a large area of ground, with immense concrete buildings..." The original structure was built of wood and it "...is unnecessary to say that all parts of the mill as formally constructed were exceedingly combustible" (Anon., 1908).

A 1907 fire destroyed these buildings in a few minutes. The lessons learned by the owners prompted reconstruction to be of reinforced concrete and brick for the walls and floors. A light wooden roof, supported by cast-iron columns, was the finishing element. The new mill contained a large band saw, three veneer saws, and a "huge" slicing machine. The company also operated a mahogany mill in Havana, Cuba, and sawmills in Kentucky. The Cincinnati bandmill was geared to flitch production. Lumber produced in the process was cut into dimension stock and sold to furniture and desk factories.

The mill was modern by contemporary standards and electric motors were used to direct-drive the veneer saws and slicer. The cut veneer was placed in racks for drying except for some "...dried over the cylinders which feed the carriages on the veneer saws." A considerable amount of Cuban cedar was cut for veneered cigar boxes.

In the early years of the 20th century, Thomas J. Moffett was president, assisted by E. W. Robbins as vice president. Published materials do not reveal the eventual fate of the company or the company officials involved.

Ohio Veneer Company

Two brothers, Ferdinand and Joseph Boskin, were the proprietors of both the Ohio Veneer Company and the Cincinnati Hardwood Lumber Company. The lumber company produced "finely figured, quarter-sawed oak veneers," mahogany lumber, and thin veneers of other species. The plant and offices were located in the 2600 block of Colerain Avenue and extended to the Miami and Erie Canal (Anon., 1904).

No data were readily available concerning the early history of the company although a turn-of-the-century date is probable. By 1909, the company was well established and growing rapidly. "The machinery equipment is of late and improved design, and consists of two slicers, one of them said to be the largest operating in the United States, and capable of cutting veneers up to 14 feet long. This machine has a daily capacity of 30,000 feet and weighs forty-five tons. The other machines are two veneer saws and one large rotary cutting machine. Besides this equipment the company has also a completely equipped saw mill" (Anon., 1909b).

The Ohio Veneer Company maintained a large inventory in a large, three-story brick warehouse of Circassian walnut, mahogany, English brown oak, quartered oak, curly birch, rosewood, white holly veneers and thin lumber.

When necessary, the Boskin brothers traveled to the log markets of Europe personally to select mahogany and Circassian walnut logs for shipment to Cincinnati. Four salesman were kept on the road prior to World War I selling the company's veneer and thin lumber products (Anon., 1913).

In a letter dated March 29, 1939, and addressed to the Pierson-Hollowell Company in Indianapolis, William J. Boskin announced the end of production: "We are liquidating stock, buildings, and machinery."

Hartzell Hardwood, Inc.

Hartzell Industries, Inc., has been a relatively small, multifaceted industrial empire. The company's base of operations is Piqua, Ohio, which oversees plants not only in Piqua but also Wilmington, Ohio; Decatur, Indiana; and Benton, Arkansas. The early beginning of this diversified business can be traced back to 1875 and to an Indiana farm boy with a dream. What happened in the next 100 years evolved in logical, sequential stages from lumber and veneer production to the manufacture of products such as airplane propellers, industrial machinery, furniture and novelty wood products, industrial fans, and plastic products. The original moving genius and entrepreneur was John T. Hartzell, a dynamic young man, determined not to stay "down" on the family farm.

With $25 borrowed from relatives (his father had refused to loan him the money), he acquired a horse and a huckster wagon and traveled from farm to farm buying butter and eggs and selling "store-bought" goods to farm wives. His contacts with farms and farmers stirred his interest in the manufacture and sale of lightning rods, which became the second phase of his successful entrepreneurial career. In 1875, he built his first sawmill, in Greenville, Ohio, and cut the local hardwood species including oak and ash. Dimension lumber soon found a dependable and growing market with customers who built wagons. This observation led to Hartzell's decision in 1885 to manufacture farm wagons himself. Hartzell wagons became well known in the Midwest. He was joined in this lumber and wagon enterprise about 1890 by his son, George, who had been born in 1875. The firm was renamed John T. Hartzell and Son (Anon., 1975).

In 1900, George bought out his ailing father, using borrowed money. The firm was moved to Piqua and the name changed to George W. Hartzell Company (Fig. 6-6). The company output was restricted largely to bandsaw cut walnut lumber produced in its modern turn-of-the-century mill. Always innovative, George W. Hartzell developed export markets in England and Germany. He also developed a system for steaming walnut so that both the sapwood and heartwood became uniformly brown. It was then possible to market sapwood walnut lumber in Europe which the company had previously not been able to sell. Although not immediately

Figure 6-6
George W. Hartzell was known for the quality of his products. His initials, G.W., came to mean "Good Walnut" in the hardwood industry.

accepted by the American lumber market, the technique eventually became an accepted manufacturing process in walnut mills selling lumber in domestic markets (Anon., undated).

When walnut lumber exports to Europe were embargoed in the early days of World War I, Hartzell's company acquired an antiquated rotary lathe and began cutting veneers in 1915. Production was initially hampered by a lack of ancillary equipment which was considered essential just a few years later. The cut veneer was dried in a heated room where it was held by finger clamps. The Hartzell company also began producing walnut gun stocks, walnut airplane propeller blanks, and later, in 1917, hand-crafted walnut propellers. This latter product was prompted by Robert N. Hartzell's boyhood friendship with Wilbur Wright who lived and worked in nearby Dayton (Anon., 1969a).

A trade journal in 1917 also recognized that George Hartzell had an eye for landscape beauty and that this was not incompatible with the attainment of increased business profits.

> Given equality in product, is a customer more likely to be in a receptive mood when he has climbed over piles of trash to a barn of an office, or when his course lies over a white crushed-stone driveway, bordered by smooth lawns and trimmed hedges, to a vine-covered building and into a home-like office, with rug-covered floors and growing plants in the windows? Mr. Hartzell... thinks it productive of sales. There is an instinctive feeling that the product partakes of the character of a neat and well-ordered factory; also, the surroundings have a marked effect upon employees. These are psychological axioms that are taken by keen men in

all trades and are exemplified by the personnel and product of the Hartzell mills" (Anon., 1917b).

After the war, George Hartzell traveled through Europe renewing business contacts and buying Carpathian elm burls and African mahogany in the form of logs, crotches, and cut veneers. It was in about 1927 that the company bought its first dryer, a Proctor, which performed well but was extremely slow in operation. It was used for many years until it was replaced by a 103-foot-long James dryer in 1942. This dryer was used for about 30 years. A Hale dryer was purchased in 1965 and another Hale jet-type dryer was installed in 1978.

In the middle 1920s, it became evident that many of their veneers, particularly walnut, were exhibiting considerable buckle which made them difficult to sell. With the company's typical ingenuity, they bought an obsolete French Oil Mill Machinery Company press designed for the automobile industry. As a result, Hartzell could sell 100 percent buckle-free veneer. This provided them a considerable competitive edge until the technique became standard for the industry. Hartzell was known affectionately in his later years as "Uncle George." He and his company were also widely known for the initials, GWH, which meant in the trade: Good Walnut Hartzell.

Over time, the company acquired a reconditioned 12-foot slicer from the Dayton Veneer Company (after that firm was destroyed by fire), a new 17-foot Capital slicer in 1948, and a new 13-foot Capital staylog rotary lathe installed in 1952 or 1953. The new lathe permitted the company to cut better quality rift white oak than was possible with the old lathe. Their last veneer slicer was acquired when the equipment from the defunct Mitchell Veneer Company of Greensburg, Indiana, was sold. The veneer machinery of the 1980s consisted of the 17-foot and two 12-foot Capital slicers and a modern 12-foot staylog lathe. Together these machines were capable of producing 100 million square feet of quality veneer annually.

In 1928, George Hartzell relinquished sole ownership of his company when he brought in his son, Robert N. Hartzell, as vice president and the business was incorporated as Hartzell Industries, Inc. Robert N. Hartzell, born June 3, 1896, was 22 when he began his career in the wood products business with his father. Robert became president upon the death of his father in 1933. Miriam Hartzell, George's widow, assisted her son as vice president until 1952. In that year, the second George W. (Bill) Hartzell, Robert's son, became company vice president.

A 1928 report describes in detail the careful process of selection which enabled the Hartzells to produce fine walnut veneers. The best logs were segregated from the lumber logs as the rail cars were unloaded so that only the "...large, choice logs with clear, finely-figured grain are reserved to be manufactured into veneer." These logs were given further inspection to determine whether they were more suitable for sliced or rotary-cut veneer. Stumps and burls were processed on the stump lathe.

At all stages of production, efficiency was stressed by utilizing labor-saving techniques and machines. This was coupled with a policy of reducing waste by remanufacturing veneer clippings which became inputs for secondary manufacturing companies. This strategy was also employed by the sawmill department (Booth, 1928).

The tenure of Robert N. Hartzell as president was an eventful one during which a number of new divisions and product lines were created. World War II contributions included development of: gluing techniques in glider construction, propeller blades, aircraft veneers, and wooden parts for aircraft. The early experience gained from manufacturing wooden propeller blades was utilized during World War II when government orders expanded their output. With the loss of military business after the war, the propeller division began manufacturing controllable metal propellers for light aircraft. Larger propellers were made for commercial aircraft, beginning in 1948. Industrial fan production based on airfoil technology, which had begun in 1927, was continually expanded while Robert Hartzell was president. A hydraulic products division was added in 1961. Other areas of the business were created as new related products were developed and found marketable. Several business acquisitions were also made. Robert Hartzell died at the age of 72 on December 11, 1968. The momentum continued after George William (Bill) Hartzell took over as president of the firm in 1969.

Bill Hartzell came to this position with a wide range of practical experience and technical training. When asked about this, he responded by saying that his first lessons were learned in the veneer crating room of Hartzell Industries at age 15. He left the firm just long enough to serve his country during World War II and acquire a degree in mechanical engineering at Purdue University, graduating in 1947. Returning to Piqua, he spent several years in the fan division before transferring to the lumber division in 1971. While there, he invented and perfected the Hartzell carriage and produced other practical and advanced innovations in the field of hydraulic sawmill equipment. As president of Hartzell Hardwoods, his long-term devotion to the face-veneer and related wood products is apparent from his many contributions to the veneer and walnut trade associations and by his aggressive actions to expand company divisions and acquire firms which integrated well with his firm.

Decatur Industries, Inc., located in Decatur, Indiana, was acquired in 1969. This firm produced wood novelties for pipe smokers and various consumer gift items such as wall plaques and book ends. This division provided a good market for Hartzell walnut lumber by increasing the utilization of end trimmings and short lumber. The company also acquired the Arkansas Face Veneer Company at Benton, Arkansas, in 1971. This firm was started by Paul Mooney in 1956 and grew to become one of the largest hardwood veneer "face" plants in the country producing faces for stock panels and the furniture trade. Hartzell's Piqua veneer mill supplied about 10 percent of the face plant's requirement of a carload of veneer each working day in normal times (Burrell, 1972).

Although up to 25 percent of Hartzell's veneer was once sold in foreign markets, in 1982 almost all of it was being sold domestically through commission salesmen. Sales were also made through a showroom located at the veneer mill in Piqua. Roughly two-thirds of the veneer production was walnut and pecan. Maple, ash, oak, and cherry made up the balance.

American Woodcrafters, organized in 1975, was the last division to be formed in the corporate organization. This division designed and marketed solid and veneered walnut furniture sold by mail in kit form for the do-it-yourself home craftsman. This division also provided an integrated market for the lumber and veneer divisions.

The expansion of the company into its several divisions was accompanied by decentralizing manufacturing capabilities and management responsibilities. Thomas H. Hartzell and James R. Hartzell, brothers of Bill Hartzell, provided family leadership in these decentralized units. As the company's centennial brochure points out: "Hartzell Industries is unusual in its history of having continued with a closely knit harmonious family management through multiple generations."

George William Hartzell retired as chairman of the board of Hartzell Industries at the age of 61. His brother, Thomas H. Hartzell, succeeded him as president and chief executive officer of Hartzell Industries. Tom Hartzell joined the company in 1953 after graduating from Ohio State University with a degree in industrial management. James R. Hartzell remained as vice president of operations for the company.

Dayton Veneer Company

The city of Dayton, Ohio, over the early years of the century, had at least three veneer mills. The Dayton Veneer Company evolved out of the Dayton Sawmill that was owned and operated by H. C. Hossafous, beginning about 1900, and was located on South Broadway at the C. H. and D. Railway. The Miami Lumber and Veneer Company, located at 1921 East 1st Street, was out of business by about 1911. The third was the J. W. Frye Lumber and Veneer Company, situated at the rear of 702 South Broadway, probably the site of the Dayton Veneer Company. The Buckeye Veneer Company which sliced quartered oak was a new firm in 1912. The Buckeye plant was located 800-810 West Bolender Avenue.

Russell T. (Bud) Miller, closely identified with the Curry-Miller Veneer Company in Indianapolis, began his working life in 1910 at age 14 at the Cincinnati Veneer Company. This was a short-term job because the mill burned a few months later. Armed with this experience, he moved to Dayton in 1912 to work on the Buckeye Veneer Company mill which was under construction. One of Miller's jobs was to mix, by hand, the concrete for the slicer. Upon completion, he measured and inspected the first veneer produced by the company.

In 1915, Miller switched companies and worked for the H. C. Hossafous organization until called for military service. During the wartime period or shortly thereafter, Hossafous sold his company to J. W. Frye who took in other partners to form the Dayton Veneer Company. Upon returning to Dayton in 1919, Miller secured a job with Dayton Veneer as an inspector who also measured the veneer output. Within two months, he was appointed plant superintendent when his predecessor was sent to High Point, North Carolina, to sell the company's veneer in an expanding market area. Miller remained with Dayton Veneer until 1930 when he moved to Indianapolis. J. W. Frye bought the Buckeye Veneer Company about 1922 and ran it and the Dayton Veneer Company as separate companies. Within a year, Buckeye Veneer had burned and was not rebuilt.

In 1925, the Dayton Veneer Company was "...pleased to announce the completion of a splendidly modern new sawmill..." This new sawmill was equipped with a seven-foot Clark bandsaw, "all the most modern accessories," a new 102-inch rotary in addition to an older 96-inch rotary, and a 13-foot slicer. Walnut lumber and veneer were specialties of the firm although it produced significant quantities of white oak and mahogany as lumber and veneer (Anon., 1925a).

The new facilities and equipment coincided with a general reorganization of the company. S. C. Brown and his partner, J. E. Harris, reported in 1927 that "The Dayton Veneer Company produces at least as much American Walnut as any firm in the country...[and that] sales have increased more than one hundred percent in those two years..." (Anon., 1927). Ownership of the company passed into the hands of J. Hampton Hoult and C. F. Rice in 1937. Prior to this time, these men had been located in Grand Rapids, Michigan, where they were "...well known in the furniture and hardwood industries" (Anon., 1937a). The last listing for the company in a Dayton City Directory was 1944.

ILLINOIS MILLS

C. L. Willey Company

Chicago's largest veneer dealer in 1895 was Cameron L. Willey. His sales company, founded in 1890 at Thirty-fifth and Iron Streets, was not the only one. J. Rayner, at Fulton and Morgan Streets; and Adolph Sturm, at 55 West Washington Street, fleshed out the "Big 3" in Chicago. All three handled domestic and exotic veneers. Much of the domestic and all of the foreign veneers were cut in New York City. It is known that C. L. Willey furnished 300,000 square feet of Tabasco mahogany veneer for the Marquette Building and, prior to 1895, the English oak veneer and lumber for the Chicago Public Library. Orders such as these whetted Willey's interest in producing veneer himself in Chicago (Anon., 1895).

Unlike Rayner and Sturm, Willey proceeded in 1897 with a large-scale investment in a building site, buildings, and equipment. The plant was located on a 12-acre parcel at 2558 South Robey Street and was served by the C. B. & O railroad and by water transportation from its own slip on the south fork of the Chicago River. By 1909, the C. L. Willey Company was "...the largest producer of veneer in the world, and one of the foremost handlers of mahogany and other imported high-class woods" (Anon., 1909b). Veneers were produced using three slicers, three rotary lathes, and 11 veneer saws. Most of the production was sold to the Pullman Company at Burnside, Illinois. In addition to the Chicago mill, Willey had an extensive band sawmill and veneer operation in Memphis. The Tennessee plant began production in 1907 with two rotary veneer machines and a double band sawmill with a capacity to cut 40,000 to 50,000 board feet of hardwood timber per day. Willey's son, Charles B., served as general manager of the Memphis "department" and J. D. Laskey was manager of sales (Anon., 1907b).

C. L. Willey died on November 28, 1916, after an outstanding business career. Annual veneer cutting capacity ultimately reached 100 million square feet and was coupled with equipment capable of drying 200,000 feet daily. Willey "...came to be recognized as a man of almost magic[al] judgment of logs and how best to cut them" (Anon., 1922c).

The business was incorporated in 1917, at which time Charles B. Willey became president. He remained president until his death in 1920. Clarence Ingalls succeeded Willey but resigned in 1921. A. N. Cordell, a friend of C. B. Willey and former banker, took on the presidency after Ingalls left the company. An old-time employee, H. B. McClain became vice president and production manager. However, by January 1923, Charles A. Snapp had become president (Anon., 1923).

Throughout this period of company turmoil, Mrs. C. B. Willey remained as chairman of the board of directors and may have been at the center of the problems the company encountered. Ross Kreamer, who knew of the situation, gave this insight:

> Charlie [Willey] had a good mill. He was a big operator. He died young and his wife tried to take it over and didn't do too well at it. She was a heavy drinker and married her favorite bartender. Ingalls at the time was their general manager and superintendent. [Ingall's] son told me that his father came back to the office after being away on company business and found the bartender at his desk with his feet propped up on it. [Ingalls] just picked him up and knocked the living hell out of him. The man ended up on the ground floor from the second floor. Mr. Ingalls just packed up and left and went to Nashville where he started up a veneer mill with a single slicer.

It is known from an advertisement that the C. L. Willey Company was still doing business in 1928. The company appears to have gone out of business soon thereafter.

R. S. Bacon Veneer Company

Robert S. Bacon was born in Hannibal, Missouri, on April 16, 1872. He is known to have been a Quincy, Illinois, druggist who decided "...to trade his pills and potions for logs and lathes" when he moved to Chicago in the early 1890s. However, the "logs and veneer lathes" followed his employment with the Dayton Book and Paper Company of Chicago. He resigned from this firm in 1898 to become a veneer merchant with some other Chicago businessmen. In 1900 he incorporated his veneer activities as the R. S. Bacon Veneer Company. Incorporators were R. S. Bacon, Julia S. Bacon, and Charles G. Comstock; and the company had a capitalized value of $15,000. In 1908, Bacon bought a 12-foot Capital slicer and installed it in a small building on Anne Street in downtown Chicago. The firm prospered rapidly and was later moved to Lake Street. The company imported logs and manufactured veneers from mahogany, Circassian walnut, rosewood, and other quality tropical hardwood logs. American black walnut, oak, and birch were also used for veneer purposes (Anon., 1921c).

Success in this initial venture led to the organization of the Bacon-Underwood Veneer Company at Mobile, Alabama, with a sawmill and factory located at Stockton, Alabama. Bacon spent three years in Alabama establishing and developing this enterprise. This was typical of Bacon's approach to the veneer business. In 1913 he went to the Caucasus mountains of Russia to obtain first-hand information on the Circassian walnut resource and to establish direct market connections between it and his Chicago veneer mill. On his return trip to the United States, he sought out and developed European markets for his Chicago-produced veneers. However, these efforts were ultimately frustrated by the outbreak of World War I in 1914 and the Russian Revolution (Burrell, 1965).

The business was well established when Bacon sold out to Frank H. Scott in 1921. Scott had joined the company in 1901 as a stenographer and bookkeeper but rose rapidly in the firm as he acquired stock and more important responsibilities. Under his direction a new plant was built specifically for veneer production at 4702 Augusta Boulevard, just off Cicero Avenue in Chicago. The decision to stay in Chicago was an alternative to buying a company which later evolved into the Amos-Thompson Corporation. The asking price of $23,000 was less than the land cost of the Augusta Boulevard site. Bacon was evidently influenced by his perception of small town life in Indiana and is reputed to have said: "Who would ever really want to live in Edinburg, Indiana?."

The offices and warehouses of the company were moved to the Augusta Boulevard site in January 1918. The cutting room was said to be well lighted and equipped with a 16-foot Capital slicer, a Capital 100-inch lathe, and a Proctor and Swartz veneer dryer. The two private offices were "elegantly paneled and sumptuously furnished..." (Anon., 1928b).

Scott owned the firm until his death on October 28, 1928. The company was subsequently sold by Scott's family to Joseph F. Mertes on January 10, 1929. He had come to the firm in 1914 as a salesman, having previously been assistant to the superintendent of the Tell City Furniture Company in southern Indiana. On the death of R.S. Bacon in 1921, Mertes became a partner in the veneer company and was elected vice president. Mertes was active in the Veneer Association and served as its president on two occasions (Anon., 1952). A former employee remembers Mertes as "...a gruff, but soft hearted gentleman." In 1954, Mertes was presented with an honorary law degree from Georgetown University. Although semi-retired at the time, he continued in the capacity of president-treasurer of R.S. Bacon Veneer Company until 1963. He died in 1966.

In March 1929, Henry P. Walsh was hired to fill the vacated vice president position. Walsh had gone to work in the veneer industry at the age of 14 when he was hired as an office boy by the Chicago Veneer Company on South Robey Street for $1.50 for a week's work of 60 hours. This was later raised to $2 after successfully completing his first month on the job. In contrast, his boss, the mill superintendent, was making the princely sum of $12 per week. When hired by the Bacon company, Walsh was president of the Veneer Manufacturers Company which he had helped organize in 1910 (Anon., 1929c).

George F. Wilhelm joined the R. S. Bacon Company in 1937. Wilhelm was employed first in the sample room before moving into management positions. Having survived the worst years of the Great Depression, R. S. Bacon was advertising in 1937 that the company had "...over 60 varieties of fine fancy foreign and domestic veneers ready for immediate shipment" from their Augusta Boulevard location and their Fayetteville, Arkansas, plant (J. P. Brower Veneer Company) acquired in March, 1937. Brower Veneer Company had been long noted for the fine color of its walnut products. "Here we are specializing in the production of walnut veneers and lumber of the same superior quality and workmanship we have always maintained in our Chicago mill." A large warehouse facility in Portland, Oregon, was also acquired "...for the convenience of our Western customers" (Anon., 1937).

The Chicago mill, in the late 1930s and into the 1940s, specialized in mahogany veneer. When African mahogany logs became difficult to obtain because hostilities in Europe had spread to the sea lanes, the Bacon Company began buying mahogany logs in Honduras. Transportation was still a complex problem which could only be solved with patience and ingenuity. This is illustrated by the efforts in 1941 of R. J. Wilzbach, a vice president for log procurement, who was able to secure the charter services of a 2800-ton ship to transport "...a cargo of Honduras mahogany logs that had been carefully selected for their particular suitability for making veneer." After a successful landing in New Orleans, "The transportation problem again bobbed up but, through considerable maneuvering, necessary freight cars were finally secured" and the logs were delivered to "...the company's private switch at Chicago" (Anon., 1941). The Bacon Company continued to import Central American mahogany logs during World War II which were manufactured into aircraft veneers and veneers used in constructing PT boats (Fig. 6-7).

After being a vice president of the firm for many years, Wilhelm and four other company officials bought the company from Mertes on March 1, 1963. Wilhelm became company president; James G. Cotter, vice president (veneer division); R. Donald Elliott, vice president (lumber division); John Jensen, vice president for log procurement; and Clara Casey, treasurer (Burrell, 1965).

Although the firm manufactured mahogany veneers almost exclusively in the early 1950s, the company had ceased to cut this species by 1960. Production of walnut veneers and walnut lumber then began to dominate the company's line of products. During the "mahogany era," an 8-foot band mill was built in 1947 in Sunman, Indiana. The mill was originally designed to produce mahogany lumber for a large midwestern mahogany

Figure 6-7
The R. S. Bacon Veneer Company and other similar veneer firms manufactured mahogany veneers that were used in PT boats. The Higgins Boat Company of New Orleans built many of these plywood-hulled patrol and torpedo boats.

market which existed at that time. The sawmill also cut flitches which were trucked to the company's veneer mill in Chicago. This continued until 1951 when the sawmill was abandoned.

The most unique contribution of the R. S. Bacon Company was the conception and manufacture of grooved decorative, hardwood plywood wall panels. This type of panel product was "invented" by George Wilhelm in his basement in May 1952, using a table saw and small plywood panels with which he could easily experiment. The idea worked although the full potential was not immediately apparent.

> In 1953, we became one of the first firms in the nation to go into the manufacture of grooved hardwood plywood paneling on a large scale. From the outset, our goal for the product, called Panawall, was to create new markets for wood and to get fuller utilization of our veneer logs. We accomplished this in two ways: (1) by use of all sound veneers regardless of character markings, and (2) by use of random width, sliced veneer strips in the manufacture of our 4- x 8-foot and 4- x 7-foot paneling. As recently as seven years ago [1950], industry experts contended the public would never accept wood products unless the veneer was "clean." We believed the public would buy and appreciate the innate character of wood as represented by the burl, swirl, and pin knots. So we made a deluxe paneling on this theory and production doubled and redoubled in each 18-month period. We've never had a complaint on the quality of our face veneers (Anon., 1957).

The finished panels, when installed, looked similar to random-width, solid plank paneling—but at a fraction of the material and labor costs of the traditional product. The grooves at the "joints" were, at first, hand routed before a more efficient machining technique was developed. By 1957, R. S. Bacon employed 125 people and produced 15,000 square feet of paneling per day. These quarter-inch-thick panels were offered in a variety of domestic and tropical species plus knotty pine.

The grooved plywood panel industry developed rapidly. Within six months, Dulaney in Louisville was in production. The two firms had a virtual monopoly until Roddis Plywood in Marshfield, Wisconsin, came out with its own line a few years later. The Roddis organization had earlier been involved in selling the R. S. Bacon product after their Chicago representative had viewed the first panels and stimulated the interest of Roddis executives.

In the early days, Wilhelm recalled, that all he "...had to do was little more than pick up the telephone and call the one distributor with the best financial rating in a given market area." A chain of wholesale outlets was built up for Panawall in this way. Initially, the panels were unfinished and limited to three species, one size, and one thickness. These constraints were relaxed as competition increased. Prefinishing was added about 1954 when Roddis Plywood introduced this innovation. Character-marked paneling was introduced in 1956. Unlike the many competitors which joined the market, Bacon's Panawall product was always grooved along the splice between adjoining sections of face veneer. As a result, there was no fixed pattern from panel to panel. Production capacity was limited to about 800 custom made panels per day as late as 1965 when Ed Beebe, a former Roddis employee, was made plywood production manager.

In order to gain needed space for the lay-up and prefinishing of their wall panels, the firm shifted veneer production to Dubuque, Iowa, during 1964 and 1965. The Dubuque operation became the Hubbard Walnut Division. The strategy was to produce veneer close to the point of log origin and reduce freight costs. The move involved the transfer of 12-foot and 16-foot Capital slicers, a Proctor-Swartz dryer, plus supporting equipment. A 12-foot Capital lathe and Proctor-Swartz dryer from the dismantled Freiberg mill in New Orleans were installed in the new Dubuque plant prior to the transfer so that veneer production would not be interrupted.

For some, this was an historic milestone as it represented the final retreat of veneer mills from the major cities of the United States. One veneer executive summed it up this way: "This does not mean to overlook such fine cities as Louisville and Indianapolis that do have mills, but rather to recall the days of sometime back when there were a number of mills in cities such as Chicago and New York" (Stem, 1965a).

The panel industry that R. S. Bacon Company had created grew beyond all expectations. By 1965, the industry had national wall panel sales of $120 million and individual firms were producing as many as 40,000 panels on a daily basis. Lauan panels shipped in from the Far East were another competitive factor. Having elected to remain with its historic tradition as a veneer producer, the Bacon Company, perhaps wisely, never joined the ranks of the gang-grooved, mass producers which entered the market in 1957. However, the company did enter into a licensing arrangement with Denny, Mott & Dickson of London, England, and displayed its Panawall products at a handsome showroom at Adelaide House in London (Stem, 1965b). The British firm also produced Panawall at its plant at Kings Lynn and was responsible for worldwide distribution outside the Western Hemisphere. In 1968, John J. Walsh was elected vice president of a newly created Panawall Division. This division was short-lived as all of the plywood manufacturing equipment was sold in 1969 to the Weber Veneer and Plywood Company of Shawano, Wisconsin, for its affiliate, Panawall Company, Inc. The Augusta Street property was sold and the veneer warehousing function was moved to Dubuque (Anon., 1969b).

In 1981, a large existing building at Grundy Center, Iowa, was purchased and veneer and warehousing operations were moved to that location. The sawmill at Dubuque continued to be used to cut flitches for the Grundy Center veneer mill where walnut, butternut, white oak, red oak, and cherry veneers were produced. Sales offices remained located at the company's Chicago area office in Hillside, Illinois; Calgary, Alberta; and

High Point, North Carolina, where warehouses were also maintained.

Robert Maricich joined the company in 1973 to become manager of the Dubuque production division. The move to Grundy Center was his responsibility. Management responsibilities there were taken over in 1982 by John Stasi. In 1984, the sawmill was moved to Grundy Center, thus severing the "Dubuque connection" of R. S. Bacon which had existed for more than 30 years.

Of Wilhelm's company officers in 1963, only John Jensen and Clara Casey were with the firm in 1985. James McCracken, Wilhelm's nephew, was made executive vice president in 1972. He had come up "through the log yard" for about 20 years. McCracken was responsible for developing the Architectural Division of the company which imported specialty veneers from world-wide sources for a diversity of design and architectural applications.

Gone are the glory days when one-third of the company's veneer went into the production of Panawall. Wilhelm said that he was comfortable with a smaller-sized business and liked "...to think that we are merchandisers who work veneer to its best end utilization. We are selective in what we buy and when it comes to veneer we set a fair price and keep it." On December 13, 1985, James McCracken became president of R. S. Bacon Veneer Company with George Wilhelm remaining as the firm's chief executive officer.

Dean Company

Richmond Dean was the patriarch of the Dean family. He came to Chicago from St. Louis to become a senior vice president and general manager of the Pullman Sleeping Car Company at the turn of the century. In St. Louis he had been associated with a hardwood lumber business. It is likely that one of his customers had been the Pullman Company which was the largest consumer of fine hardwoods in the United States at the time. His major responsibility, the procurement of quality wood products at a reasonable cost, was firmly spelled out to him by Pullman company president, Robert Todd Lincoln.

Richmond Dean had four sons: John Richmond, Thomas A. (Tom), Philip D., and Charles D. Dean. John R. Dean was the oldest son, having been born in 1891. Tom Dean smiled as he described John "...as a fun-loving, high-spirited young man" which may have led to his early "graduation" from Notre Dame University. In order to redirect his energies in more positive ways, John's father invested in the Black Brothers Lumber and Veneer Company located at 22nd Street and South Crawford Avenue in Chicago and put the young man to work. John Dean was not entirely without veneer experience as he had worked for the C. L. Willey Company where he had acquired practical knowledge of the business in the mill and as a salesman. The Black Brothers, John and Horace, had started their lumber and veneer business in 1905. Two segment saws had been used to produce veneer in addition to a 12-foot Capital slicer which was added later.

The Black Lumber Company had originally been located at the corner of Blue Island Avenue and Robey Street when it was known as the Fred W. Black Lumber Company. Fred Black was president and treasurer and Horace Black was company secretary. The mill utilized only southern hardwoods and also operated sawmills in Mississippi and Tennessee (Anon., 1909c). The firm name was changed to the Dean-Spicker Company when Julius T. Spicker became a partner in the business in 1917. Spicker had come from Louisville where he had been employed by the C. C. Mengel & Brother Company for 19 years. There, Spicker had gained import/export experience and had been sales manager. The company he joined was deemed "...so thoroughly modern and so efficiently arranged that it offers interesting possibilities for those interested in [veneer] production.." (Anon., 1917c).

Thomas A. Dean had been employed by the Dean-Spicker Company at various times as he was growing up and during summer vacations when not attending Georgetown University. Upon graduation in 1920, he joined the firm as a full-time employee. Spicker resigned from the company in 1921 to form the Ingalls-Spicker-Ransom Company in Nashville, Tennessee (Anon., 1921b). This Nashville firm specialized in cabinet woods with particular emphasis on mahogany, walnut, and oak lumber and veneer. The veneer production capacity was estimated to be between 60 and 75 million feet per year. In 1923, he became associated with the C. L. Willey Company in Chicago as a vice president. Spicker was eulogized when he died in 1928 as "... one of the best authorities on foreign and domestic veneer" and as a "...man with a forceful personality" (Anon., 1928c).

Following Spicker's resignation from the Dean-Spicker Company, John R. Dean became president and Thomas A. Dean, vice president. In 1924, Spicker's name was dropped and the name, the Dean Company, was adopted (Anon., 1922). The Dean Company moved its primary manufacturing facilities in 1927 to a 15-acre site on Elm Avenue in Portsmouth, Virginia. The plant that was constructed was "...conceded to be one of the finest veneer plants in the South." It was well equipped with a Coe rotary lathe, two veneer saws and a slicer. The new firm was organized as the Dixie Veneer Company, a division of the Dean Company, with John R. Dean, president; Thomas A. Dean, vice president; and Philip D. Dean as secretary and treasurer. (Anon., 1928a). The Chicago veneering facilities were totally abandoned several years later in 1932. However, company headquarters remained in Chicago with John R. Dean continuing as president until his death in 1940.

Between 1940 and 1944, Thomas A. Dean served as president of the firm before becoming chairman of the board. He held this position until he retired from the firm in 1963. Just after he became president, Tom Dean was approached in 1941 by the Army Air Corps (Dayton-Wright) to produce precision-cut Sitka spruce and Douglas fir for use in the construction of certain types of aircraft. Given some assurances that the federal government wouldn't subsidize another veneer mill next door, Dean Company officials headed West, locating an old automobile plant at Gresham, Oregon. Between August 1941 and plant start-up in October 1942, a used 156-inch Capital slicer, clippers, and dryers were located and shipped to Gresham. Once installed in the completely renovated building, the plant quickly began producing veneers for use in military aircraft, boats, and Signal Corps equipment. In 1943, a 17-foot slicer and a Capital staylog lathe were added. Joe Goldston was the first plant superintendent of the Olympic Manufacturing Corporation and was responsible for much of the Herculean effort in physically assembling a producing veneer mill in 14 months. Olympic converted to peacetime production in 1945 under the supervision of James S. Vaughan, general manager. Mill capacity by this time was over 200,000 square feet per day. Annual production in the early post-war years was in the range of 25 to 30 million square feet. The mill continues to operate as the Olympic Manufacturing Corporation Division of the Dean Company and concentrates on West Coast hardwood species (Anon., 1943a).

Less fortunate was the decision, in 1946, to produce walnut veneers at a plant located in the flood plain at Kansas City, Kansas. The plant fell victim to the Big Flood of 1954. The site was inundated with water and buried with silt and logs from other upstream veneer establishments. A valuable inventory of crated veneer was flooded as well. The redried veneer was found to have absorbed silt particles which caused the Dean Company no end of problems when furniture manufacturers attempted to apply finishes to these veneers. The loss of manufactured veneers would have been much greater if Tom Dean had listened to the assurances of the Corps of Engineers that the ominous flooding situation at Kansas City was under control. As a precaution, Dean had ordered that all warehoused veneer was to be shipped immediately if it had already been sold and was pending a request for shipment by the purchaser. This reduced the loss considerably. Nevertheless, the Kansas City plant was closed because of the flooding uncertainties.

Tom Dean was succeeded as president by Philip D. Dean in 1944, a position he held until his death in 1952. Charles D. Dean, the youngest of the Dean brothers, was elected vice president in the same restructuring of the organization. Charles later became executive vice president of the company prior to his retirement in 1960. He died in Florida in 1962 at the age of 56 while returning from a fishing trip.

Upon the death of Philip D. Dean in 1952, Ted J. Connelly, who had been with the company since 1933, became president and remained so until 1973 when he became chairman of the board. He held this position until June 30, 1981, when he retired from the company. He is credited with introducing a progressive pension/profit sharing plan which was the first of its kind in the hardwood face veneer industry. He had "...started at the bottom of the mill...at Norfolk" a few years after graduating from Loyola University in Chicago. Several years were devoted to learning the production end of the business before returning to Chicago and the main office of the Dean Company. From then on it was a steady rise from being the president's secretary to company secretary, sales manager, and executive vice president (Anon., 1953). James S. Vaughan followed Connelly as president until Edward V. Horton became president in 1978. William A. Costoplos was named executive vice president in 1978.

In 1957, Dean Company production was shifted to the Norfolk Veneer Company, owned by the Thompson Mahogany Company, where the two firms shared facilities on a rotation basis. Relationships were generally cordial at the higher administrative levels although frictions were known to occur at the production level.

The Dean family interests were sold in 1963 to several of the company's officers and managers, including Ted Connelly and Edward Horton. Soon thereafter, a decision was reached for the Dean Company to disassociate itself from Norfolk Veneer Company and to build a new plant at Princeton, West Virginia. The location was selected so the company would be able to manufacture veneers from domestic species in the Appalachian area while being able to continue manufacturing veneer from exotic species shipped in from East Coast ports. The proposed site was also in close proximity to the firm's largest customers. An attractive, modern and spacious veneer mill was built there in 1965 (Anon., 1965a).

Veneer production got under way the following year using two Capital slicers and one staylog lathe. This plant was incorporated as the New Dixie Veneer Company Division of the Dean Company. Executive offices were moved from Chicago to Princeton on September 1, 1971. Two years later, William A. Costoplos was transferred from the Chicago sales office to become director of sales at Princeton. Other actions of the board of directors included the promotion of Connelly to chairman and chief executive officer; James S. Vaughan, president; and Edward V. Horton as executive vice president and secretary (Anon., 1968, 1973).

The Dean Company maintained a sales and marketing office in High Point, North Carolina, and was also involved in international markets. Bork-Dean, an import company, specialized in teak and other exotic veneers manufactured by Bork-Singapore. Danish interests were involved in the Bork-Singapore operation.

The "old" Dean Company (Dixie Veneer Company) site on Elm Avenue and the hardwood lumber business were sold to Charles D. Dean, Jr., and Lester Fletcher in 1963 when the veneer production enterprise and the Dean name (the Dean Company) was sold to Ted Connelly and his associates. The lumber business operated as the Dean Hardwood Lumber Company and dealt essentially with mahogany and teak lumber and dimension stock. After Lester Fletcher died in 1981, this business was carried on by the surviving partner (Bowman, 1982).

On February 9, 1984, tragedy struck the Dean Company when a chartered plane crashed, killing William A. Costoplos, company president; Kermit Howell, vice president of manufacturing; and two other Dean employees. These men had the year before been primarily responsible for putting into motion a $4 million modernization program. It was a tribute to them and the men that succeeded them that the plan became a reality in 1985. A. Herbert McClaugherty summed up the scope of the renovation as follows:

> We started at one end of the mill and went through to the other. We could see that in 10 years we wouldn't be competitive if we didn't make some changes. It's a big step for any operation, especially in our industry (Donnell, 1985).

McClaugherty had joined the Dean Company in 1983 as a director and was promoted to company president following the death of Costoplos. The key to the plan was the installation of two continuous production lines featuring Babcock-BSH dryers equipped with humidity and computerized process controls. The automatic drying lines are identical except the working width of one is 18 feet, installed behind a Capital slicer, and the second is 13 feet, used with a Capital staylog lathe. Seventy-seven employees working an eight-hour shift produced 8 million square feet per month. The modernization increased output by an additional 1.5 million square feet per month. Cherry accounted for 40 percent of the cut in the early 1980s, followed by red and white oak, ash, maple, birch, poplar, and walnut. In 1985, only about five percent of the production was exported; the balance went to major furniture markets.

Rounding out the new team at the Dean Company were Don Brohard, secretary-treasurer; Bill De Ruise, plant manager at Princeton; and Jim Vaughan, senior vice president and general manager of West Coast operations, at Gresham, Oregon.

MICHIGAN

Veneer mills in the Lake States have cut a variety of species which reflect the composition of the northern hardwood stands. Yellow and white birch and hard and soft maple have been favored species; they account for more than 50 percent of the veneer produced. These species, along with basswood, do not have internal grain patterns which can be enhanced through slicing. Oaks, elm, and ash species provide the contrast, having strongly differentiated grain structure that is associated with ring-porous species. Veneer saws were commonly used for veneering these species. This was true for the Grand Rapids Veneer Company and the Munising Wood Products company. However, most Lake States veneer mills, past and present with few exceptions, have been strongly oriented toward rotary veneering operations.

Birds Eye Veneer Company

One of the most interesting Michigan veneer companies is the Birds Eye Veneer Company of Escanaba in the state's Upper Peninsula. It has roots extending back to the Strong Veneer Company in Gerry, New York. The New York company was organized in 1888 by J. Strong, B. E. Strong, and Charles T. Tuxford, a young man born in 1863. Their purpose was to take over the operation of the historic mill of J. Strong & Sons. Soon after this was accomplished, Charles Tuxford left on a selling trip "...carrying only ash burl and wampa ash samples and only made enough to cover his expenses...Later the company began the manufacture of birds eye maple veneers" (Anon., 1921a).

Leaving the company in 1897, Tuxford opened a "jobbing" business in Jamestown, New York, where he specialized in birds eye maple. This specialization took him to northern Michigan in 1905 where he purchased logs having these small, circular distortions in the wood. Four years later, he organized or helped to organize the Escanaba Veneer Company in Escanaba. He operated this mill for two years as general manager before resigning because of ill health and returning to Jamestown as a sales representative for the company. The following year, 1912, he became associated with John C. Kirkpatrick and, on January 25, 1913, they organized the Birds Eye Veneer Company (Anon., 1933). The company officers elected were Kirkpatrick, president; Dr. G. W. Earle, vice president; and C. C. Patterson, secretary-treasurer. C. T. Tuxford was listed as vice president and general manager. There is some confusion in the literature about the company's birth. A second version follows: "The Birds Eye Veneer Company was organized in Escanaba, Michigan, in 1912 with Juel Lee as president and general manager" (Graham, 1956). It is very likely that Juel Lee was involved in some way as he is known to have been general manager in 1935 and could have been at an earlier date (Anon., 1935).

Tuxford apparently stayed on as general manager until 1918 when he returned to Jamestown in a sales capacity. President Kirkpatrick immediately appointed Charles J. Byrns to be general manager. Byrns acquired a large financial interest in the company prior to Kirkpatrick's death in 1922 (Anon., 1922b). Born in 1861, in Allegheny, Pennsylvania, Kirkpatrick became a leading public figure in Escanaba and the Upper Peninsula of Michigan. He was president of several companies including the Pittsburgh & Lake Superior Iron Company, the Escanaba National Bank, the National Pole Company, the Escanaba Morning Press Company, and the A. J. Kirstin Company. Active in political circles, he also served as vice president or director of a number of other regional businesses. The organizational vacuum left by his death created the need to reorganize the Birds Eye Company in 1925. W. P. Schuldes, a Kirkpatrick business associate, was elected president; C. T. Tuxford, vice president; and C. J. Byrns was elected secretary-treasurer in addition to his duties as general manager.

Byrns had been born on a farm near Fort Henry, New York, in 1860 or 1861 and was 58 years of age when hired by Kirkpatrick as general manager. The Byrns family had moved to Ishpeming, Marquette County, Michigan, when he was 18. Employed first as a bookkeeper in a retail lumber business, his "...forcefulness, intelligence and conscientiousness of service brought him the full responsibility of managing the retail lumber yards of his employers at Ishpeming and Marquette." He subsequently was made retail business manager of the Deer Lake Lumber Company. He was only 24 when that company discontinued its retail business but it financed Byrns so he could continue the business under his own name. Other executive and organizing challenges honed his management and marketing skills prior to his appointment as general manager of the Birds Eye Veneer Company. This accounted for his "...astounding ability to get things done and done right." Above all, he was reported to be honest and exceedingly just and fair in dealing with associates and his work force. "But let no man who esteems his friendship, or who desires to retain favor in his eyes, attempt to take advantage of his fairness" (Anon., 1925b).

The physical plant for which Byrns became responsible was substantial. Its main, two-story brick building was 100 x 368 feet and contained the cutting room, dry kiln, drying rooms, shipping department, and company offices. Two other buildings housed the power plant and six reinforced concrete boiling vats. The cutting room alone occupied 8,000 square feet and included the two 102-inch Capital lathes, veneer clippers, and other ancillary equipment. The mill was known for the amenities provided for workers. A well-appointed lunch room and locker facilities, complete with showers, are examples of the arrangements made "...for the comfort, health and safety of the employees." Outside, the plant was softened with a "beautifully" terraced lawn, trees, shrubbery and flowers (Anon., 1919).

Byrns was elected president, treasurer, and general manager of the company in December 1929, at the annual meeting of the board of directors. Tuxford remained as vice president and W. P. Schuldes was elected secretary. Prior to the annual meeting, Byrns had been secretary-treasurer and general manager. As "... evidence of their appreciation of his service during the past year, this year having been the most successful the company has ever had...," the board of directors presented Byrns with a "Buick Roadster" (Anon., 1929a).

One month later, Byrns announced an ambitious expansion plan to add two new buildings to accommodate a veneer dryer in addition to the one in use, a new 88-inch lathe, clipper, another boiling vat, and additions to the ventilating equipment to provide "...ideal working conditions by eliminating excess moisture, heat and dust, and providing free circulation of fresh air." The new dryer eliminated the need of the rack drying system and this space was allocated to the veneer clippers during the rearrangement of equipment. The outdoor beautification program "...which has made the Birds Eye plant unique among industrial plants..." was also continued. The expansion program resulted from the growing demand for birds eye veneer for furniture and cabinets. Under the direction of Byrns, this Escanaba firm was producing 75 percent of the birds eye veneer produced in the world (Anon., 1930c). In 1930, the Hardwood Record trade publication interviewed C. J. Byrns to inquire about the factors responsible for his company's prominence in the specialized production of birds eye veneers. A portion of that interview follows:

> Birds eye maple's introduction and growth to a favored position among cabinet woods is a comparatively recent occurrence. "In fact," said Mr. Byrns, when questioned in this connection, "at the time I assumed the management of our company, now some twelve years ago, birds eye maple, although in use for many years, was given but little attention among the figured woods of the world. My observation convinced me, however, that this beautiful wood lacked only a sponsor having the moral courage to place it prominently and rightfully among the choicest treasures of the forests." This recognizance by Mr. Byrns of the possibilities of birds eye maple and his subsequent steps in verifying and developing his ideas have become known as one of the most outstanding examples of keen business judgment and initiative in the hardwood industry.

The interview continued by exploring Byrns' investments in research to discover tensile strength of the product, appropriate stains, and the operations required to produce beautiful veneers and finished products. This interview could well have been his swan-song because he died two years later, on July 10, 1932, at the age of 72, after two years of failing health. An extensive obituary records the many novel and efficient methods and ideas introduced by this remarkable man (Anon., 1932).

Charles T. Tuxford was promoted to president of Birds Eye Veneer Company when Byrns died. His tenure as president was brief; he died the next year, in Jamestown, New York, on November 8, 1933, at the age of 70. Tuxford was reputed to be the oldest rotary-cut veneer man living, having started his career in 1889 (Anon., 1933).

The void created by the deaths of these two experienced executives was apparent at the 1935 annual meeting. The widow of J. C. Kirkpatrick was elected president and E. R. Empson, secretary. Juel Lee was elected general manager, and this is the first indication in the literature that a new era had begun (Anon., 1935a).

Advertisements in the trade papers clearly indicate the shift from specialization in birds eye maple to a diversity of northern Michigan species, including birch, brown ash, red oak, elm, and basswood. In addition to lathes, a veneer saw and slicer were used in manufacturing veneer. Sales representation had been expanded from midwestern furniture centers to New York City; Santa Monica, California; and Portland, Oregon.

In 1939, Birds Eye leased the "green mill" of the Escanaba Veneer Company. The plant consisted of a staylog lathe which was operated as a full rotary, another Capital rotary lathe, and a veneer dryer. Outright purchase was made in 1950. Fire completely destroyed the mill in the spring of 1954. To maintain production levels, the main Birds Eye plant operated 24 hours per day using one slicer, four lathes, a segment veneer saw, and three veneer dryers (Anon., 1954). The 1909 plant built by Tuxford was never replaced. However, Birds Eye did acquire the idle Evans Products Company veneer operation at Marquette, Michigan.

Juel Lee was born in 1894, so he was a mere lad of 18 when he became associated with the Birds Eye Veneer Company. Working his way through the company to general manager and to president some time after 1933, he continued in this post until his death in 1961. His son, J. Russell Lee, born in 1924, provided information about the involvement of other family members in the business.

Following his graduation from Michigan Technological University in forestry, J. Russell Lee joined his father as an assistant and became the company's executive vice president in 1959. When his father died two years later, J. Russell became president and general manager of the corporation. Juel's second son, Warren J. Lee, became active in the firm about 1950 and was named treasurer in 1959. During the 1961 company restructuring, he was elected executive vice president. Juel's grandson and son of Warren Lee, Marcus, became a vice president in 1981. Marcus Lee, like his uncle, graduated with a forestry degree from Michigan Technological University and joined the family firm after graduation.

The Marquette division of Birds Eye ultimately became surrounded by the campus of Northern Michigan University which prevented expansion of this facility. The alternative was to move equipment from Marquette to Escanaba and to expand production at the main plant. This was accompanied in 1964 by mill modernization, including a new bridge crane, new veneer dryer, lift trucks, and conveyors to increase labor efficiency. A 24,000-square foot warehouse was also constructed along with a new sample and sales room. When the dust had settled, the veneering equipment consisted of four rotary lathes and two slicers. Birch and maple remained the primary species cut, and lesser amounts of veneer were cut from oak, cherry, elm, and basswood. The annual capacity of the mill was projected to increase from 50 to about 65 million square feet per year. Birds Eye Veneer Company, in a recent year, wholesaled veneer and plywood and sold birds eye maple logs to domestic mills and the export market. Production of veneer was limited to that cut on a single slicer in Mercer, Wisconsin.

The annual Birds Eye Venison Banquet originated in 1965 and continued for about 20 years. Held in New Albany, the annual gathering attracted scores of veneer manufacturers and secondary manufacturers from the eastern one-half of the country. It was a veritable Who's Who of the industry and an excellent means of industry-wide communication. The demise of these banquets probably reflects the significant changes taking place in the Birds Eye business operation as well as throughout the entire industry. Shifting from traditional, family-owned veneer businesses operating in a close-knit fraternity of "friendly," but aggressive, competitors, the industry is now dominated by foreign nationals operating in the global economy.

Chapter 7

VENEER FIRMS ON THE FRONTIERS OF THE HARDWOOD HEARTLAND

From the eastern mountains to the western prairies and from Canada to the Midsouth, hardwood face-veneer mills produce quality veneer even though they lie outside the Central Hardwood region. These mills take advantage of regional species, whether they are maple and birch of the North Country, western walnut of Missouri and Kansas, oak stands of the Appalachians, or the hardwood trees found on the alluvial soils of the Mississippi delta, stretching from Missouri to Louisiana. Although fewer in number, many of these "frontier" firms have been crafted to be at the cutting edge of innovative technology. Owners and executives have in their times provided exemplary leadership for the industry.

MISSOURI

Missouri Valley Veneers

It is not easy to understand fully the complex of companies involved in ownership of Missouri Valley Veneers. Basic ownership is vested in Walnut Products, Inc., and Iowa-Missouri Walnut Company, which jointly own Missouri Valley Walnut. Missouri Valley Veneers is one of the three departments of Missouri Valley Walnut. The Christgen family owns Walnut Products, Inc., and Dan Duncan owns Iowa-Missouri Walnut Company.

Kenneth H. Christgen entered the industry in 1936 as a secretary in the office of the American Walnut Company in Kansas City, Kansas, while he was enrolled in a Kansas City business college (Fig. 7-1). When he was interviewed, Christgen said that this company was essentially a sawmilling concern specializing in walnut lumber and walnut gunstocks. In 1937, American Walnut and the Frank Purcell Walnut Lumber Company combined resources to form Puramco Veneer, Inc. Puramco was subsequently acquired entirely by the Purcell organization. The veneer plant had an annual capacity of 20 million square feet, produced by a 10-foot Capital slicer, and a 125-inch Capital lathe.

The Frank Purcell Walnut Lumber Company had operated for many years as producers of walnut logs and as manufacturers of walnut lumber. Frank Purcell, Sr. (Fig. 7-2), founded the company and he was succeeded by his son, Frank, Jr., who began veneer production under the name of Frank Purcell Walnut Veneer Company. Headquarters were in Chicago but the mills and factory were located at 12th Street and the Kaw River in

Figure 7-1
Kenneth H. Christgen has been associated with Kansas and Missouri forest products businesses for more than 50 years. Black walnut products have played prominent roles in several of his business ventures.

Figure 7-2
Frank Purcell, Sr., was the founder of a company which produced walnut logs and lumber for many years in Kansas City.

Kansas City. Frank, Jr., died in 1946. The company was not only one of the old, established manufacturers of walnut products in the center of the country but also a most important factor in the domestic and foreign export business in the late 1930s (Anon., 1937). The 1951 Kaw River flood ended the operation but not before it had been owned by the Dean Company of Chicago and later by Cleo Whorton of Dubuque, Iowa.

Christgen parlayed his business training and experience into a close working relationship with Robert V. Duncan of the Iowa-Missouri Company. Duncan earlier had been president of American Walnut Company. In order to expand production capacity, Iowa-Missouri moved in 1938 from Kansas City to St. Joseph, Missouri, and formed the Iowa-Missouri Walnut Company with Ken Christgen as secretary-treasurer. Christgen resigned from the company in 1949 although remaining associated with it "...one way or another ever since."

Before establishing himself permanently in St. Joseph, Christgen owned and operated a sawmill in Kansas City (Walnut Products Company) and another sawmill in West Des Moines, Iowa (Des Moines Walnut Company). Both of these companies survived and were closely interrelated in business with the Iowa-Missouri Company and the Missouri Valley Walnut Company. On a single large property in St. Joseph, Iowa-Missouri is on one end, the Missouri Valley Company is in the middle along with Walnut Products, and the Des Moines Walnut Company is on the other end doing business as St. Joseph Woods.

The veneer operation started with a trade publication's advertisement in 1963 about equipment for sale by the Freiberg Mahogany Company in New Orleans. Christgen recalled that this ad prompted his taking a trip to New Orleans, not realizing that the purchase of equipment was a very small part of the total cost of a veneer business. "We...bought the 17-foot Capital slicer, knife grinder, and clipper and were in operation by May of the next year after purchasing a new Hale dryer."

> In 1977, we went to Italy and negotiated the purchase of a vertical slicer. We installed that here and started operating it at the end of 1978. It was the first Cremona in the United States and the first vertical slicer Cremona had sold outside of Europe. We made a very good deal because Cremona wanted to get one they could show in the United States. We operate it plus the [1920] Capital slicer and other Cremona equipment.

"Veneers make a triple pass through the oil-heated Cremona dryer on about a 2.5-minute cycle. The Hale dryer makes a single pass-through on a 7-minute cycle" (Lee, 1981). In 1984, the length of the Cremona dryer was doubled. Additional veneering equipment was purchased when the Turri veneer mill (International Timber) in Springfield, Missouri, went out of business about 1986. The Springfield mill had been equipped with a Cremona rotary lathe and a Cremona horizontal slicer.

Dennis Christgen joined the family business in 1964 and was general manager of Missouri Valley Veneer Company in 1985, responsible for veneer manufacturing and sales. Ken Christgen, Sr., was president; Dan Duncan, owner of Iowa-Missouri Walnut Company, was vice president; Ken Christgen, Jr., was vice president; and Hugh Welch was treasurer. Ken, Jr., came into the business in 1958. A third son, Ed, joined his father and brothers in 1968. In 1985, he was sales manager of Walnut Products and general manager of St. Joseph Woods.

Together, these entrepreneurs produced a large variety of products, including hardwood lumber, turning squares, high quality gun stocks, planks, and veneers from timber harvested for the most part (80 to 85 percent) within 600 miles of St. Joseph. Competition for logs appeared to be keen. In addition to other domestic producers, buyers representing German, Japanese, and Korean companies were often very active in the region. Species in the region include walnut, red and white oak, ash, sycamore, basswood, hickory, and soft maple. Between Walnut Products, Inc., and Iowa-Missouri, their 42 Moore dry kilns had a capacity for 1.25 million board feet per charge, the largest in the United States.

Detailed records were kept on the purchased logs which were segregated into three categories: lumber logs, veneer logs, and gunstock logs. These were then allocated to the company or department which could most efficiently utilize them. As a company brochure stated: "Each log we handle is recorded and documented in our in-house computer bank. This means we can provide products to our customers with specific environmental and geographic information" (Anon., 1985c). The system also permitted a check on plant and log buying efficiency. For example, flitches for the veneer mill were numbered when cut in either of the two sawmills. This also provided the key to the complete history of all the veneer cut from them. "With just its number, [veneer] can be traced back to the log source, to the supplier, where it was bought, what it cost, who flitched it, when it was sliced, and which shift dried it..." As Christgen put it: "We know what each [log producer] is doing because in the end we look at each lot. If there's been a credit, the lot has made a profit. If there's a debit, it's a loss and it is put against his record. It's just that plain and they know we know."

Missouri Valley Veneer Company veneers were sold domestically by sales representatives in North Carolina; Louisville, Kentucky; Los Angeles; and St. Joseph. Because of excessive competition in some of these markets, the Christgens thoroughly explored foreign sales opportunities. Ed Christgen, as sales manager, typically worked through export agents in overseas countries and shipped to Western Europe, Africa, the Middle East, the Orient, Australia, South America, Canada, and Mexico. A full time Korean company representative handled sales in South Korea. Employees at St. Joe were encouraged to become conversant in German, Spanish, and French, languages spoken in some degree of fluency in the firm's offices.

In the last half century, it can be seen that the Christgen family enterprises experienced tremendous growth and success based on the timely introduction of new technology, a commitment to produce quality products and service. Nonetheless, the St. Joseph complex underwent major financial reorganization during 1988 and 1989.

VIRGINIA

In addition to the Norfolk Veneer Mills, Inc., and the Virginia Log Company, which are treated in Chapter 8, two other Virginia firms—the Lane Company and the Erath Veneer Corporation of Virginia—are of historical interest. A third, Pulaski Veneer Corporation of Pulaski, Virginia, was owned by Chick Richardson and was new in 1926. It represents mills in the state that eventually failed to survive competitively. The Pulaski mill was equipped with a Capital 16-foot slicer, two rotary lathes, and a veneer saw. The mill initially cut walnut veneer "...gathered in from the blue limestone coves and valleys in the neighborhood. A large future supply is in sight, also much fine maple and oak" (Traveler, 1927). It was estimated that the annual cut would be in excess of 50 million square feet. The veneer mill served as a captive mill for the Pulaski Furniture Company. Veneers were sold to other local furniture plants as well. Eventually, the veneer mill was moved to Dublin, seven miles away, because of an expansion of company furniture operations. Sometime later, furniture company officials determined that it was less expensive to purchase their veneer requirements on the open market; consequently, they phased out the in-house veneer plant. Ross Kreamer bought the slicer and one of the rotaries for the Burnside Veneer Company, located in Burnside, Kentucky.

Lane Company

The Lane Company, located at Altavista, Virginia, was founded in 1907 by John Edward and Henry Lee Lane. They worked together as partners in Lane Brothers Company, General Contractors, a firm which constructed railroads, bridges, and other civil works in the rebuilding of the "New South" after the Civil War. Altavista was conceived by them after they learned that the Virginian Railroad (later merged with the Norfolk and Western Railroad) was going to intersect and cross the main line of the Southern Railway at this point in the Virginia Piedmont. Two thousand acres of land were purchased by the Lane brothers for the town and Altavista began to grow as a rail hub. In March 1912, John Edward Lane attended a public auction at Altavista and, for $500, bought a bankrupt packing box plant owned by J. W. Bobbitt. In June of that year, Lane suggested to his 20-year-old son, Edward Hudson Lane, that he leave the family home near Charlottesville and go to Altavista for the purpose of manufacturing cedar chests in the box plant. John Edward Lane had got the idea from A.T. Quick that this might be a profitable venture. Quick had headed up the Consolidated Furniture Works of Altavista. Although this firm had gone out of business, Quick told Lane that cedar chests had been their most profitable item (Lane, 1963).

The beginning of the business was not auspicious. The purchased building was only 30 feet wide and about 60 feet long. It had no side walls and the machinery was unsuitable for manufacturing cedar chests. The usable equipment consisted of only two small dry kilns, a boiler, and a small engine. Before production could start, another $50,000 had to be invested in machinery considered necessary to manufacture cedar chests. Lane said later: "I found the [machinery] company that would give us the longest credit terms possible to pay for the needed machinery...and ordered it" much to the displeasure of his father whose original $500 investment had ballooned to $50,500. Starting with five workers, the company produced simple chests ornamented with copper strips at the rate of five to 10 per day. The sales volume the first year was less than $50,000. Consistent progress was made as the company expanded and as a well-trained, in-house sales organization developed new markets for the cedar chests (Lane, 1962).

Although Edward Hudson Lane, born on July 4, 1891, had graduated from Virginia Polytechnic Institute and had taken woodworking courses, he "...had no manufacturing experience and had never heard of a cedar chest!" The first 10 years were particularly difficult for the company. It had been named the Standard Red Cedar Chest Company rather than the Lane Company because of uncertainties about the ultimate success of the business and the possibilities of damaging the reputation of Lane Brothers, General Contractors.

During World War I, the company shifted to manufacturing pine ammunition boxes. Money was lost on the government contracts but the company did learn a valuable production lesson. A rudimentary assembly line was developed which evolved into a moving conveyor system for assembling, finishing, crating, and transporting units to the warehouse or shipping dock.

In 1922, the name was changed to the Lane Company and a national advertising program was launched. Cedar chests were produced "...with fine cabinet exteriors such as mahogany and walnut to match or harmonize with other fine household furniture." Involved was the creation and promotion of an attractive trade mark. This required aggressive selling to capitalize on the money being spent for advertising. It also involved the development of an appeal for beauty and utility and a demand based on these qualities rather than price. Glamorous Hollywood stars were frequently featured in the company's advertising to promote this concept. At the same time, the Arthur D. Little Company was enlisted to conduct a research program to determine whether cedar chests really did kill moths.

Based on the research findings, manufacturing processes were adjusted so the chests would be aroma-tight. The degree of success can be measured by the fact that the National Union Insurance Company of Pittsburg was able to underwrite moth insurance policies covering the contents of Lane chests. This also was fed into a national advertising campaign promoting the fact that Lane was the first company to guarantee moth protection backed by an insurance policy included in the sale of each chest (Anon., 1961).

In March of 1925, Edward H. Lane was elected president of the firm to recognize his business successes. Lane firmly believed that any success he had had was directly related to name identification, research, constant improvements in product quality, and advertising. By 1930, the plant was manufacturing 250 to 300 chests per day. Sales were not only increased by an imaginative advertising program and by product research but also by sociological studies of potential customers. The Lane Hope (Love) Chest became an American tradition for prospective brides. Even today, Lane's hope chest remains the "flagship" of the company's diversified furniture line which has developed in recent years.

Edward Lane had, at an early date, decided that his company should saw its own red cedar lumber rather than depend on outside suppliers. The five percent saved paid for the national advertising and also set a raw-materials policy for the company which it continues to follow. When the Lane Company found that its expanding production was impacting the supply of red cedar timber, the company shifted to red cedar veneer to conserve the resource; and it began slicing its own veneer. The slicer, a horizontal Capital slicer, was purchased in 1945 from Chester B. Stem, Inc. This machine was the "famous" 165-inch European style, underslung slicer built in 1897 by the Capital Machine Company of Indianapolis. The slicer, known affectionately as "Sally Open Bottom" when it was owned by the Stem Company, has been refitted a number of times to upgrade its production efficiency.

The slicer produced a better quality red cedar veneer than a vertical slicer because the veneers were less inclined to buckle when cut on the horizontal machine. Since the machine was equipped with a special set of hydraulic dogs and a hydraulic flitch table, the operator could mount the flitch on the slicer without additional manpower assistance and do it faster than the similar task performed on a vertical slicer. The venerable slicer can still accurately cut approximately 200,000 square feet of veneer per week within the exacting tolerances required by the company. It is the oldest piece of operating veneer slicing equipment in the United States (Anon., 1977).

During World War II, the Lane Company contributed to the war effort by producing aircraft plywood, tail assemblies for airplanes, dog sleds, and glider wings. All of these products were made utilizing wood and the wood-working skills of the company's employees. The company also continued to produce cedar chests, primarily to service the market created by overseas members of the armed forces who bought the chests for delivery to wives and the "girl back home." This market was carefully fostered by the Lane Company which skillfully used national advertising in conjunction with the cooperation of "hometown" furniture stores.

After the war, the Lane Company purchased a lumber and panel plant in 1946, located at Smyrna, Tennessee, to position the firm in one of the most productive aromatic red cedar regions in the country. The purchase included a large supply of cedar logs, cedar lumber, and standing cedar timber. This Smyrna mill has continued to supply the Altavista plant with red cedar logs suitable for veneering as well as red cedar lumber and panels.

The late 1940s were prosperous years for the company. Recognizing the inherent weakness in depending on a single product for long-term business success, the Lane Company instituted a program of product diversification. The first major shift occurred in 1950 when Lane obtained a contract with General Electric to build TV cabinets. This was followed in 1951 with a line of occasional tables which were shown at the April Furniture Mart at High Point, North Carolina. The success of this product line led to the production of a diversity of tables, backed up once again by a dynamic sales force.

After the Lane Company began producing other types of furniture, company executive Edward O. Shelton said the firm found a need for tulip poplar cross-banding veneer. In 1956, the company purchased a new Capital 165-inch vertical slicer for the Altavista plant. This machine produces about 800,000 square feet of veneer per week. The firm also bought the Bald Knob furniture plant at Rocky Mount, Virginia, in 1956. Rotary cut tulip poplar and oak veneers are produced there. Since then, Lane has grown to become a furniture industry giant, with 21 plants in four states. Much of the fancy face-veneers used in "face design" layouts are purchased on the open market. The firm in recent years produced and purchased a total of about 60 million square feet annually.

The Lane Company has always been known for its innovative approaches to business, including its product development research. An interesting example was its pioneering development of particle board cores laminated with veneer. Initial studies were begun on particle board research as early as 1915 as an end use for wood waste. When synthetic resin glues were developed in Germany and became available in the United States in the 1930s, the product became feasible to produce. Because of the interruption of World War II, the company did not produce commercial quantities of particle board again until 1947. Its extrusion-type press was built on the principle of a small experimental press developed by Lane Company engineers at the Altavista plant. The manufactured particle board was first used successfully in the tops of two cedar chest models. Other applications for particle board as core stock have since been demonstrated and used.

Cooperation and teamwork have been important to the success of the Lane Company. The continuing integration of raw materials production into an increasingly complex manufacturing capacity has also been important. However, the long tradition of excellence in product engineering, knowledgeable managers, and aggressive marketing practices provide the essential elements that have distinguished this company.

Erath Veneer Corporation of Virginia

The Erath Veneer Corporation mill at Rocky Mount, Virginia, was originally known as the Helms Veneer Company. Mr. and Mrs. R. J. (Jack) Helms constructed the plant. Both had linkages to the Bassett furniture organization. Jack Helms had been employed by Bassett American and his wife's mother had been a Bassett prior to her marriage. Mill production was initially directed toward the manufacture of black walnut veneers. This presumed potential never fully materialized because of the low quality of the black walnut resource in the region.

Helms started with the purchase of a small plywood plant utilizing one old lathe and a small press. This facility was soon abandoned in 1965 and was virtually lost in an expanded layout of new buildings and equipment. Two rotary lathes, two slicers, a staylog lathe, and three dryers were part of the $2.5 million capital outlay which occupied the 48-acre site. Annual plant capacity was estimated to be 90 million square feet of crossbanding and 72 million square feet of face-veneer. When it was operating on a one-shift basis, 70 people were employed in the crossbanding plant, 67 in the face-veneer operation and 13 in the sawmill. Besides walnut, the other species cut for face-veneer were white oak, red oak, maple, cherry, and pecan. Overseeing day-to-day operations was Posie Collins, general manager and vice president (Anon., 1965).

In 1968, the company filed for bankruptcy under the Subchapter S provision of the law. When the mill reopened later in 1968, it was under the management of George S. Erath. With a partner, he assumed management control and provided additional capital investment. Mike Lefever joined the firm at this time as an assistant manager. He had gained valuable experience working with his uncle in a furniture company's face-veneer plant. In 1973, he was promoted to vice president for manufacturing at the Rocky Mount mill which had been reincorporated as the Erath Veneer Corporation of Virginia. He was later appointed general manager of the company.

Before becoming involved in the manufacture of hardwood veneers, George S. Erath had been one of the more astute veneer salesmen located in High Point, North Carolina. He had received excellent training with the dean of High Point veneer salesmen, Jimmie Mitchell, while employed by him. The veneer mill was a "backward" integrative step which assured Erath a continuing flow of quality hickory, white oak, cherry, red oak, poplar, and ash veneers for his sales organization. Erath also determined that the key to the successful operation of the mill was cutting Appalachian hickory veneers. Hickory was plentiful locally and hickory/pecan veneers were a "hot" item for furniture and panel products. Seventy to 80 percent of the 50 million square feet of the veneer cut in 1982 was hickory. Erath also cut and exported "white" poplar to Italy as Italian walnut. Some persimmon was also cut. His keen eye for markets, along with salesmanship with a highly qualified production team, proved to be a winning combination. The emphasis continues to be on quality: "When you think quality, think Erath." Erath also believed in expert advice. Howard Flock, former president of Wood Mosaic in Louisville, was a regular consultant four weeks per year at the plant. He worked on production strategies and provided counsel as a member of the company's board of directors.

In 1970, George Erath and John Erath, his brother, along with several other investors, purchased another "distressed" veneer mill in Martinsburg, West Virginia. This mill was later sold to his brother who bought out the remaining partners.

The production equipment of the Virginia mill consisted of two Capital slicers and a 12-foot Capital staylog lathe. Erath considered this to be a small operation and believed the only way for domestic firms to survive was to stay in specialized production areas and out of the large-scale veneer business markets. "There is no room for middle-size operations which must compete in both markets. The properly motivated veneer mill management can afford to provide the kind of service and close technical support to domestic face plants and furniture firms which other veneer firms are unable or are unwilling to provide." He was also concerned about the decline in traditional American values, including values associated with the ownership of fine furniture in a mobile society. Erath was also well aware of the pitfalls of the marketing of veneers in foreign markets where language barriers, cultural differences, and unfamiliarity with marketing patterns and institutions may trap the unwary American veneer producer.

While sticking to domestic markets for veneer sales, Erath purchased fancy and exotic veneers from a number of foreign sources. The Danzer organization in Germany custom cut veneers for Erath, and he had novel arrangements with other foreign producers. Believing in the growing importance of South American woods and veneers, Erath obtained in 1975 exclusive agreements in Brazil with Laserma S/A at Nova Iguacu near Rio de Janeiro and Continental Veneers in Cuiaba, state of Mato Grosso. This may have been "...the first such large-scale venture between an American sales organization and Brazilian manufacturers" (Moore, 1975). It was expected that Erath would take a major share of these companies' production and market the veneer in the United States. Erath's sales organization has been the southeastern sales representative for B. L. Curry and Sons, Inc., of New Albany, Indiana, which specializes in pecan veneers.

The entire Erath organization exuded confidence in the future. This was based on a dedication to producing the finest veneers possible and marketing their veneers through an alert sales organization which has the capacity for marketing a complete line of domestic hardwood veneers in addition to the fine veneers produced from African, Asian, and South American timber species.

WEST VIRGINIA

Breece Veneer Company

The Kenova and Portsmouth Rim and Spoke Company, established in 1904, was the ancestor of the Breece Veneer Company located in Kenova, West Virginia. Two brothers, John T. and George W. Breece, were partners in a business supplying wheel parts for the wagon building industry. In 1910, the company began to produce sawn veneer in competition with the Three States Manufacturing Company which produced quarter and plain sawed oak veneer and thin lumber in Kenova.

The company switched totally to veneers in 1917 and changed the name of the firm to Breece Manufacturing Company. In addition to using veneer saws, the Breece brothers purchased an eight-foot Capital slicer and a 76-inch full rotary lathe. A 16-foot Capital slicer was purchased in 1924, reflecting a shift to larger mahogany logs. Both the eight-foot Capital slicer and lathe were retained along with four veneer saws. By this time, the Breece brothers also had established themselves in Portsmouth, Ohio, with a plant manufacturing figured plywood and hardwood doors. This establishment maintained the Breece Manufacturing Company name. The Kenova operation became the Kenova Veneer Company and was managed by Harry T. Breece, son of George W. Breece (Anon., 1925).

Eventually, the veneer company was inherited by the two sons of Harry T. Breece: Harry, Jr., and William (Bill). The sons had been involved with mill operations beginning in the 1940s. The mill was gradually converted to a custom-cut operation. In 1965, a new Capital 165-inch slicer was installed. The older 16-foot slicer and the staylog lathe were sold about 1967. By 1972, the mill was nearly a 100-percent custom-cut plant and had principal customers in Coldwater, Michigan (World Wide Woods, Ltd.), and Winchester, Kentucky (Timber and Veneers, Inc.). Eighty-five percent of the veneer cut was white oak.

The Breece brothers sold their mill in 1985 to Theodor (Ted) P. Franklin, owner of Pacific Rim Hardwoods, Inc. Franklin had first installed a slicer near Mauckport, Indiana, where it occupied space within the Norstam Veneer Company building before he moved his business to Kenova. Pacific Rim specialized in expensive and rare veneers such as myrtlewood, Pacific oak, Oregon ash, Monterey cypress, Pacific yew, and Koa. Franklin maintained sales headquarters and a warehouse in High Point, North Carolina. The company has since gone out of business.

Erath Veneer Corporation

The Erath Veneer Corporation was located on the site of what was once the Auburn Wagon Works in Martinsburg, West Virginia. It was in these abandoned buildings that Max Wagenberg began operating the Martinsburg Veneer Company in the 1930s, using a 1915 Capital slicer. The mill was purchased in 1951 by Boyd J. Cummings in association with Thorton H. Gogoll. As the son of E. M. Cummings, a well known manufacturer of veneer faces in New Albany, Indiana, Boyd had literally grown up in the veneer trade. In purchasing the mill, Boyd took in several partners, including two Germans. One of these, Hubert Gretsky, later was killed at the Atlantic Veneer mill in Beaufort, North Carolina. He was scalded to death after mysteriously falling into a vat of hot water used to treat flitches prior to their being sliced. Boyd is remembered in Martinsburg as having a flamboyant, outgoing personality and as being a well-liked, sociable individual. He led the good life and was the chief veneer salesman for the firm—on and off the golf course. Always short of cash, Boyd had the reputation of selling veneer at discount prices which was an irritant to the rest of the industry. Nevertheless, he was well respected for being a good veneer manufacturing technician (Anon., 1952).

A fire in 1965 completely destroyed the mill. Cummings was initially unable to rebuild the mill because it had been underinsured. Financing was ultimately arranged through the Economic Development Administration. The EDA was part of President Johnson's "Great Society" program designed, among other things, to increase the prosperity of the Appalachian region and to aid the impoverished black population. Since Cummings had hired a number of black workers during a period when not many minority citizens were industrially employed in Martinsburg, it was an easy matter for EDA to target funding for rebuilding the mill. Unfortunately, the new structure was overbuilt and could not support the imposed loan repayments and at the same time provide the necessary cash flow to operate the mill. In the spring of 1968, the mill was taken over by four former employees of Litton Industries, a large multifaceted conglomerate.

The group consisted essentially of men who had been successful in various marketing and manufacturing ventures and saw promise in the rebirth of Cummings' rebuilt mill. Gordon Jacox presided as president and general manager and was assisted by Joe Kolb, vice president and manager (Anon., 1968). With the exception of Kolb, who had been an officer in the previous organization, the four outside investors had no previous experience in the veneer industry and promptly lost $1,000,000 in the next 10 months. In January 1969, the business reverted to the EDA in Washington, D. C.

Since George Erath had recently been able to reorganize and revitalize successfully the Helms mill at Rocky Mount, Virginia, EDA officials immediately contacted Erath about taking over their defunct mill in West Virginia. This occurred in February 1969, after Erath had made several trips to Washington, D. C., to renegotiate the terms of the repayment plan. The modern veneer mill that Erath took over was equipped with two Capital slicers and a staylog lathe. A Hale dryer and a Proctor-Swartz dryer completed the essential equipment. All of the machines were purchased in 1966 or 1967 for the reconstructed mill. In 1973, George's younger brother, John, moved to Martinsburg to manage the firm. In 1978, John Erath, born in 1940, became sole owner and president of the plant when he bought out his brother. That same year he hired Smedley W. Lynn as general manager.

Lynn was a good choice as manager although he had had no previous veneer manufacturing experience. However, he brought to the firm his business acumen acquired as a comptroller for Aetna Life Insurance Company, as an owner/operator of two fast food restaurants and as the independent owner with his wife of an ICC-certified trucking business. Lynn indicated that the firm prided itself on being intimately aware of customer needs. He believed that it is essential for the company's sales people to understand customer concepts of product manufacture so that the Erath corporation could effectively assist them in their production process and problem resolution.

The Erath operation was continually being reviewed for the possible introduction of new technology which would be cost saving and would enhance product quality. Production lines were automated and a conscious capital reinvestment policy was pursued.

Taking advantage of the Interstate highway system, the firm specialized in overnight truck deliveries to a wide range of East Coast and Midwest localities. Erath also introduced the concept of marketing clipped, palletized veneers to domestic customers as opposed to the traditional method of selling crated veneers. Veneers were palletized by grade or to customer specifications. A given pallet could be assembled from as many as seven or eight logs. Lynn estimated that 450,000 square feet of palletized veneer could be shipped on a single truck, compared with 250,000 square feet if the veneer were crated. Unloading was also facilitated.

In addition to veneers cut for their own markets, the Erath Veneer Corporation began in 1975 to cut veneers for a "resident" custom cutter, Fritz Bleidel, who exported his veneers to European markets. The Erath export operation was handled by Dieter Johnsen in the early 1980s. Veneers were sold in European, Asian, and South American markets through sales representatives.

The Erath philosophy was to sell top-of-the-line veneer cut from purchased top-grade veneer logs. White and red oak, ash, and walnut veneers were cut with specific customers in mind as well as their specific requirements for thickness, veneer "style," and veneer packaging. The company purposely sought out particular market niches to serve.

Lynn, looking to the future, believed that the export market for veneers to Europe would continue flourishing although it would be impacted by economic cycles from time to time. His reasoning was based on Europeans' propensity for expressing wealth in fine furniture in their homes rather than external ostentation and on the probabilities that European and Asian furniture manufacturers would increasingly market furniture in North American markets. With this scenario, containers used to ship veneers overseas would be returned with palletized and disassembled furniture.

In 1983, Lynne Erath, wife of John Erath, replaced Smedley Lynn as secretary-treasurer and Frederic Buehler, hired as an office manager, was named general manager. John Erath and Dieter Johnsen remained president and vice president, respectively. The work force consisted of 121 officers and other employees. Erath Domestic International Sales Corporation, a wholly-owned company, sold the veneer manufactured by the parent corporation. For reasons not entirely clear, Erath Veneer Corporation ceased operations in 1989.

TENNESSEE

Because of Tennessee's long East-West axis, a number of different forest resources have been exploited by veneer mills. The mountainous areas around Knoxville have attracted entrepreneurs interested in oak, yellow poplar, ash, maple species, and walnut. One of the earliest was Perlie B. Raymond who had gained his veneering experience with the Williamsons in Indianapolis. With his brother-in-law, George F. Adams, he established the Knoxville Veneer Company in 1910. It was reputed to be "...one of the newest and best types of veneer plants in the country, and Mr. Raymond is adding to his reputation as a high-class veneer manufacturer by the stock he is making at the Knoxville plant. The supplies of oak, ash, black walnut, and other high-class woods come from logs grown in the vicinity of Knoxville" (Anon., 1911). The Brunswick-Balke-Collender Company was also established in this era and continued to cut veneer until it sold its slicer to Sam Mengel in 1922 when Foreign and Domestic Veneer Company built its veneer mill in Knoxville. Further west, notable veneer companies were built in Jackson and Memphis, utilizing species adapted to the Mississippi River valley.

Ashby Veneer and Lumber Company

The Ashby Veneer and Lumber Company was founded in 1913 by Thompson VanMeter (Tom) Ashby. Born in 1875, he was the son of Robert Ashby (1854-1929) who was a farmer, grain elevator operator, and owner of retail lumber yards in the Ladoga, Indiana, area. Robert, a strong-willed individual, insisted that Tom either attend Wabash College or go to work on the family farm when he reached college age. Tom refused to do either and ran away from home to Montana where he was a freight hauler and worked in logging camps and sawmills for 10 years.

When Tom returned to Ladoga, his father gave him a grain elevator and a lumber yard at Bainbridge after he married a local girl. Determined not to spend the rest of his life in a town of 300, he joined Harry Daugherty in the veneer business. Daugherty, also a native of Ladoga and a boyhood friend of Tom, operated a veneer mill in Ladoga. Together, as partners, they formed the Hoosier Veneer Company in Indianapolis. They were later joined by Tom's brother, Frederick Forsythe Ashby (1872-1963), who had graduated from Purdue University with a degree in Civil Engineering. By mutual agreement in 1912, after some discussions about who was going to be the "boss," Tom sold out his interest to Harry Daugherty and bought the Blatchley Veneer Company at Jackson, Tennessee, which had recently burned. By the time Tom had reworked all of the equipment, he had no money left to operate the plant. The machinery was then sold to a group of Arkansas businessmen who intended to manufacture steamer trunks. Tom went along with the machinery to Little Rock as a salaried plant superintendent. When that venture failed, he was able to buy back the machinery for half the price he had sold it for.

Returning to Jackson with his machines and some working capital, Tom began operating in 1914 as the Ashby Veneer and Lumber Company. His older brother, Fred, was recruited in 1918 from Hoosier Veneer in Indianapolis to help run the expanding business. In several respects, Jackson was an ideal location for a wood products business. The community was served by five railroads which ran from the town like spokes from a wheel hub. Logs were brought to the mill and the products were then conveniently shipped nation-wide by this efficient rail transportation network. The forests of the region were extensive and both upland and bottomland species were easily obtained (Ashby, 1957).

Between 1914 and 1929, the principal products sold were rotary cut veneers for furniture and plywood firms and hardwood lumber which was widely distributed. The veneer plant was considered to be one of the most modern in the industry, operating with a 102-inch Capital Machine Company lathe and two smaller machines, a 65-inch Capital lathe and a 56-inch high-speed Coe lathe. The drying operation consisted of a 1926, 11-section Coe roller kiln with a 14-foot opening and a Proctor and a Proctor-Swartz textile dryer. A fire during World War II necessitated the purchase of a new Coe dryer and a reworking of the textile dryer. Equipment to produce berry baskets and bushel hampers was added during the 1920s to supply containers for the strawberry and potato-growing regions of Kentucky, Tennessee, and Alabama. Tom Ashby handled production and sales responsibilities for the firm while Fred devoted his efforts toward the purchase of logs and timber lands. The company still retained ownership and managed 4,000 timbered acres in the Hatchie River bottoms near Jackson in the late 1980s.

During the mid-1930s, the company began manufacturing flat veneer products (ice cream sticks and tongue depressors) to offset declining sales in furniture veneer. The company also extended its distribution and sales of vegetable hampers to California and, for two years, operated a container assembly plant at Calipatria, California, utilizing cut-to-size veneers shipped from Jackson, Tennessee. An office/warehouse for distributing containers was maintained in Colorado until 1946 when the entire basket-making and distribution operation was sold to a firm in Paris, Texas.

By 1932, Tom Ashby's three sons were entering the business: George R., a 1931 graduate in mechanical engineering from Purdue University; William L., a 1932 graduate from Indiana University in business administration, with a major in marketing; and Tom, Jr., who graduated from the University of Tennessee in 1938. All three had worked for the company during summer vacations, beginning in their high school years, and were thoroughly familiar with all departments of the business. They were made full partners in 1944 and took over the management responsibilities for the firm. In 1953, after the death of Tom Ashby, Sr., Fred Ashby sold his interest in the business to his nephews.

The Ashby brothers divided the management of the company business and each assumed responsibility for a department of the overall operation. George was responsible for raw material procurement, forest management on company-owned properties, and lumber production and sales at Crutchfield, Kentucky, and Town Creek, Alabama. Tom, Jr., was in charge of sawmill operations at Jackson and the production of dimension lumber.

William (Bill), a natural-born salesman and organizer, directed the veneer end of the business and was in charge of the production and sale of all veneer products. He also established sales offices in the principal furniture production centers and commissioned a network of jobbers throughout the United States to sell the company's line of woodenware products. Bill Ashby also was active in promoting quality hardwood products for the industry and served as president of the Fine Hardwood Association for a two-year term between 1954 and 1956. In 1957, the firm employed 200 workers, had an annual payroll of $450,000, and processed 7.5 million board feet of hardwood logs annually. Total annual sales were about $1.5 million (Anon., 1958).

The Depression years were difficult for the company. It was necessary to retrench and some operations were lost, including sawmills and a plywood venture in Canada which the company operated jointly with Hoosier Veneer of Indianapolis. During World War II, the company recovered economically by being involved in the production of aircraft veneer for gliders and DeHavilland Mosquito bombers and by producing veneer for PT boats built by the Higgins Boat Company of New Orleans. Most of this veneer was 1/10-inch sweetgum, although some sycamore veneer was also furnished. After the war, the company specialized in producing the "skins" and cross-bands for hollow-core doors. Veneer pieces were edge-spliced and cut to door-size pieces which were then bundled and shipped to door manufacturers. When outcompeted for this market by imported lauan veneers from the Far East, the Ashby Company concentrated on veneer parts for piano and other specialty items in which veneers were cut to size for easy "lay-up" when processed by customers.

Tom, Jr., died in 1958 at the age of 41. George, who was the chairman of the board, died on November 3, 1981, at age 71. Bill retired from the firm in 1974 to pursue other business interests. Robert C. Ashby, George's son, took on additional leadership responsibilities for the firm at the time of his father's death. Robert had grown up in the business and had served as company president prior to 1981. His preparation for the job included an undergraduate degree in industrial engineering from Georgia Tech and an MBA from the University of North Carolina. He entered the business on a full time basis in 1962. J. Gordon Fields assisted him as executive vice president.

In recent years, the principal species cut were poplar, sweetgum, and sycamore. The facilities were equipped to meet most demands for rotary veneers in terms of lengths and thicknesses (1/75- to 1/4-inch) required by the furniture market and the musical case industry. Specialty flatware items were still manufactured as well. Sales were mainly handled through manufacturers' representatives in major market areas such as Grand Rapids, Michigan; Louisville-New Albany; Los Angeles; New York City; and in North Carolina.

Saddled with obsolete machines and aging facilities, the company was sold and the buildings dismantled in 1988.

Nickey Brothers, Incorporated

There has been a succession of lumber and wood products enterprises owned or controlled by the Nickey family, dating back to the year 1866. The corporate existence of the present company began on April 18, 1907, when it was incorporated as the Nickey-Dickson Lumber Company of Gallaway, Tennessee. By amendments to the charter, the corporate name was changed to Nickey Brothers Hardwood Lumber Company on January 13, 1910, and to Nickey Brothers and Bass on June 24, 1913. Since October 6, 1915, the company has operated under the corporate name of Nickey Brothers, Incorporated (Cobb,undated).

The Nickey family odyssey with the hardwood industry started in 1831 when Dr. Nickey, a physician in Staunton, Virginia, brought his family by covered wagon to a densely forested wilderness homestead about 12 miles from Fort Wayne, Indiana. Addison B. Nickey was born in 1846 and remained on the farm through his early adult years, raising cattle. In 1865, he invested in a portable sawmill with a man by the name of Gandy at Churubusco. This partnership lasted about five years, with much of the 6,000-board foot daily output going to the Wabash Railroad in the form of dimension timbers.

Addison B. Nickey's son, Samuel M. Nickey, was born on September 9, 1868. He was later joined by two sisters and a brother, William E.

Nickey. In 1882, Addison Nickey contracted tuberculosis and was advised by his doctor to move to Arizona. This he did for a year, leaving Sam, at age 14, to manage his farm and to "...maintain the family fortune back in Indiana." The sawmill was closed down during this period. Sam had gained experience in handling livestock during this interim and stayed in the livestock business until he was 20 years old, buying and selling cattle, hogs, and sheep locally and shipping them to East Buffalo, New York. After Addison Nickey returned from Arizona, he expanded the Churubusco sawmilling operation and built a bandmill at Auburn, Indiana, in partnership with Gandy.

On the advice of his father in 1888 ("You can't make a living on the farm. You'd better get in the lumber business."), Sam went into partnership with a Mr. Perine at Auburn and wrote in his memoirs that "...for one or two years we had a pretty hard time getting along, but we finally got so things looked pretty good." At this point, Addison Nickey suggested that he and his son buy out their respective partners and merge their resources. Sam Nickey (1953) continued:

> We made money every year for three or four years, mostly because we were cutting export oak. We did our own logging. I think we had 25 logging teams. We bought our own timber just around Auburn. They'd load the logs onto the wagon and bring them into the sawmill. These men lived at home with their families. You see, 12 miles was about the limit of our getting logs. There was enough timber to keep the sawmill going at that time for three, four or five years. While it was slow picking, still we made a little money, and we ran that mill for about ten years, till about '98. But the timber that we were getting, especially the oak, was all full of these black spot worms. We couldn't do anything with it except put it in railroad ties. It got so bad that my father said, "We've got to move out of here some way or another."

The Nickeys ultimately located a new supply of old-growth timber near the Wabash Railroad in the vicinity of Princeton in the southwest corner of Indiana. John S. Dickson was a local log producer and had an intimate knowledge of the area's timber resource. Dickson guided Addison Nickey on a tour of the region while Sam returned to Auburn. Sam recalled that later in the week he received a telegram from his father which read: "I've seen more timber than ever. Finest in the world. Make arrangements to close our business immediately. Will build a sawmill in Princeton."

The sawmill, built in Princeton about 1896, was operated under the name of A. B. Nickey and Sons although Dickson was a partner in the business. The mill produced dimension lumber for the railroad until it was realized that the quality of the timber warranted a higher degree of utilization. They began to "quarter-saw" the logs instead of plain sawing them, air dried the lumber, and sold the lumber to cabinet makers and furniture manufacturers. Utilizing the network of railroads in the area, logs were bought within a radius of 40 miles from the mill. Within 10 years, the firm had accumulated a cash surplus of $500,000.

Seeking investment opportunities, the Nickeys looked South beginning in 1901 along the Illinois Central Railroad in Tennessee. They bought 8,000 acres at $5.13 per acre in an area south of Memphis, at Pritchard, and 3,000 acres near Marks, Mississippi. The land and timber were bought under the name Green River Lumber Company which the Nickeys had incorporated in 1905 in Memphis. A. B. and Sam also journeyed to the West Coast and bought 1,800 acres of Douglas-fir timber on the Columbia River, 40 miles upstream from Portland, Oregon. The parcel was purchased under the name of A. B. Nickey and Sons, for $60,000, and was sold three years later for $250,000.

In 1907, Sam's wife died after an extended illness. Later that year he and his daughter left on an epic trip around the world on one of the first cruise ships to do so. A stop in the Philippine Islands proved to be providential for the firm. Sam was greeted there by an old friend who was Consul General of the Islands. The Consul General encouraged Sam to evaluate the great quantity of quality mahogany timber standing in the Philippine forests as an economic development potential and suggested that a boat trip around Mindanao could be arranged for this purpose. The cruise schedule made this impossible, but Sam had been impressed by what he had seen near Manila. Years later, immediately after World War II, Sam sent Russell C. Stadelman to the Philippines to arrange for logs to be shipped to Memphis. Logs could be purchased and shipped to Memphis at less cost than tulip poplar logs shipped to Memphis from North Carolina.

Sam was married for the second time in 1911 to Lois Metzger, his first wife's sister. Lois was the mother of Samuel M. Nickey, Jr., and Lois Nickey Eason. Eleanor Nickey, daughter by his first wife, died in New Mexico in 1926.

The Nickeys were inventive and constantly seeking new opportunities. With one 16-foot slicer (the first in the South) and two lathes, the company began producing veneer in 1913. These items of equipment were in addition to four veneer saws located in an adjoining department. The drying room contained the "...widest textile dryer ever built." The veneer departments were planned and installed by Walt G. Bass, a Nickey partner at the time (Quigley, 1915). The slicer and lathes cut mainly walnut, figured red gum, and quartered oak, while the veneer saws ran "...to the highest capacity cutting oak, ash, and gum" (Anon., 1916). In 1928, another slicer, three lathes, two Coe dryers and a Proctor dryer were added. Production of plywood began in 1947.

Sam had actually made 25 or 30 hardwood plywood panels in 1917 with which he constructed buildings on his cotton plantation. This early experience, along with his continuing study of glues, paid dividends when the company began to produce plywood commercially several years later.

In 1920, the Nickeys formed the Capilano Timber Company and bought 5,000 acres, including one billion feet of Douglas-fir timber, for $180,000. The land was located in the watershed for the city of Vancouver, British Columbia. At a cost of $400,000, a rail line was constructed around a mountainside to reach the timber. Before it was completed, a major rock slide rendered it useless. The company was also plagued with labor troubles fomented by employees who were members of Woodworkers of the World (IWW). An arson fire which destroyed 50 million feet of cut logs and a high-production sawmill also discouraged them from continuing. Sam concluded that: "It's a long way out there. Too far away [and] that was part of the trouble." It was estimated that Sam and his brother, W. E. Nickey, and other family investors lost over $1 million in this venture.

Sam Nickey, Sr., developed a strong interest in forestry and woodland management on the company's 17,000 acres of timber and was one of the first wood products executives to provide assistance to other private landowners. By 1953, an additional 20,000 acres were in these privately-owned tree farms, under the supervision of a professional forester. Sam wrote: "These people that have these [tree farms] are tickled to death with the [idea]. We have no claim on their land in any way. The only claim that we have is that they won't sell their timber unless we get a chance to bid on it...[for] the services that we render...we don't charge them anything." Sam was also first to manufacture Presto-logs in the South to utilize mill waste, one of the first southerners to produce hardwood flooring in 1922, and among the first to recognize the qualities inherent in sweet gum trees for furniture veneers and lumber. Oak flooring became a significant product line until the Depression when the company had sales offices and distribution centers in Los Angeles, Des Moines, and Seattle (Nickey, 1953). During the 1930s, the company had logging operations in Mexico and Guatemala and maintained offices in Buenos Aires and London. A large percentage of the products was exported.

Sam's passion for new ideas extended into agriculture. He built one of the first alfalfa dehydrating plants on his farm in Arkansas and planted rice on river bottom lands he owned in Mississippi. He remained active in civic affairs in Memphis and industry associations until his death at age 90, on April 5, 1959. At the time he died, the Nickey Brothers processing operations at 2700 Summer Avenue covered 30 acres and employed 800 workers. They were supported by raw material sources from all over the world, including more than 80 tree farms managed by the company in the Memphis region (McCarthy, 1953). Sam Nickey, Jr., took control of the company in 1959, the same year he became president of the Hardwood Plywood Institute.

Development of the firm had not been sustained without setbacks. The family had been weakened financially going into the Depression by losses in the British Columbia venture. On January 9, 1940, the Nickey lumber mill was destroyed by a $350,000 fire. Five lumber drying kilns and four million board feet of lumber were destroyed. The mill was rebuilt in time to contribute to the World War II effort, however. An Army-Navy "E" Award was presented to Nickey Brothers, Inc., on June 15, 1943. It was presented in recognition of the firm's production of "feather thickness" veneers which were used in construction of components for the Mosquito bomber. A second "E" Award was presented in December 1943. The Nickey's war effort was augmented by the efforts of Sam Nickey, Jr., who served on the War Production Board.

Sam's children had joined the business as they completed their educations. They were integrated into the management structure as they gained experience and maturity. By 1956, Sam Nickey, Jr., was executive vice president in charge of the day-to-day activities of the firm. His sister, Lois, was company secretary and her husband, W. Jeter Eason, was vice president.

After World War II, Nickey Brothers increasingly focused attention on the timber resources of the Philippine Islands. The firm is credited with introducing and obtaining market acceptance for lauan plywood. Three veneer and lumber mills were ultimately established in the Philippines. One was near Zamboanga and the two others were on the North Coast of Mindanao.

One of these latter two mills operated as the Mahogany Products Company on the Aguson river in northern Mindanao. This business was established at the suggestion of Carl Wheeler in 1956. He was having a difficult time getting Georgia-Pacific to finance a core stock mill which would supply its Savannah plant which he managed. "I said: Look, if you guys will put up a core stock mill, I'll give you a purchase order for half the output. When [Nickeys] went into another venture, we again took half of their output. We had a very close liaison with the Nickeys." Russell Stadelman, Sr., Nickey's log buyer, made at least 40 trips to the Philippines for the company. Stadelman later became general manager of the veneer division before leaving to establish his own business.

By the late 1950s, the firm had phased out of the domestic veneer business. Sam Nickey III entered the company at about this time when the competition from Japanese plywood was becoming very keen. Sam III summarized the impact when asked about this: "The Japanese killed us," somewhat lamenting that the Nickeys had, after the war, played host to a Japanese delegation which had visited Memphis to study the Nickey operation in depth. Labor problems also plagued the firm from time to time in spite of forward-looking management programs. A major strike began in August 1948 when 700 workers affiliated with the International Woodworkers of America (CIO) struck the firm.

Following an extensive analysis of the business by a management consultant and an examination of losses it had incurred in 1960, a decision was made to begin curtailing certain operations and to sell some assets to cover bank debts. Inventories were reduced and overhead expenses were cut back substantially. Lumber operations were sharply curtailed in 1962. Another fire which started on February 2, 1964, in grass along the railroad, destroyed seven buildings and warehouses.

Nickey Brothers, Inc., has remained a viable company selling exotic veneers. In 1981, Sam III became president of the International Hardwood Products Association; and, in 1973, Nickey Brothers, Inc., joined forces with an architectural engineering and planning firm. The new company, Nickey-Bellante International, has a sawmill at Itacoatiara on the Amazon River. The major species sawn at this mill is banak. A high proportion is exported to England and the Continent (Anon., 1973).

CANADA

Besides General Woods and Veneers Ltd., in Longueuil, Quebec, and the Interforest veneer mill in Durham, Ontario, both owned by persons of European origin after World War II and discussed in Chapter 8, the only other hardwood face-veneer plant in Canada is Pannill Veneer Company of Kitchener, Ontario.

Pannill was organized in 1939 at Lindsey, Ontario, for the purpose of manufacturing furniture quality veneers. New facilities and improved operational methods were key elements when the operation was moved in 1942 to Kitchener, in the heart of Canada's furniture manufacturing industry. Although Pannill was designed originally for face-veneer production, equipment was added during World War II to produce birch veneer for military aircraft. This enabled the company to produce commercial veneers, after the war, that were suitable for crossbanding, core stock, and rotary-cut face-veneers. A Coe 110-inch lathe was used for preliminary barking and "rounding up" the blocks, and three on-line machines produced the veneer: a Capital 125-inch staylog lathe, a Coe 110-inch rotary lathe, and a 12-foot Capital slicer (1940 model). About 1.5 million square feet of plain and fancy face-veneer were produced per week in the mid-1950s (Graham, 1956).

Pannill was purchased in 1966 by W. A. Baechler and his son, Glen. W. A. had previously been a vice president and general manager of the company. Among improvements made by them were the acquisition of a hydraulic 12-foot slicer and a hydraulic staylog lathe. The slicer was in addition to a 17-foot slicer already owned by the company. Purchase of a new Proctor-Swartz dryer was also a part of the plan to increase efficiency (Anon., 1967). The company was later sold to Edward F. Yott.

Chapter 8

NORTH AMERICAN VENEER FIRMS WITH FOREIGN INVESTORS OR OWNERS

European hardwood veneer entrepreneurs came to the United States following World War II for a number of reasons. Among them were the constraints and corruption associated with former Colonial African timber sources, the freedom to export any and all logs from the United States, fear of the Russians, the stability of our governments, and the potential to seek new economic opportunities. Interviews with these men revealed the dedication and "drive" that they gave to revitalize the North American hardwood face-veneer industry. It explains, in large measure, why many of the traditional family-owned American firms have suffered from the competition or gone out of business in the post-war period. To more fully understand a significant portion of the foreign-owned component of the industry as it approaches the year 2000, it is necessary to understand the veneer industry in Europe as it existed in the years following 1900.

Paris was long the focal point of activity for the industry. Between 1900 and 1939, French veneer companies had perfected the art of veneer manufacturing to the extent that the group competed among themselves while protecting their individual market niches. None were threatened seriously by foreign firms. Another reason for this dominance was the industry's access to vast overseas Colonial holdings of tropical forest resources. Both were requisites for the exquisite furniture styles developed in France.

In the rest of mainland Europe, perhaps because of personal preference and more modest living standards, simpler furniture styles were customary in the early years of the century. As economic conditions changed, many young veneer men of Central Europe tended to gravitate to Paris. This was particularly true of the southern Germans, the Swiss, Austro-Hungarians, and Italians. Probably, the best training opportunities were offered by Marcel Miguet's veneer-cutting firm in Paris. Miguet was known widely as the "King" of the veneer business. The company had a large installed veneering capacity, had access to many exotic species, and produced excellent quality veneers. Like some other counterparts, Miguet had a network of trading companies and merchandising techniques suitable for dealing with discriminating customers.

World War II disrupted and largely destroyed the industry in the Paris area. Michael Boel, in commenting on those tragic times, commented in an interview that:

> Many of the best people in the veneer industry and in the veneer market were lost or expelled...because they happened to be Jewish. For us younger people, it is difficult to understand. Just to imagine the concentration of highly intelligent and highly trained and specialized and successful people on that many square miles in the city of Paris. They were the best and [some] were survivors... that had the foresight...to escape the Nazis in good time and [said to themselves]: "We will see to it that we will have some money, we'll see to it that our children can come." They were the survivors.

Of course, not all of the 10 or so European entrepreneurs who established beachheads in North America had this Parisian experience. Neither were all of them successful in their quest. However, when aggregated, they became the major factor in the hardwood face-veneer scene in Canada and the United States. Some elected to buy and rehabilitate existing plants while others chose to start new companies beginning with bricks and mortar. Much of what is known publicly about them and the firms they acquired is presented in the following pages. The first four of these veneer mills have been owned in recent years by men who knew each other in Paris.

The most dramatic individual in this group of Europeans was Karl Danzer who chose to buy The David R. Webb Company, a venerable Indiana firm with a long history of producing quality hardwood face-veneers.

THE DAVID R. WEBB COMPANY

The cutting of hardwood veneers in Edinburgh, Indiana, began in 1895 at the Muloda Veneer Company. The location of the veneer mill was ideal because the best walnut timber was located to the north and the best oak grew to the south of Edinburgh. Organized by Roscoe Mutz, A. J. Loughery, and Cash Davis, the firm's name was taken from the first two letters of each last name. The veneer company was operated by them until 1909 when the business was sold to David R. Webb. He had been a well-established hardware merchant in Edinburgh for many years prior to the purchase of the veneer business. Webb promptly changed the name of the firm to the David R. Webb Company. Oak veneer was the principal product and most of it was sold to the furniture industry in Grand Rapids, Michigan. Six years later, in 1915, the company was inherited by Webb's daughter and son-in-law, Ruth Webb Brodbeck and Percy Ryan Brodbeck (Otto, 1972). Following World War I, the Webb company enlarged its operations and installed new and more modern equipment which included larger slicers and modern-type dryers. With a return to walnut furniture, Webb Veneers became internationally known for the production of quality architectural veneers.

The company remained in the Webb-Brodbeck family until 1941 when it was sold to the Grunwald and Grossman families. These families had long been associated with the European veneer industry and had conducted business in France, Great Britain, Switzerland, and Austria before seeking refuge in the United States and England prior to World War II. Ferdinand Grunwald was the first member of these related families to emigrate to the United States and probably the first European to come to the United States to ultimately own and operate a hardwood veneer mill.

Interviews with several men who knew Grunwald are the source of the following composite biographical information:

> Even as a young man, he was a dynamic Hungarian with intelligence, talent and a thirst to succeed in business. After the first World War, Grunwald became a salesman for the Grossman veneer firm in Vienna. The Grossman family was very wealthy. It was "old line" Austrian and very conservative. Grunwald went to Japan in the 1920s and obtained a major portion of the supply of Japanese ash. On the strength of this success, he demanded and was granted a partnership in the business. He, later, travelled to Russia where he obtained a near-monopoly on the supply of walnut burls. These burls were sent to France where they were veneered. Without a doubt, Ferdinand Grunwald was one of the finest veneer manufacturers and businessmen of the twentieth century.
>
> When the second World War broke out, Grunwald was in Paris and remained there until he was arrested by the Nazis because of his religious heritage. He was released from prison for reasons associated with his Hungarian citizenship. Grunwald, soon thereafter, made his way to the United States via Portugal. He was able to establish business connections in Canada and, after about a year or so, bought the Webb factory in Edinburgh, for about $50,000.
>
> The mill was an obsolete and depreciated plant when purchased. To help operate the factory, he hired Max Spoerri, a Swiss citizen, who had worked for him prior to the war in Europe. Spoerri's principal functions were to price and to sell veneer for Grunwald although he performed other functions as well. In addition, a distant relative of Grunwald's, Michael Joelson, was hired.
>
> Joelson, an engineer, knew nothing about the veneer business and relied initially on the help of associates for their knowledge of woods and veneers. He learned his new craft very quickly and was soon embarked on a project to renovate the plant. Progress was monitored by Grunwald on his periodic trips to Indiana.
>
> Grunwald died when he was about 62, in a Swiss hospital. After his death the business was managed by Joelson and Spoerri and directed by Victor Grossman from England and Mrs. Ferdinand Grunwald in New York.

Eventually the business was sold to the Rutland Corporation in 1966, which, until 1964, had been known as the Rutland Railway Corporation. As a result of selling its railroad interests, the corporation had a large amount of cash and a large tax loss (tax credit) carry-over. In order to shelter the company's cash position from being taxed and to take advantage of its tax credit position, the Rutland Corporation purchased the veneer company. Michael Joelson was retained as manager. Unfortunately, the David R. Webb Company was not able to generate enough profit to offset the carry-over tax credits. Because of this unsatisfactory financial situation, the Rutland Corporation was merged in 1968 with the Walter Reade Organization, Inc., a theatrical company based in New York City.

In 1971, the assets of the David R. Webb Company were isolated and placed in a separate Webb Veneer Division of the Walter Reade Organization. The next step in this scenario was to sell the veneer division which had been a ship on the turbulent seas of the financial world for the preceding six years. This last act occurred on December 22, 1972. The urgent need for cash was given as the reason for the sale although the conglomerate mix of a theatrical firm with a hardwood face-veneer business in the hinterlands of Indiana surely must have been a secondary story line in the Reade Organization's decision to sell (Anon., 1975).

The purchaser of the David R. Webb Company this time was the Karl Danzer Company of Rutlingen, Germany. In 1975, this company was recognized as the 1183rd largest firm in Europe. It was also part of a Swiss holding company known as ANBE. Danzer's other North American business interests, including Webb Export, Inc., located in Edinburgh, and another face-veneer company in Durham, Ontario (Interforest), were also business elements of ANBE.

Karl Danzer has been somewhat of a shadowy character in the international veneer scene, beginning prior to World War II. Webb employees were reserved when interviewed about the man in a desire to respect his right of privacy. However, three men who knew him well as business partners have been able to shed some light on his prewar and postwar years in Europe. One of these was Carlo Iseli of Zurich, Switzerland, former owner of Universal Veneer of Newark, Ohio. Iseli knew Danzer, first in Paris when Danzer was a representative of German veneer interests, and later as a business partner in Switzerland. When asked about Danzer, Iseli replied:

> We started at almost the same time in Paris. Danzer was about 19 when he came to Paris. We were brokers and learned the veneer business at the same time. I think we all got excellent backgrounds there. Danzer went very often to Paris during the war when France was occupied by the Germans. They bought veneer in France because German production was down. It was very easy for him to travel to France because the German government assigned him to do that during the war. I know that Karl Danzer did quite a lot to defend the interests of the French veneer manufacturers and always had good relationships with the French people.

These observations were confirmed by the other two individuals who were personally acquainted with Danzer. Percy Simmons added:

> During the war when the Germans occupied Paris, Karl Danzer occupied a position of importance. His exact title I don't know, but it was something akin to being in charge of timber or the veneer industry. At that time, one must say without any acrimony whatsoever, we're talking about the Germans being in Paris and he had an administrative post. A German then was a German.

Leon Groscott, a Parisian veneer manufacturer before and after World War II, was one of the Jewish entrepreneurs who stayed in Europe during the war. He and his family first fled to the Vichey region of France when the country was partitioned by Adolph Hitler. After the Germans took over administration of this area from their puppet, Marshall Petain, Groscott went "underground" in the forests of southern France. When interviewed in Paris in May, 1983, Groscott spoke kindly of Danzer, minimizing the importance of Danzer's war activities and indicating his continuing high personal regard for the man.

After the war, Danzer was associated with Carlos Wessell, a Hamburg businessman. Iseli made this observation:

> Wessell was not originally in the veneer business. After this partnership broke up in 1962 or 1963, the Danzer group became very important. In my opinion, it is the largest and most important group in the veneer business today, with veneer plants not only in Germany and Belgium but also in Africa, the Ivory Coast, Zaire, Brazil, and the United States.

When the David R. Webb Company, Inc., was purchased by the Danzer organization in 1972, it was organized under the laws of Delaware. A program was immediately instituted to make Webb the largest, most modern veneer plant in the United States. Ground was broken in October 1973, and construction was completed exactly one year later. The new facilities provided about 67,000 square feet of useful production space. Two Capital slicers were transferred from the old plant and added to the two new Capital slicers purchased for the facility. Webb's Capital staylog lathe was also moved to the new plant. The old plant was then converted to a warehouse. Combined floor space of the two plants totaled 111,000 square feet when the expansion program was completed. The flow of materials and product was designed for increased efficiency using conveyers and other forms of mechanized handling to reduce labor costs.

John A. Grunwald was retained as president of David R. Webb when Danzer purchased the company. Grunwald had been encouraged to come to the United States in 1950 when he was 15 years old, by his uncle, Ferdinand Grunwald. Born in Budapest, Hungary, young Grunwald was a personable, intelligent teenager who quickly adapted to all aspects of American life. He was educated in New York, Massachusetts, and Indiana schools before embarking on a career in the wood industry. Grunwald first spent several months in the furniture shops of Saginaw, Michigan, and with the Cadillac Plywood Company in Cadillac, Michigan, before being given a year's assignment to buy logs in Europe. He joined the Webb Company in 1958 and worked in a variety of production and sales positions. Grunwald became general manager after Spoerri's death in 1969. He was later named the corporation president and quickly developed into a dynamic industrial leader.

Per R. Kalstrup, born in Denmark, was brought into the Webb Company in 1968 after serving as a Webb representative for a year in the

Chicago and Wisconsin areas. He had started his business career with the Dean Company in Chicago and earlier had worked for a lumber company while attending Loyola University in Chicago. He utilized his previous sales experience and eventually was promoted to vice president in charge of sales for both veneer and lumber. His responsibilities were later carried forward in the Danzer organization as executive vice president. Other experienced key personnel also continued, including Michael J. Tuccio, treasurer, and Lowell E. Green, timber manager.

Since 1972, the company has grown from 150 to 400 employees and has expanded into modern brick facilities for its office complex and production centers. The firm's four Capital slicers and one staylog lathe have an annual capacity to produce up to 300 million square feet of high quality veneer. One and one-half million feet of logs are required annually by the Webb plant. These logs come primarily from the Midwest and various other locations in the eastern United States. The company also owns 1,000 acres of its own timber land and provides landowner assistance and educational programs for timber growers interested in growing quality trees for the veneer market (Anon., 1979).

Webb's modern bandmill, complete with laser beams to assist in the alignment of logs on the carriage, produces 20,000 board feet per shift. Although lumber started out more or less as a by-product of veneer production, emphasis on lumber began in 1977. At that time annual lumber output was 300,000 board feet. By 1986, the company was producing six-million board feet of hardwood lumber each year and about one-third of it was walnut. Additional lumber is purchased to maintain peak efficiency of the 10 company kilns. These kilns have a capacity of about 800,000 board feet.

GENERAL WOODS AND VENEERS, LTD.

Isidore (Dore) Elefant, founder of General Woods and Veneers, escaped from France with his family in 1939 and made his way to Canada at the age of 34. He had been an associate of Karl Danzer and Carlo Iseli in Paris prior to World War II. The three men worked closely together at a time when Paris was still the world-wide center for the buying and selling of face-veneer manufactured in Europe. After World War II, the three partners split over business disagreements, apparently, vowing to outdo each other in the hardwood veneer industry. Each ultimately invested his capital and reputation in veneer ventures in the United States, and two (Elefant and Danzer) began their North American odyssey in Canada. The final split between Danzer and Elefant occurred about 1972.

Shortly after arriving in Canada, Elefant established himself in the lumber trade with the Transcontinental Lumber Company which marketed structural lumber manufactured by the numerous small mills operating in the Canadian "bush." He also began custom cutting veneer in the United

States as he had previously done in Europe. This activity was the beginning of General Woods and Veneers, Ltd. Actual manufacturing of veneer in Canada did not begin until 1947 when Elefant installed an old bandsaw and an equally old Johnson City slicer in a building at Longueuil, Quebec, near Montreal. The building had been used to produce aircraft during World War II. This was the first plant to produce sliced, hardwood face-veneers in Canada. By 1968, the plant was operating with a Capital slicer and staylog lathe and a Merritt staylog lathe. Annual production was estimated to be about 75 million square feet. Custom cutting in Europe augmented domestic production as African veneers could be more economically manufactured there because of lower labor costs and reduced freight charges on imported African logs (Anon., 1968).

The General Woods and Veneers plant has increased its productive capacity several times since 1968 and has introduced a number of innovations. In 1969, with the firm prospering, "...lavish new plant offices and showrooms [were opened] in Ville Jacques Cartier on May 16th..." (Burrell, 1969). In this expansion, a separate architectural center was also inaugurated to attract this specialized user group and to feature unusual and exotic veneers and other wood products.

Isidore Elefant stepped down as president of the firm in 1971 and was succeeded by his son, J. Marcel Elefant, who had been born in 1938. Marcel's son, Steven, a business administration graduate, began full-time employment with General Woods and Veneers in 1984. Another son, Robert, was later employed by the company. Marcel, who had previously been vice president of the firm, became about as well known as his father in North American veneer industry circles.

Isidore remained active in the firm as a director of the company. He had always been concerned with raw material supplies and, consequently, spent a considerable amount of his time globe-trotting to keep pace with company requirements.

One of the features developed in 1973 for the mill was a vacuum device to hold the flitches to the flitch table. Upwards of 15 tons of vacuum "pressure" were exerted to hold the flitch in place. By eliminating retractable "dogs," standard-sized (about one-inch) backing boards were reduced to one thickness of veneer. With the vacuum process, only about 60 seconds were required to swing a flitch into place, apply the vacuum, move the carriage into place, and activate the slicer. Mill management has estimated that the technique increases veneer yield by 10 percent and overall productivity by 15 percent. At the time the new technology was installed, Elefant was quoted as saying: "The innovation is so revolutionary, so advantageous, that we prefer to offer it to all in the industry who may wish to cut costs and improve their yield" (Burrell, 1973). Further technological innovations have included the development of a specialized planer for the flitch preparation operation and the adoption, in 1976, of computer technology and electronics used in measuring clipped veneers width-wise rather than length-wise. This speeded up the measurement and bundling of veneers 50 percent while increasing the accuracy of the measurement (Anon., 1976a).

General Woods and Veneers, Ltd., is now one of North America's leading suppliers of hardwood veneers and lumber. Besides its Canadian plant, the firm has several subsidiaries, including Genwove U. S. Ltd., located in Indian Trail, North Carolina, and Swords Veneer and Lumber Company, located in Rock Island, Illinois. The Elefants also expanded into chair and furniture components in Ontario and have other branches, sales offices, and warehouses across the North American continent and in England. In 1989, the General Woods and Veneers group of companies was operating 16 Capital slicers and staylog lathes. A wide variety of domestic species was used by the company, including red and white oak, yellow pine, knotty pine, walnut, pecan, hickory, maple, cherry, and eastern red cedar.

Genwove U.S. Ltd.

In 1973, General Woods and Veneers, Ltd., bought the Atlantic Lumber Company veneer mill at Indian Trail, North Carolina. Bill Berry and Don Brown were involved in the management of the plant at the time. The Elefant organization was interested in this location because of regional furniture plants and the easy access to High Point, North Carolina, where the company located its principal sales office. The other attractive feature of this location was that the mill could economically tap into the oak and walnut areas of the Midwest, the Mississippi delta hardwoods, and the red cedar stands of the Appalachian mountains. The Indian Trail and Longueuil mills maintained a centralized log purchasing center at Winchester, Virginia. Genwove employed one additional log buyer to supply its Indian Trail mill.

Between 1978 and 1981, 100,000 square feet of space were added to the plant, creating an efficiently organized veneer mill and face plant. The number of veneer machines was increased from two slicers and one staylog lathe to five slicers and one lathe in 1982. New Babcock dryers were installed as well.

Bob Solo, a wood technology graduate of Yale University, was selected by Elefant to orchestrate the expansion and management of the veneer mill and the construction and operation of a face plant within the same complex. He was assisted by Harvey Hayes, plant superintendent, and Karl Helms who directed the maintenance program. The highly efficient and polished plant operations were starkly contrasted by the Spartan offices tucked into an obscure corner of the mill.

The Indian Trail plant bought veneers from South American sources and cooperated with other Elefant mills in mutually filling orders and completing inventories of diverse species. The principal sales office in High Point was serviced by warehouses located at Indian Trail, Montreal, and Calgary. Other locations in Canada and South American "connections" additionally served Genwove customers.

Swords Veneer and Lumber Company

Clarence Swords started his productive life as a farmer on land inherited by his wife and her sister, Mrs. Alex Morton. Swords left the farm in 1923 to cut and sell walnut logs as an independent log producer (Anon., 1946). Paul Widener, a long-time Swords employee, knew Mr. Swords and characterized him "...as basically a log man. That was his big thing. He was a very nice gentleman and was fairly decent to work for." Swords continued as a log producer until 1931 when he "backed into" the veneer business after becoming disgusted with the log prices he was receiving at veneer mills. This new venture was incorporated in Indianapolis with his brother-in-law, Alex G. Morton, as Swords-Morton Veneers, Inc. The company custom cut walnut veneers in various mills in Indiana, but primarily at Curry-Miller Veneers, Inc., in Indianapolis. Logs were procured for the most part in Illinois, Missouri, and Iowa during the first few years the firm operated. This corporation was disbanded in 1937 to reemerge as Swords-Morton Veneer Company.

In 1940, a sawmill was built in Rock Island, Illinois. The firm was incorporated in Illinois as Swords-Morton Veneer and Lumber Company. Walnut flitches cut at the mill were trucked to Indianapolis for custom cutting into veneer. During World War II, the mill produced gunstocks from the walnut logs obtained in northern Illinois, southern Wisconsin, and Iowa.

After the sawmill was destroyed by fire in 1946, a new plant was constructed, incorporating a seven-foot Sinker-Davis bandmill and assorted second-hand veneer equipment. Veneering equipment included a 10-foot rotary lathe, a 16-foot Capital veneer slicer, and two dryers: a one-way James veneer dryer and a two-way Proctor-Swartz. Alex Morton was bought out by Swords shortly after the new mill was built.

Howard Flock, a widely-known and respected veneer professional, left Mitchell Veneer in Indiana in 1950 to become the production manager at Swords-Morton. Before leaving two years later, Flock had coaxed some rather ancient machinery to respond with a new burst of productivity. By 1955, the company was annually producing one million feet of lumber and more than 15 million square feet of veneer (Gunnason, 1955). A total work force of 60 people was employed under the direction of Daniel Mack, vice president and general manager. Veneer sales were directed by Paul Widener.

Widener remembers that the original equipment was not the most modern when he joined the company in 1951.

> The flitch table on the slicer would quiver and eventually break. We would take it off and lay it on large iron cross pieces to true it up. We'd put a charcoal fire under it and a crew from Indianapolis would braze it

> back together. The rotary [staylog lathe] was so worn out that it would cut about 40 sheets of veneer and then it would spin about 20 times so the spindle could catch up. We still cut good veneer. In his later years, Mr. Swords sort of semi-retired. Dan Mack, Jerry Swords [a nephew], and I would often run the plant when he would go off to Florida. The only way we could get hold of him was with a telegram. Then, he would call us back. Jerry Swords finally left us and went to work for the David R. Webb Company in Edinburgh, Indiana. He bought logs for them until he retired. Dan Mack left, when Swords was bought out, to become an accountant in Charlotte, North Carolina.
>
> The veneer was sold primarily in the North Carolina area and in Michigan and Wisconsin markets when I joined the company. Then, we expanded down into the Indiana-Kentucky area. Soon after, we began specializing in Canada. We would cover everything from Kitchener, Ontario, to Quebec and back again. Of course, we were in direct competition with the company which eventually bought the company from Clarence Swords. When the European market opened up, we started doing business with Karl Heinz Moehring in Durham, Canada. He was one of our big customers and we sold him veneer by the car load. Most of this veneer was shipped to Germany for TV cabinets. We continued to do a lot of walnut business with him until Swords Veneer and Lumber Company was sold. Clarence Swords lived to see many changes in his company before his death in 1970.

The plant was fully acquired by General Woods and Veneers, Ltd., of Longueuil, Quebec, in 1962. This followed a period during which there was a three-way partnership among Isidore Elefant, Karl Danzer, and William Roddis, a plywood manufacturer in Wisconsin. When Roddis and Danzer left the company, Elefant shortened the company name to Swords Veneer and Lumber Company, Inc.

One of Elefant's first actions was to begin replacing worn-out veneer equipment with better machines. A 1944 model, 16-foot Capital slicer was purchased from the disbanded Ichabod T. Williams & Sons Astoria mill in New Orleans; a 16-foot, 1917 model Capital slicer was transferred from the General Woods mill at Longueuil; and, in 1963, a new 16-foot slicer was purchased from the Capital Machine Company. A Capital staylog lathe and a Capital full rotary completed the inventory of veneer cutting machinery at the Rock Island plant. The old Sinker-Davis bandmill was retained and refitted with a new carriage. Production emphasis had shifted to the export market before the mill was purchased from Clarence Swords. It remained so when Elefant became sole owner and introduced a European-style clipper line.

Gus Stetter came to work for the Swords Company as general manager in 1973 after acquiring extensive veneer experience at the Evansville Veneer and Lumber Company. He was instrumental in increasing the productivity of the mill and integrating new technology into the production system. One of the novel installations was a Caneli saw manufactured in Germany. Only three others had been installed in North America by 1985.

The saw cuts from above and, as Stetter described it, the machine:

> ...basically is used to make just two cuts to the heart of the log with the aid of five laser beams. Everything else is done on the head rig to process the log into three equal pieces. Our sawyers can saw five logs an hour and that's it. Out of the five logs we get 15 flitches.

The apparent advantages are that the veneer is "tight cut" and very smooth. The process also eliminates quarter-sliced veneer "...which the American market does not readily accept."

Red oak was the principal species veneered in the 1980s. White oak, walnut, cherry, and birds eye maple are other featured species. The red oak is primarily from Wisconsin. Stetter continued...

> At times it's real difficult to get good logs. It's like everything else. Tremendous red oak logs come from Wisconsin, but there are areas there where you can cut an entire tract and not get one decent veneer log. Our log buyers work hard and spend a lot of time going from log yard to log yard looking for good veneer logs. Procurement will get more difficult as time goes by.

The Wisconsin log buyer was augmented by a contract buyer who "worked" both Michigan and Wisconsin. Log buyers were also located in northern Missouri and southern Illinois. Standing timber was sometimes purchased and contract cut.

The small quantity of birds eye maple received was cut on the full rotary lathe. Stetter said, "Birds eye maple is a very scarce commodity. The logs are not easy to come by and are very expensive. We cut it on the full-round rotary in order to get all the grain character and to elongate the eyes."

Until recently, exported veneer was shipped to Canada where it was containerized and marketed overseas by the parent company. Containerized veneer was also sent to foreign destinations by way of the Port of New Orleans. Domestic veneer in 1985 was unclipped and crated. When sold, the veneer was frequently uncrated and palletized to reduce shipping and handling costs. Stetter concluded his remarks by saying that:

> We put the pallets in vans or on flat beds and ship them all over the country this way. Export thickness veneer is also sold to some customers on the West Coast and plywood and door manufacturers in Wisconsin. The furniture manufacturers in the High Point area have been less receptive to this trend toward thinner veneers.

Because of the strength of the U. S. dollar in 1985 and reduced demands for their veneers in foreign markets, Swords devoted more attention to the domestic markets. Evidence of this was the opening of a High Point sales office in 1984 from which two salesmen operated. This office competed to a degree with Elefant's Indian Trail mill which also maintained a sales office in High Point. However, species differences in the two regions reduced much of the direct competition one would have otherwise expected. Two salesmen used Rock Island as a base for selling veneers in the Midwest and Lake States and another covered the West Coast from Benton, Arkansas.

The officers of the company, in the mid-1980s, were Isidore Elefant, chairman of the board; Marcel Elefant, president; Paul Widener, vice president, and James Hasty, treasurer and assistant secretary. Seventy-five employees were involved in the multi-million dollar production and sales operations of the company.

UNIVERSAL VENEER MILL CORPORATION

Dedication of the Universal Veneer Mill at Newark, Ohio, took place on May 16, 1980, almost exactly a year after the foundations were poured. This new corporation was founded by Carlo Iseli in December 1978. It was just the latest accomplishment of this 69-year-old man who had nearly a half-century of experience in the veneer business when the new mill was dedicated. As Iseli said: "I still work very hard. I couldn't be a man if I was not working. I'm not ready to retire" (Dixon and Flock, 1980).

Swiss-born, Iseli began his career with a veneer firm in Paris in 1929 by debarking logs with an ax. In an interview in 1985, he recalled that the veneer business was a family tradition.

> I had two uncles who were in the veneer business. They started at the end of the last century. When I was twenty years old, I went to Paris which was at this time the most important center of veneer production. I worked at the time in the famous veneer plant in Parc-St. Maur, not far from Paris. I worked at all the different jobs you have in a veneer mill, to the vats, the bandsaw, and so on. I also worked at the slicer. Of course, the slicer was completely different than we have today. We were producing maybe nine to ten sheets a minute. Today, we have slicers that are running very fast and are producing 70 to 80 sheets in a minute. Afterwards, I worked at the dryer and the clipper. They didn't have a clipping line and we were producing at this time French walnut. French walnut is, in my opinion, still the king of all the wood species. Unfortunately, there are only a few walnuts left in France because in the years 1929 and 1956 there were big frosts.
>
> Once I finished learning the machines, I went to another firm which was also producing veneer. There, I learned how to clip and price the veneer. At that time we priced the veneer by sheets and not by square foot or square meter. A small picture can have a higher value then a very big picture or a very big sheet. It was in my opinion a very interesting time. We didn't produce as much as they do today in a veneer mill but everything was created with more craftsmanship.
>
> In 1931, I went to Bulgaria. Bulgaria at the time had wonderful walnut. I went to Greece, also to Turkey and to Russia to buy walnut. We were producing at a veneer mill first in Kazanlik which was in the heart of Bulgarian walnut country. Later on, I went to Darvenitza, near Sofia, where there was a larger veneer mill. We produced wonderful walnut.

It was at this point that he became an associate of Karl Danzer and Isidore Elefant and managed veneer mills for them in Kazanlik and Darvenitza.

> The demand was especially big in England. They had very big furniture factories in England at this time. I think England, then, was a bigger market than Germany is today. We sold to England from Sofia. The veneer was shipped through the Black Sea from Valna or Burgas to Liverpool or London. Some went also to Antwerp in Belgium and, then, the veneers were transferred to England. At this time, I had a chance to learn quite a lot about the walnut business. We bought extremely nice logs in the average of two to three tons; the biggest ones five to six tons.
>
> We were not able to bring them to the veneer mill before we cut them into several pieces. The transportation was much higher priced than the logs. The logs were very cheap but of excellent quality. I remember this time as one of the nicest times in the veneer business.

In 1933, Iseli returned home from Bulgaria and went to work in his uncle's veneer firm in Zurich. After World War II, Carlo and his brother, Bernhardt, organized a veneer mill of their own in 1946. This relationship continued until 1956 when Bernhardt left to start a door plant at Brunegg (Moore, 1976). Iseli continued his recollections by saying:

> We were veneer dealers in the beginning and, due to my knowledge, we bought quite a lot of logs in Switzerland and France. We also bought logs in the Ivory Coast, Cameroons, and Nigeria. We brought these logs to Switzerland and we sliced the logs in one of the only two custom slicing mills in Switzerland. So, I decided in 1960 to build my own plant. This was the beginning of the Zurich veneer mill [Zuercher Funierwerk, Ltd.], incorporated at the end of 1960 and, in 1961, the plant was built. We went into production in 1962. In the beginning, I had the famous Danzer Company of Reutlingen, Germany, as a 30-percent partner during the first two years. Afterwards, I bought his partnership. We made tremendous big progress in our factory and started working two shifts. About 70 percent of our production went into export to Germany and Austria and other countries such as Spain, South Africa, and Australia. About 30 percent of the production was sold in Switzerland. We also exported some to the States, especially extremely good architectural logs. Our mill was producing teak and Indian rosewood veneers among others. Later on, we started also to produce Brazilian rosewood until the time they put on the embargo in Brazil.
>
> Our mill in Zurich always had a famous name for excellent production and this was from the beginning. In my opinion, if you want to make veneer, you have to do good slicing, good clipping, and good preparation for the furniture people. Because of my knowledge, the fact is we did good work in our mill. We became a very important firm. Maybe, I can also mention that our firm was a member of the European Veneer Manufacturers Association. I became vice president and then, afterwards, for three years I was the president. This organization is still going and includes probably all the members from Germany, France, Italy, Austria, Spain, England, Sweden and Switzerland.

In 1958, I began coming to the United States on a regular basis. We first bought birds eye maple from Michigan. I was also one of the first importers of American walnut logs. We imported walnut for a long time. We started also to import oak—white oak first, and later, red oak when Europeans became more used to this species.

I finally decided to open a mill here in Newark, Ohio. I wanted to start my plant in 1975-76 with another partner. In 1978, I decided to start this plant which was built during 1979. In 1980, we went into production. We started at a time when business was difficult for us because of the 1981 recession. We also had a lot of technical problems in this factory due to the dryers. We had to change the dryers and bought German dryers which in my opinion are the best ones. Business came back in 1982 and, in 1983, we went to two shifts until the summer of last year [1984] and now we have a new recession. We produce veneer for export in the European style. Everything is sliced into European thickness; walnut is 0.55 mm and oak is 0.65 mm. They are the most important species we are producing here.

The business, of course, in Europe is very difficult for the reason that everybody wants to have only the best veneer without knots, without defects. In my opinion, that's a big mistake. That's against nature. We should use walnut veneer, oak veneer, or other species like it is grown. For instance, if you see furniture like Louis XIV, Louis XV, and Louis XVI furniture, you will see they were using a lot of walnut with defects. This kind of furniture is sold today at tremendously high prices.

You have always to be up to date with the veneer machines. There are a lot of changes with the machines. For these reasons you have to practically invest every year for new machines. About five years ago everybody was feeding their dryers by hand. We still do it here. There are some factories here that are feeding the veneer directly from the slicer to the dryer. In my opinion this hurts the quality of the veneer. In Zurich, we bought a new dryer from Babcock which not only drys the veneer but also presses the veneer at the same time. This is a very important thing. I think this new kind of dryer is the future in our business.

When a container [40,000 square meters] of veneer leaves Newark, it goes mostly to Baltimore from here. You have other places where you can ship the veneer—Norfolk, New York, and so on. Baltimore is the best place to ship the veneer. We have special rates from here to Baltimore and we could also ship through Wilmington but we use Baltimore, I would say, 95 percent of the time. The containers are usually trucked to Baltimore. They are also sent by railroad but mostly by truck. Then, the veneer is shipped to Hamburg or to Bremen; or, if we sell veneer to the Far East, it will go to different countries like Singapore. [At the foreign destination] the veneer is sold to a consumer or to a veneer dealer. The dealers mostly come here to inspect the veneer in our warehouse. Afterwards, we put the veneer on pallets. The pallets are put in a container and shipped. For Germany, the containers are trucked from Bremen or from Hamburg to the final destination usually. The dealers are wholesalers who sell to smaller furniture factories.

Iseli was convinced that log buying was the most important part of a veneer mill's operation and personally spent three to four months per year searching for the finest quality veneer logs in Asia, Africa, and in North and South America for his veneer mills.

It was his growing realization of the increasing importance of North American log sources in world trade and the impact of higher energy costs that motivated him to search for a mill site in the United States. By exporting dried veneers instead of logs, he could reduce shipping costs and the wastes associated with manufacturing and clipping.

In the European tradition, the Universal mill custom cut veneers for other entrepreneurs as well as for the Iseli organization. The veneer slicing equipment consisted of three Capital slicers and a 12-foot Capital staylog lathe. Drying was done with one Hale and two Babcock-Schilde dryers. Keller clippers and electronic measuring machines completed the essential machinery. The plant layout was designed by Hans Landert to maximize efficiency of materials handling and waste disposal. Bark, trim, and clippings were conveyed to the boiler system which produced heat for the log vats and the dryers. Back-up gas heat was available for the dryers but had not been needed. The production floor and office space were under a single roof which covered 86,000 square feet.

Iseli was able to integrate into the Ohio business community in a relatively short period of time. Local financing, along with associates such as John Schwartz and Sally Wiley, assisted in establishing the business. Never quite able to reach his expectations for the Newark mill, Iseli sold the company to Klaus Krajewski in 1988. Krajewski is regarded by his contemporaries in the industry as an enterprising representative of the younger generation of the hardwood veneer executives.

VIRGINIA LOG COMPANY

Karl Rudolph is the last of the group who received Parisian veneer experience prior to World War II. A native of Leipzig, Germany, he was not apparently a major figure in the business lives of Karl Danzer, Carlo Iseli or Percy Simmons. During World War II, Rudolph moved West to escape the Russians. Later, he established a veneer business in Minden, West Falen, West Germany, under the name of Rudolph and Studier. His was one of the first three European firms to export logs from the United States after the war. With solid business acumen, Rudolph followed his log supply back across the Atlantic and established the Virginia Log Company veneer mill in West Point, Virginia. Its business office was located in Richmond. The mill was equipped with two slicers and one staylog lathe. Rudolph continued to export logs while manufacturing quality veneers primarily for export. He was reticent about granting an interview and little, if anything, has appeared in the trade publications about his activities.

The next group of Europeans who became established in the United States received their training or experience elsewhere. Karl Heinz Moehring, whose story follows, interned in Germany.

ATLANTIC VENEER CORPORATION

The person most closely identified with the Atlantic Veneer Corporation is Karl Heinz Moehring. He and his close associates created an intercontinental veneer and wood products empire in less than three decades, often under trying circumstances. His hardwood face-veneer mills are operated on three continents in addition to commercial operations in Africa.

Moehring began his association with the veneer industry in 1939 at age 16 when he became an apprentice to a German log importer who custom cut his logs at local veneer mills in the port area of Hamburg. Moehring's diligence, intelligence, and love for his work were rewarded two and one-half years later when he was given the status of "full fledged employee." Moehring's basic philosophy about work, which never changed, is to be "... the first one to arrive in the morning and the last one to leave at night."

Always ambitious, Moehring became a veneer agent after World War II, buying and reselling veneer in 1948. Three years later, in 1951, he built his first veneer mill in Lemgo, Germany. The mill cut veneers from logs imported mostly from West Africa. From its modest beginnings, the mill grew to be one of the largest veneer producing establishments in Europe. Hans Moehring, Karl's brother, managed this operation (Moore, 1977).

Seeking still broader horizons, Moehring emigrated to Canada in 1954 and bought a large farm north of Toronto, near Durham, Ontario. Beginning in 1955, the farm served as a base of operations for log buying and log exporting. Operations were conducted out of a seven-stall, garage-type structure. On the farm were "...a lot of mature gray elm which he cut into logs and sold overseas."

Moehring also purchased veneers for export. Hamann remembers first meeting him at the David R. Webb Company in 1957 or 1958. He had been told that there would be a red-haired German coming to look at everything in the mill. "On that day, Moehring bought almost our complete walnut inventory in about 10 minutes. He simply looked at the first 50 sets of samples and said 'I will take them'." Hamann concluded that Moehring was a man of action, "...but he backs up his actions and makes them count" (Hamann, 1975).

These veneer mill visits did one other thing for Moehring. They exposed him to veneer cutting technology unknown in Europe. European veneer cutters were by tradition all horizontal slicers developed to slice the large African timbers which flowed to European ports from the tropical forest of West African colonies. These machines produced quality veneers but cut veneer at only 12 to 15 slices per minute. The Capital Machine Company slicer was a revelation to Moehring. When he decided to build his own veneer mill, it is not surprising that he equipped it with Capital veneering machines. The mill was built at the Durham, Ontario, location. Michael Boel began his long association with Moehring when he became an integral part of the Durham venture.

The business was incorporated as the Maple Leaf Veneer Company and was highly successful. In 1963, it was sold to the Abitibi Power and Paper Company, a large Canadian forest products company. By then, the plant was producing veneers utilizing four Capital slicers and a Capital staylog lathe. The complex was horizontally integrated to the extent that hardwood plywood panels were being manufactured. Six plywood presses were involved in this production process. Abitibi later sold the entire operation to the Danzer organization which changed the name of the company to Interforest, Ltd.

The decision to sell his Durham operation was prompted by Moehring's perspective of the large market in the United States for quality veneers and the opportunities to expand the production of veneer for export into the world markets. The large forested acreages of hardwood timber in the eastern states were also an attractant which could not be denied. Early in 1964, Moehring, Boel, and Hermann F. Meyer began looking for a plant site. After scouring the East Coast, they selected Beaufort, North Carolina, located "...on the eastern bank of the Newport River, just across the estuary from the deep water port at Morehead City" (Anon., 1976b). Rail lines also connected the port site to the heartland of the prime eastern hardwood forests and woodlands. Beaufort was also well connected by highways to the principal furniture centers in North Carolina and other coastal states. The close proximity to the Atlantic Ocean prompted Moehring to name his new company, Atlantic Veneer Corporation.

Following the Moehring approach to the veneer business, the initial plant was small, efficient, clean, and designed for expansion as workers became trained and markets expanded. As Hermann Meyer explained: "Start small and grow as you train your workers." By starting small, he meant that by 1965 a $3 million plant had been located on a 150-acre site which included 150,000 square feet of floor space. A docking area to accommodate 1,000-ton capacity barges and a one-mile double siding on the Southern Railway accommodating 50 cars at a time were completed in the initial phases of construction (Anon., 1965). By the late 1980s, the plant had been expanded to accommodate 10 Capital slicers and two Capital staylog lathes. In less than a decade, Atlantic Veneer became the largest producer of hardwood veneer in North America and second-largest in the Western hemisphere. Only Moehring's Brazilian veneer operations, consisting of 30 veneer slicers and staylog lathes, were larger than the Beaufort complex.

At the Beaufort mill, an eight-foot band mill reduces about 100 tons of domestic and foreign logs into precision-cut flitches each day. The flitches, heated in a battery of stainless steel tanks equipped with electronic controls, are located directly behind the slicers. The largest Capital slicers, 225-inch "supersize" machines, can accommodate two flitches at the same time which effectively doubles the productivity of a single machine. The cut veneers are dried using Schilde double-band dryers. Thirty tons of plant waste are burned daily to supply much of the firm's energy requirements. Only one 55-gallon barrel of ash is produced by the boilers in each day's operation. The aggressive nature of Atlantic Veneer in purchasing logs and standing timber throughout eastern United States can by judged by the purchase of a parcel of walnut timber in northwestern Ohio which was purported to be the finest quality walnut timber sold in recent years. As reported by the New York Times in 1977, the 18 walnut trees sold:

> ...for $80,000.01 for 18 black walnut trees, including one the hardwood industry considered the most perfect and valuable black walnut tree in the nation. When the bidding was over, Hermann Meyer, the president of Atlantic Veneer Corp. of Beaufort, N. C., put its value at $30,000. It measured 57 feet to the first limb and was 38.4 inches at 4 1/2 feet above the ground...The second highest bidder, the David R. Webb Co. of Indiana, offered $78,750.

It was later reported that this single "perfect" tree "...would produce veneers for more than 800 suites of furniture in superb rift-cut quality" (Anon., 1977).

Logs arrive at the Beaufort mill by truck, railroad car, and ship. A large log yard adjacent to the company's waterfront docking facilities is equipped with water-sprinkling devices to protect logs from insects and molds while they are being accumulated for trans-shipment to Atlantic's Vitoria, Brazil, operations. Ships leave Beaufort for Vitoria, Brazil, six times a year and return with exotic veneers and panel products.

Hermann F. Meyer oversaw the entire operation at Beaufort as corporate president from its beginning. Under his guidance the plant was vertically integrated to include a face-making division and a panel plant. Two-ply backings produced in Atlantic Veneer's Brazil factories were "faced" with Beaufort-produced veneers into quality hardwood panels.

Rainer Ingmann, another long-time Moehring employee, was company production manager for 12 years prior to 1980 when he left to run the Fairfield Veneer Company (a Bohlke company) located near Cincinnati, Ohio. Both Meyer and Ingmann expressed satisfaction about the success of the Atlantic Veneer Corporation. They believed that training had been a key element, with every employee being well-trained for specific operations. Meyer said, for example, that when a veneer knife is changed, "...each worker is at his own station, ready to do his part...every man has been trained to handle a certain part of the operation" (Ross, 1970). This philosophy is carried out throughout the well laid-out and immaculately clean buildings which are considered to be other necessary elements in producing quality finished products.

A large wholesale lumber operation was developed which features lumber pre-drying and conditioning installations in lieu of outside air drying stacks. The lumber is then kiln-dried and shipped, for the most part, to overseas markets in shipping containers. The process provides customers with "bright" lumber rather than wood that is characteristically gray in color because it was air-dried prior to kiln drying.

Hermann Meyer expressed pride in the establishment he presides over and was prompt to praise his young, energetic managers and employees. He mirrors Moehring's work ethic and is usually the first to arrive in the morning and the last to leave at night. Meyer looks to the future with confidence.

> We have a great deal of confidence in the future of the veneer and lumber business, particularly in the United States. The plant in Beaufort has become a very important link in the Moehring worldwide chain, and we are confident that the recently completed expansion of our plant will be followed by even greater improvements in the near future.

Moehring's incentive to expand his business more on a world-wide basis occurred on February 28, 1968, when the Brazilian government placed an embargo on rosewood logs. The market for rosewood veneers was at a peak of fashion at that time and much in demand for furniture and expensive wall panels. On March 2, 1968, Atlantic Veneer Corporation's Brazilian company was registered as Atlantic Veneer do Brasil, S. A., and the first construction contracts were let for a site at Vitoria, Espirito Santo, a strategic location for the economic exploitation of Brazilian rosewood resources.

Both Moehring and Michael Boel were intimately involved with the development of the project and, by June 13, 1968, the initial buildings were completed and the basic equipment in place. Construction has virtually continued unabated since that time. A recent tabulation indicated that there were 21 veneer cutting lines at the Vitoria plant, consisting of at least 17 Capital slicers and three or more Capital staylog lathes. As summarized by Plywood and Panel magazine (Moore, 1977):

> Today, the Vitoria complex includes four slicing plants; a rotary mill; the two largest band saws in Brazil...housed in their own building and complete with electronic controls; a new 272-m x 102-m furniture plant which houses two panel-finishing lines; a plywood plant which produces furniture and construction plywood; a newly completed 225-m x 25-m splicing room (large enough to house 110 splicers); various office and warehouse buildings; a huge log yard with sprinkler system and log pond; several boiler houses; furniture showrooms to display the firm's new and complete line of bedroom suites; more than seven km of paved roads; and other ancillary structures totaling 83,000 cu m under roof.
>
> It's a huge complex—so big that Atlantic has purchased 45 bicycles to enable their foremen and supervisory personnel to get around the area more easily. And top management utilizes the company's fleet of Brazilian-made Volkswagens for travel within the plant grounds—and even within the plants. The big buildings are so well laid out and so well organized that you can drive down the aisles!

The complex certainly contains the largest hardwood veneer slicing operation in the world. In addition to the cutting of veneers and the manufacture of consumer products at Vitoria, Atlantic Veneer also has built and maintains two villages for its workers. Both Boel (who left Brazil in 1979 to become president of Wood Mosaic in Louisville, Kentucky) and Moehring customarily were on the job at 4:45 a.m. and still there 14 hours later. Moehring has been quoted as saying that "...absentee management is a disaster anywhere, but especially in Brazil." He has also stressed that it is necessary to work with partners, not managers.

In 1977, a second Brazilian veneer plant was constructed on the Amazon River at Itacoatiara, about 180 miles down river from Manaus. Mike Boel was in charge of the initial construction at Plant 1 which contained the foundations for eight veneer machines. In 1982, company publicity literature indicated there were 10 cutting lines at Itacoatiara where the plant was incorporated as Atlantic Veneer da Amazonia Ltda. Logs were supplied to the mill by water and by truck after a road was pushed through the jungle to connect Manaus and Itacoatiara.

Logs from Itacoatiara are shipped by boat to Vitoria about once a month. Veneers from all of Moehring's veneer operations enter world trade through a network of sales offices, and the output of any one mill can be scheduled to complement and supplement that from any other mill in the network. Moehring is, thus, able to provide a wide species range of veneers at any sales office, whether in High Point, North Carolina, or Hamburg, Germany, and for a wide range of woodworking industries utilizing fine face-veneers in their product lines.

One advantage of Moehring's world-wide veneering operation is that unique woods can be "discovered" for the furniture trade when it is looking for something different for its jaded furniture and architectural clientele. Substitutes for existing well-known and accepted species are also sometimes found, e.g., Brazilian Madera Nova (new wood) which is now considered an excellent substitute for American pecan and hickory.

Moehring has made a large commitment to the Brazilian economy. Over 2,500 Brazilians work for Atlantic Veneer. Moehring is also optimistic that Brazil's forest resources are one of the keys to the world's future supplies of hardwood products. "Brazil will be a major supplier of wood products to the rest of the world for the next 250 years. Every third hardwood tree in the world is growing in Brazil" (Moore, 1977).

FAIRFIELD VENEER MILL COMPANY

Manfred Bohlke was another German who came as a log buyer and stayed to build a veneer mill. The Fairfield Veneer Mill was built in 1980 as the production company organized by Bohlke and his associates to complement the Bohlke Veneer Company, a veneer sales organization, also located in Fairfield, Ohio. About 80 percent of the veneer was exported in the early 1980s to Australia, England, Austria, France, and Germany where the veneers were sold to brokers under the supervision of Heinz Offenheuser and Ludwig Luehrs, Bohlke's German partners. Logs and lumber were also sold overseas through the Bohlke company. Domestic veneer was sold principally in the High Point, North Carolina, market area.

Bohlke started his American career buying logs for export. Ultimately, he began custom cutting veneers at existing veneer mills, specializing in this type of service. Thiesing Veneer at Mooresville, Indiana, among others, cut Bohlke veneer to augment the production at the Fairfield mill. The necessity for owning his own mill apparently became obvious to Bohlke in 1979 when the export veneer business was just beginning to expand explosively.

Bohlke acted as chief log buyer and often flew to locations where he had contacts with log producers. Thomas Gleason, a Bohlke vice president, was also involved with log buying as well as the purchase of lumber to be sold in foreign markets. Gleason's son, Mike, also bought logs, as did several others.

Rainer (Ron) Ingmann was manager of the Fairfield Veneer Mill in 1982. In Germany he had been apprenticed to a veneer agent (broker) for 10 years. He had also served as production manager for Atlantic Veneer at Beaufort, North Carolina. The Fairfield mill was operated as if it were a custom mill and cut Bohlke veneers as well as veneers for other veneer custom cutters. About 9,000 board feet of logs were processed during the average work day.

The mill's operation was facilitated by the latest technology available for increasing production efficiency and quality. The water vats, which condition the flitches prior to slicing, were temperature-controlled electronically to maintain proper temperature while minimizing energy requirements. Three Cremona slicers produced the veneers which were dried in Cremona dryers. The production rate and high technological level of the Cremonas were attractive features influencing their selection. Some problems with maintenance of the electronic controls and difficulty of obtaining spare parts were noted as troublesome but not limiting factors in choosing this equipment.

A broad range of eastern species was cut, with white oak, red oak, walnut, and southern pine being the predominant species. Special thicknesses could be cut for export orders—white oak as thin as 1/110 inch and mahogany 1/85. Most were being cut 1/38 to 1/50 inch for export, but veneers destined for domestic markets usually were cut 1/32 inch thick. Fairfield, in 1982, was export-shipping 16 to 20 containers per month, primarily through the Port of Baltimore.

INDIANA VENEER CORPORATION

The Indiana Lumber and Veneer Company was "...one of the most noteworthy enterprises in its line in this country" soon after its incorporation in 1892. Band-sawed lumber and sawed veneers were manufactured at the firm's location at "...Twenty-third Street and the L. E. &W. (Monon) Railroad." Quarter-sawed oak was a specialty (Hyman, 1897).

Amos K. Hollowell was the company founder and president. A native of Paoli, Indiana, he moved to Indianapolis after an entrepreneurial career elsewhere as a general store merchant and manufacturer of flour milling equipment. He later was associated in Indianapolis as an investor and corporate official with the Jenney Electric Motor Company, the New Long Distance Company, and the Harris Air Pump Company. At the turn of the century, the Indiana Lumber and Veneer Company was reputed to be the largest of its kind in the world, with products being "...sold in all parts of the United States and command a large trade in Europe and other foreign countries." Hollowell enjoyed a good reputation as a man of "inflexible integrity" and as an "...exponent of broad, liberal and generous public spirit" (Dunn, 1910). He died on May 15, 1921, in Indianapolis (Anon., 1921).

The company was initially incorporated with a capital stock of $25,000. Hollowell's son, Lynden P., was vice president; O. M. Pruitt was secretary-treasurer; and William Dickerson was superintendent. The plant covered six acres and included a two-story mill, sheds, and yards. Between 75 and 100 men were employed (Hyman, 1902). By 1908, Pruitt had taken over as company president while retaining his earlier function as treasurer. Other officers were C. L. Goodwin, vice president, and M. L. Hovey, secretary. Josephus Davis may have been among the original stockholders in 1892. It is known that Minor Davis obtained his first job in the veneer business with this firm in 1904 when he was 21 and that his father, Josephus, was a log buyer for the company.

The original plant burned down in 1913. Officers of the company estimated the loss at $100,000 and all but $8,000 of that was covered by insurance (Anon., 1913). When it was rebuilt, it was described as having:

> ...three Smith, Meyer & Schnier veneer saws, each equipped with automatic set works to insure uniform cutting... and space has been left for the installation of a fourth saw. Each saw is driven by an individual G.E. motor of 50-h.p. capacity, direct-connected by means of a flexible coupling. Sliced, rotary, and half-round veneers are cut on a Capital slicer and rotary machine. The slicer is installed in the saw room, while the lathe occupies a separate room with its steaming vats. Cut veneers are trucked to a 12- x 100-ft. mechanical dryer. Something like 3,000,000 ft. of manufactured veneer [are stored] in the well-lighted and commodious warehouse..." (Anon., 1922).

John Goodman, Sr., who may have been associated earlier with the Milan Veneer Company, was apparently the owner at the time of a still later fire. The mill operated with a 1906 Capital slicer and four veneer saws. Bud Price recalled that "...The first thing we brought up here after the fire was that old slicer. Then, when we moved the sawmill up here, they brought all the flitches from 22nd Street to here on [1121 East] 24th Street." The slicer had originally been owned by the Thiesing Veneer mill. Price also remembered that: "We had one of the last veneer saws because John Goodman [Jr.] owned the veneer plant at that time and we had an awful time trying to talk him [into] tearing the saws out and putting in a rotary. We took three out. The other one sat around for a long time and was still there when he sold the mill to Paul Sanders [in 1954] but we never used it."

Another fire occurred in October 1950, while President C. L. Goodwin was in Virginia on business. O. M. Mitchell was treasurer and Oscar Montague, secretary. Seventy men temporarily lost their jobs because of the fire. The firm's specialty was still white oak veneer, and large stocks of lumber and veneer valued in excess of $100,000 were lost.

Paul and Roscoe Sanders added a Capital staylog rotary lathe after they purchased the company in 1954. They operated the Indiana Veneers, Inc., mill until August 1971 when it was sold to Luke E. Pinsonneault and O. C. "Bo" Edwards. A mill modernization program was immediately undertaken by the new owners. The plant was enlarged to accommodate a Hildebrand oil-burning, S-type veneer dryer through which the veneers pass three times. A Hale steam dryer was also added along with computers used to prepare tally sheets in the clipping department for veneers that were to be exported. In 1977, a new 165-inch Capital slicer replaced a smaller slicer. The new slicer was equipped with the latest technological advancements, including a temperature-controlled, anti-staining system for blade and bar. The 10-foot Capital staylog lathe purchased by the Sanders was retained to complete the veneer making capabilities of the firm.

Luke Pinsonneault had been plant manager at the David R. Webb company in Edinburgh. He had gained his earlier experience with Weyerhaeuser Company at Sault Ste. Marie, Ontario, and with Jamestown Veneer and Plywood in Poland, New York. At Indiana Veneer, he was president of the company and O. C. Edwards was executive vice president. Edwards had started in the veneer business with Wood-Mosaic in Louisville before shifting to the David R. Webb veneer mill in 1969.

The intent was to allocate one-half of Indiana Veneer's production capacity to the manufacture of its own veneers and the remaining capacity was to be devoted to custom cutting for others. This had been the production pattern established by the Sanders. Bud Price continued as plant superintendent. The Pinsonneault and Edwards business arrangement lasted until June 1978 when the mill was sold again. Pinsonneault and Edwards moved on to Coldwater, Michigan, where they helped to organize International Veneer Company, Inc., a veneer production company using Cremona (Italian) slicing equipment and dryers (Anon, 1971).

Veneer brokers Werner Lorenz and Herbert Kaiser, of Niedernhausen, West Germany, bought Indiana Veneer Corporation in 1978. Lorenz assumed the presidency of the Indiana company and Kaiser remained in Germany to manage their veneer brokerage business. The operational aspects of the Indianapolis veneer firm were assigned to Francis (Bud) Price as plant manager and to Lutz Machon, general manager. The Kaiser-Lorenz firm had been the principal market for Indiana Veneer's veneers for several years previous to the change in ownership and had accounted for almost 85 percent of its export trade. The primary function of the mill after the sale was to furnish veneers for Kaiser-Lorenz Gmbh which were resold to European furniture, door, and paneling manufacturers. Indiana Veneers, Inc., continued to earn the reputation for producing "top-of-the-line" veneers and particularly for the company's white and red oak veneers. The company was also reputed to be the largest exclusive manufacturer of quartered oak veneer in the world.

Werner Lorenz had originally come to Indiana to buy logs for export. The Indianapolis company he formed to handle these transactions was LVL (Lorenz Veneer Logs). He began buying veneer from Indiana Veneer Corporation in 1974 for Kaiser-Lorenz. When Pinsonneault and Edwards offered to sell him the business in 1978, the German partnership bought the mill. Another round of mill modernization began with an improved system for the internal flow of materials. Two new Capital slicers and two Keller-Hildebrand dryers doubled the production capacity of the mill.

George Minde, a salesman and engineer employed by Kaiser-Lorenz since 1972, supervised the plant in 1978. The desirability of marketing the mill's veneers in the United States and importing European veneers for domestic sales soon became apparent as a means of increasing total sales. By 1982, 35 percent of the 65 million square feet of veneer produced in the Indianapolis mill was sold domestically. Veneer sales in 1989 were managed by Jim Shobe.

The mill produced veneer with three slicers and a staylog lathe. Logs were purchased from a number of sources, including sawmills, log producers, and tracts of standing timber. The cut was about 60 percent white oak, 30 percent red oak, and 10 percent walnut, cherry, and pine. About 25 percent was purchased in Indiana. As Lorenz observed: "Indiana is a place where every log buyer and exporter is sitting. Indiana oak is that good and also expensive." No commercial lumber was cut and lumber logs were resold, if acquired, to stave mills or sawmills. Incidental lumber produced while sawing logs into flitches was used in crating and industrial blocking (Perry, 1978).

PLEASANT HILL VENEER CORPORATION

The Pleasant Hill Veneer Corporation is located in a sturdy brick building which once housed the American Scale Foundry. Pleasant Hill is about 40 miles southeast of Kansas City, Missouri. The foundry building was converted into a veneer mill in 1961 by Fred Persinger to take advantage of black walnut and pecan resources in nearby areas of Oklahoma, western Missouri, and the northwestern corner of Arkansas. Enterprise Corporation was the name originally given to the firm. The business may have been purchased later from Persinger in total or in part by Ralph

Snyder, a High Point, North Carolina, hardwood veneer salesman, with all production being sold through his sales office in High Point.

Little is retained in the collective memory of the Pleasant Hill business community about the early years of the veneer company, and the local newspaper archive for this period was destroyed by a flood in 1982. Newspaper clippings filed in the historical society extend only to 1960. It is known locally, however, that the mill had practically no sawmill equipment and was very dilapidated and run down by the middle 1970s.

In 1975, the Dichmann Furniture Company, of Kelkheim, Germany (near Frankfort), purchased the mill and changed the name of the company to the Vario Veneers Corporation. Substantial investments were made to improve the mill, including a new seven-section Babcock-Schilde dryer. The mill was cleaned up and workers were trained to produce quality veneers for the European market. Further renovation took place after a tornado in the spring of 1977 ripped through the plant. Roofs, siding, and some walls were either stripped away or destroyed. The extent of the damage can be imagined by the fact that the plant was without electrical service for four weeks. Careful planning and rebuilding of the mill was accomplished under the direction of Rudi Cesarz, vice president and production manager of the firm, and his production foreman, Sam Clark (Flock, 1979). The mill was equipped with 17- and 13-foot Capital slicers and used an older Proctor-Swartz dryer to complement the new Babcock-Schilde dryer. White oak and red oak were the principal species cut, although some walnut and other species were sliced periodically.

Unfortunately for Vario Veneers, the parent company's furniture plant in West Germany burned in 1981 and the firm was forced into bankruptcy. To satisfy creditors, the Vario Veneer Corporation asset was placed on the market. Dieter Ripple, Vario manager, closed the mill on February 1, 1982. Union Veneers Ltd., a subsidiary of Aaronson Bros. Ltd. of London, England, subsequently purchased the plant. The Aaronson company was started by Solomon Aaronson in 1939 as a privately-held firm. In 1962 it was converted into a public stock corporation. The firm later grew to include a number of subsidiaries which produced and marketed wood-faced building products in world markets.

The Pleasant Hill Veneer Corporation began operating under English ownership with about 50 employees on September 7, 1982. Company representatives expressed confidence in this venture, citing a strong demand for veneer in the face of adverse economic conditions throughout the world. Several weeks were spent retooling and modernizing the plant prior to its opening. Brian Landes, president, and Dieter Schaefer, executive vice president, anticipated selling their veneers in both domestic and foreign markets. Orders for additional equipment had been placed in September 1982 for a major expansion of production capability (Anon., 1982).

Landes had been affiliated with the parent company in London for 10 years before he came to the United States in 1982 to become president of the Pleasant Hill Veneer Company. Schaefer had a successful career in Germany working for Stoecker and Company and the Kling Furniture Company before becoming general manager of the Vario Veneer Corporation in 1974. He later held jobs with Swords Veneer and Lumber in Rock Island, Illinois, and Amos-Hill Veneer and Lumber Company in Edinburgh, Indiana.

INTERNATIONAL VENEER COMPANY

International Veneer Company (IVC), at Coldwater, Michigan, is jointly-owned by foreign and American entrepreneurs. Its location, on the northern fringe of the timber resources it utilizes, was determined by the earlier business activities of R. D. Jones, one of the company's founders. He had been a large-scale exporter of walnut logs in the 1960s. Although continuing to export logs, he later shifted his attention to custom cutting veneer at the Indiana Veneer Company in Indianapolis and at the Breece Veneer mill in Kenova, West Virginia. Veneers sliced at these mills for Jones were transported to his Coldwater warehouse for clipping and sale in the export market. Through these operations, he became acquainted with Sergio Colombo of Seregno, Italy. Colombo, born in 1939, owned SOC Transtadio of Milano as well as a large veneer brokerage warehouse in West Germany.

The result of these related business interests was a decision in 1978 to build a veneer mill in Coldwater using veneer equipment manufactured by Angelo Cremona & Figlio of Monza, Italy. Initially, the Cremona Company held 50 percent of the stock and provided favorable credit terms for the enterprise. Jones and Colombo were the other stockholders.

Luke Pinsonneault was hired to engineer the construction of the plant and to install the machinery. The plant started up in 1979 with two 3.7-meter vertical Cremona slicers. A 3.6-meter Cremona "staylog" lathe was acquired for rift cutting. In 1984, a third Cremona slicer was added. Seven-section, three-stage Cremona dryers were directly linked to the slicers by conveyors.

Bo Edwards, born in 1935, said in an interview that he had become friends with R. D. Jones "over the years" and had done custom cutting for him at the Indiana Veneer Company. After Edwards and Pinsonneault sold their plant in 1978 to Werner Lorenz, Edwards started his own business exporting logs to Europe. Within six months, he and Jones had merged into a single log-exporting company which "...ended up buying all the Cremona stock in the veneer plant...and we bought it before the mill ever went into production. We started to produce veneers about October 1, 1979." Colombo, the major stockholder after Jones was bought out, was initially the chairman of the board and company president. Edwards, as vice president, ran the operation while Colombo "...showed up in Coldwater...probably, once every four to six weeks" (Anon., 1985). William E. Craycraft, formerly affiliated with the Breece

Veneer Company in Kenova, West Virginia, for 30 years, had more recently worked with R. D. Jones in his log export business before becoming a vice president and secretary of IVC. Jerry Diehm, the youngest member of the management team, functioned as treasurer. Diehm, a German, had previously owned a warehouse in Italy before moving to Coldwater and applying for U. S. citizenship.

Originally, management's marketing strategy was to service mainly the European market by custom cutting 100 percent of the output for it. About 50 percent passed through Colombo's warehouse in West Germany and much of the rest went through an associate, Lutz Neukirchner in Hamburg.

By starting the business as a custom cutting mill, management could be freed from concerns about log procurement and the marketing of veneers. These functions were handled by the Timber Veneers Company and World Wide Woods, based at IVC. Edwards said "...all the owners of these two companies are also major owners in IVC. Much of my efforts are in Timber Veneers because that is where the logs are. The obvious advantage to start with was that the mill had nothing to worry about except producing a good product. We could put them all together which we probably will one day."

Although domestic business was limited when IVC began, the company decided in 1984 that it had grown enough to serve markets in the United States and Canada. At the time, IVC operated two nine-hour shifts and produced about two million square feet of veneer per week. This consumed about 200,000 board feet of veneer logs per month (Knowles, 1984). Edwards confirmed that their plans were "...to get into the domestic market, probably to the tune of 35-40 percent of our production. We've gone through the growing pains...so we are now ready to run at full capacity." To assist in accomplishing this, a salesman was hired for the High Point, North Carolina, market. When reminded of the distance between Coldwater and High Point and all the competitors in between, Edwards responded: "Production skills and new equipment won't overcome these disadvantages. We hope we can sell them a service and that is what we intend to do with well-manufactured products and service. We are very quality conscious. We are with our European customers and we are going to be the same down there. I have never been in an area of the world where that doesn't work." The firm's expansion plan also included the addition of 35,000 square feet of showroom and warehouse space for domestic veneers.

Two years later, Edwards observed: "As of now, we feel like we have had good acceptance in the domestic and export veneer markets, and we feel like this is based mainly on two things: the quality of logs we use and our quality of production" (Knowles, 1986).

The management team changed with the consolidation of company stock into the hands of four major stockholders. Sergio Colombo remained as chairman of the board; Bo Edwards, became president; Stephen E. Mitchell, executive vice president; Lutz Neukirchner, vice president; and Jeff Nichols, secretary-treasurer. Donovan Steffey and Randy Taylor handled sales, and David Counterman was in charge of production. By 1989, Steffey had replaced Neukirchner as vice president and David B. Johnson had become secretary.

On November 29, 1988, IVC dedicated a new, second veneer mill in South Hill, Virginia. South Hill Veneers, Inc., represented the latest "state-of-the-art" in veneering technology and was expected eventually to employ 120 people. The IVC management team operated both mills.

South Hill Veneers, Inc.

This company, located in South Hill, Virginia, is the wholly-owned subsidiary of International Veneer Company of Coldwater, Michigan. The $10 million plant began operations in 1988. It has three Capital slicers and Keller dryers which utilize an ironing-type system of drying. This minimizes buckle of the veneer sheets in the drying process.

More than 75 percent of the company's veneer was shipped to Europe in 1989. Although O. C. Edwards, president of the parent company, believes that this market will remain steady, "...Far Eastern markets [will] continue to grow stronger and this is one of the areas where...the company [is expected] to gain ground, along with expanded domestic markets" (Stevenson, 1989).

If Edwards has a concern about the future, it lies with the reluctance of the federal government to limit the flow of increasingly scarce hardwood veneer logs to foreign countries. However, this has not been a major factor in his planning for the future, which includes adding three more slicers as the company expands into its 147,000-square foot facility. Ultimately, this will increase the number of employees from about 80 to 150. Werner Paul was responsible for managing the South Hill plant in 1989.

KEYSTONE VENEERS, INC.

Keystone Veneers of Williamsport, Pennsylvania, became another hybrid of American and European enterprise. Luke Pinsonneault had long been active in Indiana veneering enterprises before being hired by the International Veneer Company (IVC) to build the plant for them at Coldwater, Michigan. O. C. Edwards was quite aware of Pinsonneault's abilities: "He installed the machinery and did the engineering at IVC. That is his specialty—production. He left in about 1983...and went into a 50/50 partnership with Werner Lorenz." Lorenz, owner of Indiana Veneer Corporation, had purchased the Indianapolis company from Pinsonneault and O. C. Edwards in 1978 so the business relationships were long-standing.

The mill built at Williamsport was of modern design with sawmill, soaking vats, and other subsidiary equipment all located in the same

85,000-square foot building with the slicers and dryers. Keystone Veneers began operations in about July 1984, with two Cremonas and companion dryers. Although much of the production was intended for overseas delivery, the company was confronted by the strength of the U. S. dollar with respect to European currencies. Fortunately for the company, Lorenz was well-connected with European markets.

Firm management in 1984 consisted of Luke Pinsonneault, president; Dan Pinsonneault, his son, vice president; and Werner Lorenz, vice president, based in Indianapolis.

PENN-BEAVER VENEER CORPORATION

Veneer production at the Penn-Beaver mill began on January 2, 1990. Its construction had begun during the waning months of 1988. Rolf Schnable, chief executive officer of the corporation, was responsible for planning and designing the mill. He was born in Germany in 1936. As a young man, he had gained veneer production experience at the Atlantic Veneer mill in Beaufort, North Carolina, where he was employed between 1966 and 1975.

The Penn-Beaver mill is located in Darlington, Pennsylvania. That location was chosen primarily because the region has good timber resources and because the regional infrastructure in the Greater Pittsburgh area is excellent. The interstate highway system is used for transporting manufactured veneer to the ports of Philadelphia and Baltimore where it is exported in containerized form. The road system also makes it possible to transport logs efficiently to the mill from the Appalachian forests and from as far west as Indiana.

At present, there are three veneer cutting lines in the plant—two Capital slicers and a Cremona staylog lathe. Schnable takes pride in having purchased only the most technologically advanced machines and equipment for this operation. A Sanborn mill is used to cut the flitches, electronically operated Babcock press dryers dry the veneer, and a high-tech Keller clipping line prepares the veneer for the export market.

All veneer is currently sold overseas through the Feldmeyer organization which has extensive market connections throughout Europe. Future plans call for an expansion of veneer production capabilities and the potential to sell veneer domestically in the United States. Jan VanGils is president of the corporation and has reponsibility for operations. Schnable and Jan-Hinrich Stoll, director of financing and company treasurer, are based in Germany. Robert C. McCartney is secretary and legal counsel for the firm.

LAKE JERICHO VENEER COMPANY

Incorporated on December 9, 1982, by Michael Boel, the Lake Jericho Veneer Company began operations on September 15, 1983 (Fig. 8-1). The

Figure 8-1
Michael Boel is among the influential foreigners who have reshaped the veneer industry in the United States. He was a close associate of Karl Heinz Moehring and served as the president of Wood-Mosaic before he built the Lake Jericho Veneer Company at a site in rural Kentucky.

company represents one of the crowning achievements of this individual who has combined innate abilities, experience, and ambition throughout his career in the veneer industry. His odyssey through his professional life, more than anything else, illustrates several of the reasons why traditional family-owned American firms find or have found it difficult to compete with foreign-owned and managed veneer companies.

Boel arrived in Montreal, Canada, from Germany at the age of 19, seeking an opportunity to "...gain some experience, to do an extended apprenticeship, and to learn about newer and larger types of operations on the West Coast." In Germany, he had been trained in skills related to veneer utilization and marketing. Headed West, he traveled only as far as Durham, Ontario, where Karl Heinz Moehring was planning to build a small, rotary-lathe mill as an adjunct to his log exporting business. Boel hired on as a laborer. By the age of 24, he had become superintendent of a plant equipped with four veneer slicers. Moehring sold this company, the Maple Leaf Veneer Company, to the Abitibi Power and Paper Company on July 13, 1963.

Out of a job, Boel returned to Germany and was a private entrepreneur for a short time until Moehring contacted him about the possibility of working on a projected mill to be built in the United States. Moehring had become increasingly concerned in 1963 about a possible log embargo prohibiting the export of black walnut from the United States. The resolution of this potential threat was to be the creation of the Atlantic Veneer Corporation in Beaufort, North Carolina. After the purchase of the property had been negotiated by Boel, he flew to Germany to meet with Moehring and to receive from him a check representing the start-up capital of the new venture. As Boel related:

> I flew back to Morehead City with the check in my pocket. The next morning I went to the First National Bank of North Carolina. It was one of the old banks where the bank manager sat behind milky-glass windows in a little box opposite the main contact area. I went to the new accounts counter and said that I wanted to open a new account and would like to speak to the manager. The manager that day was the former manager who had retired and filled in when the manager was on vacation. He was a very fine, distinguished gentleman. He heard my story and said, "Well, maybe we have to open the account for the time being in your name for the account of Atlantic Veneer Corporation. That way, it stays identified to the Company account but is not the company account until you give me the documentation. Let's draw up the papers. How much do you want to deposit to get the account started?" When I told him that I had a check for one million dollars with me, his reaction was indescribable. Never in his years of being with this bank had anyone opened an account for that much. He said to the girl, "Please show the gentleman to my office." I was ushered from the counter to his office and [offered] coffee, coke, or whatever I wanted. From there, I went across the street to a lawyer and formed the company. This was the 5th of May 1964.

When his given assignments were completed in late 1964, Boel returned to Germany and bought his father's veneer and plywood retail business and expanded it between 1964 and 1967. He also provided the logistics and organized the construction of a veneer mill in Ghana, returning to Germany in 1968. Once again, he was contacted by Moehring who asked for 90 days of assistance in establishing a mill in Vitoria, Brazil, because the Brazilians were about to embargo rosewood. Boel accepted the job offer and was actually employed in Brazil for about 15 years as a mainstay and partner in Atlantic Veneer do Brazil which became the world's largest hardwood veneer company.

Having a desire to return to the United States, Boel accepted a short-term challenge of resuscitating the Wood Mosaic plant in Louisville, Kentucky, for the new German owner, Katz of America. Boel agreed to stay only for three months unless the owner had plans to expand and to build a new plant. "They gave me sort of an amber light...not a green light..." to investigate alternative locations in Kentucky and West Virginia while the old Wood Mosaic plant was being given "...bandaid-and aspirin-type therapy." The death of Mr. Katz and the decision of the family to sell the company prompted Boel to resign in June 1972, although he stayed on until the middle of September at the request of the Katz family. "Then came the week of lawn mowing. I kept waiting for the grass to grow so I could go over it again with my lawn mower."

During this week of lawn mowing, he received a phone call from the foreign owner of a modern midwestern veneer mill who was willing to sell Boel a 51-percent interest in his mill. Those negotiations ultimately led to Boel's decision to build his own mill on a farm property he had previously acquired. While the mill building was being completed and three Capital slicers were being installed, Boel aggressively sought out former Wood-Mosaic customers and others not previously serviced by the company. To provide veneers for these customers in the interim period, he purchased logs which were custom cut at four midwestern and southern mills. The veneer was warehoused and distributed from a Louisville location until facilities had been completed at his Lake Jericho site.

In speculating on the prominence of foreign-owned mills in the United States, Boel said in a 1984 statement that:

> There must have been a vacuum into which foreign companies could introduce themselves because it is not a powerful [foreign] industry...or a powerful money takeover that changed the American veneer industry. But, why was there a vacuum? [Several mill people] must have made a decision not to invest in any more new machinery [and] not to build new plants. We will look back to this period and say, "This was a period during which the very respected, traditional, purely American veneer manufacturing [industry] laid itself bare of its defenses and allowed foreigners to come in and take it over..."

Chapter 9

FINE HARDWOOD VENEER MARKETS AND MARKET FUNCTIONS

Marketing consists of those activities and functions which lead to the exchange of goods for money in a market setting where buyers and sellers meet for this purpose. In the theoretically perfect market, where all parties are equally knowledgeable and there are many buyers and sellers, all parties in a given consummated transaction will be equally satisfied with the exchange. Veneer markets tend to be something less than perfect because of the nature of the product, the unorganized nature of the market, and the common use of agents who are not salaried employees of the producing firm. Furthermore, the markets may be located several hundred miles from the site of the exchange. The relative bargaining strength of buyers may also adversely affect the possibility of obtaining a fair market price for the seller in the exchange process.

THE MARKETS

In the early days of the Hardwood Veneer Age, practically every community had its own cabinet maker. These cabinet makers provided the market for the commercial manufacturers of veneer. John Strong, of Gerry, New York, was one of the first who catered to this market opportunity. It was his practice, sometime before 1850, to cut up a wagon load of veneer from some "desirable blocks" using his three-foot slicing machine.

> Then Mr. Strong would take this veneer in his wagon and go on what we would now term a peddling trip among the different towns and sell it to the cabinet makers. Sometimes he would make a sale outright, at other times he would accept in exchange for his veneer, pieces of furniture, musical instruments, live stock, flour, meal and any and all kinds of material that he could make use of. This is one feature of the early work that would recall some of the talks of our fathers in days when furs were made use of as a medium of exchange and when there was much more trading in the materials themselves than there was cash marketing (Anon., 1908).

In more recent times, the furniture, musical instrument, and panel manufacturers are the ones who have supplied the markets for specialized hardwood veneer entrepreneurs. Ross Kramer, a long-time veneer manufacturer and salesman between 1920 and the 1980s, recalled that:

The veneer industry flourished in Indiana and Kentucky. The veneer manufacturers were there because of the timber, and this was the same reason that the furniture plants were largely in the area, at least, not far away. Rockford, Illinois, in the early days had 27 furniture factories making top grade furniture. I would go up there and spend maybe a month calling on furniture people. There were a lot of Swedes who settled there and they were excellent woodworkers. They made some of the best furniture we had at the time. Then, I would go on to Grand Rapids, Michigan, and there were dozens of good plants there. And, on to Wisconsin where you had the Germans. Sheboygan and Two Rivers were good furniture towns. In the Wausau area there were two furniture plants, one specializing in using only walnut stump veneers. They all used a certain species of wood and concentrated on either bedroom, dining room, or occasional furniture. This was all back in the 1920s.

At Columbus, Indiana, the Orinoco Company made the highest priced and the best furniture in the world. We sold them a lot of veneer and I remember one time asking their procurement man what the company was looking for. He said: "We have to have quality. We don't care what it costs. We have to have quality." At one time the Showers Brothers Company in Bloomington, Indiana, was the largest furniture manufacturing plant in the world. And, the Singer Sewing Machine Company was the biggest veneer cutter in the country, slicing down as much as 10 million board feet of oak, walnut, and sweet gum, and they had to buy large amounts of veneer. The musical trades, such as the piano industry, were centered in northern Indiana and Chicago. Bluffton, Detroit, and Cincinnati and Muskegon all had piano factories. I don't know how many there were in Chicago. I used to call on all of them. Except for the few furniture plants that managed to survive the Depression or moved South, they're all gone. One Rockford plant would shut down in the 1930s. Then another. It was just pitiful.

William C. Linn, a retired executive with Hoosier Veneer in Indianapolis, sold his company's veneer line in Virginia and North Carolina in the 1940s. Looking back at this time in his life, Linn said: "The market was tremendous down there. It was a place where we could sell just about anything we cut. They manufactured all kinds of furniture on a mass production basis. They took the trade away from Jamestown, New York, Rockford, and Grand Rapids. The reason they could do it is simple to explain. There were no labor unions, wages were twenty cents an hour, and there was an ocean of hardwood timber close by."

The shift in the major veneer markets from the Midwest and East to the South was, in certain periods, abrupt and at other times less perceptible. Local historians have chronicled these furniture eras for several of these former "furniture capital" cities (e.g., Lydens, 1966). For a time, Grand Rapids was the hub of the industry and was noted for its "furniture mart" which attracted thousands of buyers from around the country. It was superceded as the major center by the Chicago Furniture Mart after losing a competitive struggle with it.

The Chicago Mart finally lost out in the 1960s to the Southern Furniture Market in High Point, North Carolina. Although other furniture markets exist in Dallas and on the West Coast, High Point is considered to be the "Nation's furniture center." Within a 200-mile radius of High Point, 60 percent of the country's wood bedroom furniture and more than 52 percent of the wood-based dining room furniture items are manufactured. The furniture industry in this part of the South began about 1900 as the southern economy slowly recovered from the impacts of the Civil War.

The first High Point furniture market building opened in 1921 to provide permanent exhibition space to manufacturers. There are now at least 12 such buildings as well as 27 factory showrooms in the area. It is estimated that buyers representing 85 percent of the purchasing power for home furnishings attend the High Point Market. With at least 90 area companies utilizing wood in the manufacture of furniture, hardwood veneer manufacturers are almost universally represented in High Point. Markets also exist at the veneer mill locations where most display, sometimes in elaborate facilities, veneer samples for prospective customers who come directly to them.

Traditionally, individual flitches are sliced, dried, clipped to remove "ragged" edges, crated, and warehoused. The keys to this marketing step are the two to four samples, usually three, which are "drawn" from near the top, bottom, and middle of the veneered flitch. This is in the same sequential order as it was when cut from the flitch.

Palletized veneer, destined for overseas markets and some domestic furniture manufacturers, differs from crated veneer in several ways. After being dried, the veneer is export-trimmed and clipped to eliminate defects. Edges of bundles, consisting of 24 or 32 sheets of veneer, are trimmed so the edges are essentially parallel. Each bundle is secured with plastic ties and then placed on a pallet which may contain similar veneer from several flitches. Palletized veneer is normally inspected bundle-by-bundle by prospective buyers at the veneer mill. Once the veneer is sold, pallets are loaded into containers for shipment to foreign destinations. The overall purpose is to minimize transportation cost and maximize the delivered amount of useful veneer. This method of marketing clipped and packaged veneer has proved to be attractive to some domestic manufacturers of veneered consumer products because of cost-saving features.

Exported veneer usually passes through additional marketing steps once it reaches a foreign port. An American-based veneer mill typically markets directly through the owner's foreign-owned merchant-wholesaler sales company or a sales agent located overseas. In at least one instance, the veneer is sent directly to a foreign furniture manufacturer. As related elsewhere, foreign-owned mills in the United States are strongly integrated into their overseas operations.

MARKET FUNCTIONS

Selling Veneer

Since the earliest days, independent agents have been important elements in the selling of veneer. The Veneer Association compiled a list of 220 commission salesmen in 1939 (Table 9-1) at a time when the industry was emerging from a severe and lingering, industry-wide depression. New York, Illinois, North Carolina, and Michigan were favored places to locate veneer sales agents. As the furniture industry moved to the South, the interest in High Point as a market center grew. In 1989, High Point had more than 80 veneer sales representatives working out of 34 agencies and at least 15 sales offices which assumed the name of the "parent" veneer company located elsewhere.

No individual veneer manufacturer appeared to handle sales in the same way. This is illustrated by three well-managed veneer manufacturing companies which have significantly different production capacities: Curry-Miller Veneer Company, Pierson-Hollowell Company, Inc., and Atlantic Veneer Corporation.

Curry-Miller, of Indianapolis, is a small operation with one slicer and one staylog lathe. The firm is known throughout the industry for the quality of veneer it has produced for about 60 years. Bill Miller, company president, described his sales organization as follows:

> We have been with the same customers for years and have had a good relationship with them. We are talking about Bassett and some of your other big furniture companies. They are using a lot of veneer and are buying from other veneer manufacturers. Most veneer companies, I would say, do have their pet customers that they stick with and depend on.

Table 9-1
Geographic Distribution of the 220 Hardwood Veneer Commission Salesmen Active in North American in 1939

New York		Michigan		Pennsylvania	
Jamestown	19	Grand Rapids	27	Philadelphia	4
N.Y. City area	14	West Coast		Other cities	8
Other cities	8	Los Angeles	11	Indiana	7
Illinois		Other cities	4	Ohio	7
Chicago area	27	Canada		Kentucky	5
Rockford	12	Ontario	12	Other States	23
North Carolina		Quebec	3		
High Point	20				
Other cities	9				

Source:
Veneer Association, unpublished list of Commission salesmen, March 10, 1939.

> We have one representative, you might call him a manufacturer's representative, down in the southern market at High Point and one with very low volume up in Grand Rapids. And we have one in the Chicago area who takes care of a lot of architectural trade up there. But that is low volume. Most of ours goes into the High Point area. The Chicago representative handles two or three or four different companies. The same would be true for the Grand Rapids representative because he is also low volume. He would starve to death if he was depending on us. But, in High Point it is different. We are mainly an oak business and that is the only account that he sells. For example, he sells knotty pine and we don't produce any knotty pine. He also sells foreign wood and we don't cut any of that. As far as your domestic woods, mainly oak and walnut, we are the only one that he represents. There are some other firms that have their own salaried sales employees down there.
>
> That is one of the advantages of being small. On a big operation they are going to have to look for markets. They will hit the West Coast and Canada and other places where we never have. When I mention "line," this is the type of veneer that different furniture companies look for within a species like white oak. Some companies want a narrow heart, some want cathedral, and the different grain characteristics. Some of them demand or need longer cutting. Some of them are using short cuttings which is a lower grade wood, and others might be looking for a knotty grade. There are all different grades of veneer and that is what our representative down there looks for in sorting the samples (Fig. 9-1). By knowing what the customer wants, the sales rep will put together samples representing 100,000 feet, maybe 3, 4, or 5 hundred thousand square feet of veneer at a showing. We grade the veneer here in Indianapolis and rely on the rep to make up the lines and price it at the market in High Point.

Pierson-Hollowell Company, Inc., is an upper mid-range veneer manufacturer with three slicers and one staylog lathe. The company is capable of producing more than 100 million square feet on a yearly basis. Export veneer is produced for overseas customers who visit the plant to inspect and purchase veneer for their foreign wholesale or furniture businesses. Domestic veneer is marketed by several exclusive and some independent sales organizations. Two sales agents in High Point seek out customers in Virginia, North Carolina, South Carolina, and Georgia. A Louisville-based agency "covers" Mississippi, Arkansas, Louisiana, Tennessee, and Texas. A Chicago agent sells in Illinois, Iowa, Minnesota, and Wisconsin. Another sells in Pennsylvania, New York, and Maryland. Indiana, Ohio, and Michigan are the responsibility of a Pierson-Hollowell representative. The company's sales offices and veneer showrooms at the Lawrenceburg, Indiana, mill location complete the veneer sales organization. Altogether, field agents and company officials are responsible for sales in 19 states and several foreign countries (Miller, 1984).

The largest veneer plant in the United States is the Atlantic Veneer Corporation facility at Beaufort, North Carolina. The veneer mill, in 1989, was operating 12 Capital slicers and staylog lathes. A thoroughly

Figure 9-1
Three representative sheets of domestic veneer are usually selected from each flitch. These samples represent the entire flitch and are used in grading, pricing, and selling the entire veneered flitch. A "line" is a set of samples representing several flitches which have a similar appearance. The line is created for potential customers who have known preferences for certain surface characteristics.

modern facility throughout, it is an important link in the Karl Heinz Moehring worldwide chain of veneer and lumber mills, often vertically integrated with plywood and furniture production facilities. Moehring is now capable of supplying veneers, lumber, plywood, and related products to world-wide markets (Thompson, 1988?).

Atlantic Veneer coordinates its marketing activities with a vice president for sales. He manages a staff of sales representatives who cover the High Point market area as well as all the eastern states and Canada. Through the many sales outlets, the full line of domestic and tropical hardwood veneers is sold to manufacturing firms throughout the midwestern states (Anon., 1976).

Market Realities and Institutions

During the 1920s and 1930s, veneer manufacturers were prone to complain about their independent sales agents in places like High Point. Bud Miller, of Curry-Miller Veneers, remembered those days when he was interviewed in 1981. He indicated that many Indiana producers believed that "...all those veneer sales people did in High Point was to walk around in fancy golf knickers, ride in convertibles, and sell the veneer for less than it cost to manufacture it." As a point of fact, many old-time veneer men probably didn't know their true manufacturing costs because of inadequate accounting systems. This lack of information certainly contributed to the problem and to excessive cut-throat competition. Taking advantage of the situation, furniture procurement managers were known to call for a "convention" of five or more sales people who represented mills producing veneer of similar quality and species.

Percy Simmons, arriving from England about 1934, was amazed that veneer manufacturers were at the mercy of "a handful" of furniture companies and that the mills could never muster the courage or had enough market strength to fight back.

> It went something like this. The veneer manufacturers' salesmen would each lay down their samples and quote their prices. Then, the furniture manufacturer would sharpen his knife and play one against the other.

Simmons continued by saying that very few veneer manufacturers in Europe sold directly to furniture companies. Merchant wholesalers functioned as the middlemen and kept the veneer manufacturers from being exploited. He also noted that the American veneer producer in the 1930s did not control the selling process if an agent broker was used.

> Call him a salesman. That salesman received a 10-percent commission. He had no other involvement in the business of the veneer company. Invariably the salesman sold the best veneers and the manufacturer was left with the junk, which was invariably his profit which he couldn't sell. It was worthless as an asset, whereas the mill man should have paid him five percent for selling the good stuff and twenty-five percent for selling the low quality veneer. This was a terrible weakness which was subsequently corrected. Today, most of the domestic veneer manufacturers have their own sales offices and control their sales.

This earlier "...marketing weakness also applied to the terms of the sale where a potential customer [would outline] to several veneer salesmen [the terms of sale] before they were requested to show their samples. These terms called for the usual discount, not for cash, but for a trade acceptance without interest for sixty or ninety days." This 1929 writer concluded that it was the commission method of paying salesmen "...peddling face-veneers [which] is largely responsible for the chaos now existing" (Anon., 1929).

The term "salesman" is used somewhat ambiguously in the trade literature and this makes it difficult to differentiate between a salesman who represents his mill as a salaried employee and a salesman who sells on a commission basis. However, a "jobber" is clearly an independent middleman in the marketing chain. He buys and sells veneer to specialized customers as a wholesaler. In the early decades of this century it "...seemed as though the mill men considered the jobber as their enemy and I believe if they would give this matter careful consideration, they would find that they are not only their friends, but that their interests are mutual and the jobbers are necessary to the best interests of the veneer business from the mill man's standpoint" (Walscheid, 1913).

Generally located in the largest veneer markets, customers were able to secure veneer needs on very short notice if their requirements were small or if unanticipated manufacturing emergencies occurred. The jobber also provided a service to the veneer manufacturer by usually buying rail car loads and by buying large amounts of random stock and accumulated "cut-downs" which the mill man would find it difficult to market otherwise. By buying veneer in different forest regions, the jobber was also able to warehouse a great assortment of species and veneer grades for a variety of customers. Large city jobbers supplied similar services to contractors, architects, and others with immediate needs for veneer or veneer samples. By 1927, the jobber-wholesaler was recognized in modern marketing terminology as having "...an assembling and distribution organization as well as a sales organization [which] is growing steadily and is a factor of importance in the veneer and plywood business...because of the services rendered and the consumer's developing habit...of buying more on the hand-to-mouth basis and looking to some nearby source of supply." The great wholesale houses of the J. J. Bonneau Company, J. H. Monteath Company, and The House of Constantine in New York City are representatives of this era (T. C., 1927).

Of these, The House of Constantine stands out because of its historical beginnings. Family documents record that John Constantine was a well-known cabinet maker in Leeds, England, before embarking for New York City in 1793 where he continued his craft. His son, Thomas, who had been born in England, also excelled in creating furniture pieces while still a teenager. He established his own business in 1812 at the age of 21 at 60 Vesy Street. Five years later, Thomas moved to 157 Fulton Street where he was a near neighbor of Duncan Phyfe. His business, by then known as Constantine and Company, relocated to 503 Pearl Street in 1824 where he built cabinets for many years. He and his brother, John, also maintained a mahogany log yard where they measured and inspected the shiploads of mahogany which were being shipped to the Port of New York.

When Thomas died in 1849, his son, Thomas W. Constantine, succeeded him in the cabinet business and also engaged in domestic timber operations. On one of the trips he took to develop a dependable source for black walnut, young Thomas founded the town of Constantine, Michigan. Albert Constantine, son of Thomas W., entered the firm when he was 21. His son, Albert, Jr., joined the family firm in 1920 at the age of 18. The company name was then changed to Albert Constantine and Son, Inc. The firm had earlier begun to offer customers a wide variety of domestic and exotic veneers and cabinet woods as well as woodworking tools and other materials at the wholesale and retail level. This was after machine-made furniture began to dominate the market about 1877. The business is currently owned and managed by Albert's granddaughter.

One market observer detected the trend that reduced the importance of the veneer wholesaler in the marketing process in the 1920s. He concluded that it had been brought about by veneer and plywood producers who performed the "distribution function" themselves. They set up their own distribution warehouses with attached sales organizations. Improved transportation and communication systems very likely hastened the process.

Pricing Veneer and Veneer Prices

When Bill Miller, of Curry-Miller Veneer Company, was asked how face-veneer was priced, he replied:

> With grains such as wheat and corn, there is an organized market. You can get some real feel for price because yesterday's prices are published daily. A sales representative gets that feel of the market from his experience as to whether he can charge a little bit more or charge a little less. Veneer prices stay pretty steady, except through inflation or as the price of logs and manufacturing costs go up. The price of veneer has increased over the years like everything else. But, from day to day, it stays fairly consistent. At some point price is keyed back to the demand for the furniture. If there is a boom in the market in the furniture trade, then presumably there would be some price augmentation. The other factor is the demand for veneer in Europe. If the demand is down over there, we find a lot more price competition over here from the foreign-owned mills located in the United States.

Several managers of American-owned hardwood face-veneer plants have grumbled that export-oriented mills "dump four-cent veneer for two cents" on the domestic economy in such circumstances.

The limited amount of price data available substantiate Miller's observation that average veneer prices are generally "sticky" over time when inflation and fluctuations in European demand are accounted for (Tables 9-2 and 9-3). The average price of hardwood veneer can be misleading. Such a price does not indicate the considerable range in price for the veneer, depending on its quality (Table 9-4). This appeared to be particularly true of white oak and walnut in 1980. Production cost data for three of the most important hardwood species cut in 1980 indicate a nominal profit margin of 12 to 14 percent (Table 9-5). Given the estimated market value per thousand square feet, the gross profit ranged between \$6.85 and \$11.50 per thousand square feet and averaged \$9 per thousand square feet (MSF) on an unweighted basis. This was less than one cent per square foot for sold veneer. Palletized veneer (clipped for export) followed a similar allocation pattern of relative production costs. Timber cost per MSF is reportedly somewhat higher because of greater quality expectations and the larger amount of veneer clipped to eliminate defects as log quality declines.

Table 9-2
Average Domestic Face-Veneer Sales Prices for Oak and Walnut, 1956-1988

Year	Species: Red and White Oak	Species: Walnut
	— cents per square foot —	
1956	3.6	3.9
1960	3.6	4.2
1964	3.6	4.3
1968	3.7	4.7
1972	3.8	4.7
1976	4.7	5.7
1980	6.9	8.3
1984	7.2	10.2
1988	8.3	12.1

Source:
Confidential industry data.

Table 9-3
Average Price of Hardwood Veneer Exported from the United States, by Species and Year

Year	Amount Exported	Species: Red Oak	Species: White Oak	Species: Walnut
	million sq. ft.	— cents per square foot —		
1978	789.8	6.3	7.2	8.0
1979	886.0	7.1	9.1	10.6
1980	1,036.1	8.4	11.1	9.9
1981	878.6	11.9	10.3	7.9
1982	771.9	8.6	8.4	9.6
1984	971.5	7.5	9.7	10.1
1985	703.9	7.9	9.2	7.5
1986	873.8	12.4	9.3	8.7
1987	1,151.8	7.6	9.5	9.9
1988	1,377.6	7.6	10.4	12.1

Source:
USDA Foreign Agricultural Service, Forest Products Division.

Table 9-4
Relative Amounts of Price-Graded Veneer Produced by Sample Mills for Selected Species, 1980

Price Grade per Sq.Ft.	Species		
	Red Oak	White Oak	Walnut
— cents —		— percent —	
0 - 4	11.2	12.9	8.6
4 - 6	38.9	23.1	29.5
6 - 8	44.3	40.4	29.1
8 - 10	4.6	13.5	12.9
10 - 12	0.6	4.1	7.0
12 - 14	0.3	2.6	6.0
14+	0.1	3.4	6.9

Source:
Confidential industry data based on the average of 14 veneer mills.

Table 9-5
Relative Costs and Revenues for Manufacturing Selected Species of Domestic Face-Veneer Produced by Sample Mills, 1980

Cost Item	Species		
	Red Oak	White Oak	Walnut
		— percent —	
Timber	32.0	38.7	43.6
Processing			
Buying logs	1.3	2.7	3.6
Manufacturing	39.9	33.2	27.1
Sales	6.6	7.1	6.7
Administrative	8.4	6.2	5.0
Total Costs	88.2	87.9	86.0
Profit Margin	11.8	12.1	14.0
		— dollars per MSF —	
Market Value	57.59	70.23	82.19

Source:
Confidential industry data based on the average experience of 14 veneer mills.

Hardwood Veneer Export Markets

The U. S. Department of Commerce did not begin to publish detailed data on exported hardwood veneer until 1978 (Table 9-6). Prior to then, oak veneer was not separated into red and white species groups. In 1978, red oak comprised 14.5 percent of the exported veneer. Since 1983, it has ranged between 26 and 43.4 percent (See Table 9-7 for data). Walnut veneers represented 16.3 percent of the total hardwood veneer exported in 1978. Since then, walnut exports have averaged about 10 percent and, in 1988, they represented only five percent of the total hardwood veneer exported. Several conclusions can be hypothesized from these relationships. White oak has been and continues to be the dominant exported veneer species. Because of its generally lower market price, red oak has been substituted for white oak in a number of applications. Walnut veneer exports have also experienced relative declines with respect to all veneers exported. This may be directly related to the declining size and quality of walnut logs and changed consumer preferences.

The major markets for the three principal sliced hardwood species have been Canada and West Germany during the past decade or so. Those countries imported 66.8 percent of the red oak, 63.5 percent of the white oak, and 59.2 percent of the walnut veneer in 1984, a representative year. West Germany, understandably, was the leading importer of hardwood veneers in all categories of hardwood veneer exported from the United States (Table 9-8) because of the integration of German-owned plants in

Figure 9-2
Veneer destined for export is palletized and then warehoused prior to shipment.

Figure 9-3
A standard-sized shipping container will hold about 40,000 square meters of export-thickness veneer.

Figure 9-4
After it is loaded in a shipping container, export veneer is trucked to a convenient port where the container is loaded on a cargo ship destined for Europe or Asia.

Table 9-6
Total Hardwood Face-Veneer Exported by the United States, by Category, 1950-1988

Year	Total: "fancy, face, figured, & special"	Black walnut	Other face-veneer	Oak: Red	Oak: White
	— million square feet —				
1950	20.8				
1955	28.6				
1960	30.5				
1965	98.9	80.6	18.3		
1967	60.3	44.8	15.5		
1969	121.0	92.4	28.6		
1971	117.1	97.7	19.4		
1973	180.0	90.9	89.1		
1975	150.1	63.2	86.9		
1977	224.7	93.1	131.6		
1978	789.8	128.9	184.8	68.9	407.2
1979	886.0	80.6	283.4	70.2	451.8
1980	1,036.1	117.8	287.0	120.1	511.2
1981	878.6	107.9	261.0	107.4	401.8
1982	771.9	78.6	180.4	141.0	371.9
1983	982.1	106.0	251.7	236.9	387.5
1984	971.5	84.4	251.0	250.1	385.9
1985	770.3	85.6	203.1	209.2	272.4
1986	932.7	72.6	220.8	236.3	403.0
1987	1,202.3	90.7	254.3	296.5	560.8
1988	1,377.6	69.9	249.0	360.7	698.0

Source:
U. S. Department of Commerce, Bureau of the Census. U. S. exports/Schedule E commodity by country. FT 410 (monthly). Washington, D.C.

the United States with West German companies. The one significant "blip" was Taiwan's importation of red oak veneer. This corresponds to its flourishing furniture industry which targets its production for United States and Southeast Asian markets. Since greater exports of furniture and furniture components from Asian countries can be expected in the future, it is likely that sales of domestic American veneers in this region will increase over time as well. For example, between 1977 and 1981, Taiwan's exports of furniture and furniture components to the United States increased about 300 percent on a dollar basis (Table 9-9). In 1985, Taiwan was reputed to have 43 furniture plants and one of them, Universal Furniture Industries, was the principal exporter to the United States.

Productivity

Production costs, referred to earlier, are directly linked to the productivity of the individual firms and, when aggregated, to the productivity of the industry. An inspection of modern veneer plants provides several indications that productivity has increased in the hardwood face-veneer industry. Current slicer models cut veneer at the rate of 60 to 80 slices per minute. The amount may depend on whether or not the sliced veneer is automatically conveyed directly to the dryer without human handling. The primary advantage of direct delivery to the dryer is the labor-saving reduction in cost since two to six jobs per slicer/dryer combination may be eliminated by the required capital expenditures.

The speed of the slicer has been gradually increased through periodic engineering improvements. Between 1930 and 1988, the compounded average rate of increase in slicer speed, as measured in slices per minute, has been about two percent. The general consensus among industry engineers appears to be that slicer speeds much above 80 per minute will prove to be impractical. Bob Brand, chief engineer for Capital Machine Company, believes that a much more fertile area for increasing slicer productivity will be the slicer's flitch-loading capability. "At the present time the knife is in the wood only about 50 percent of the time. Flitch loading can eventually be reduced to seconds rather than minutes. This anticipated efficiency will be augmented by a computer-controlled veneer slicing system which assists operators in reducing time spent in making adjustments throughout the cutting cycle."

Similar technological changes are being developed to improve the clipping and drying processes. Veneer drying has progressed from the use of air-drying techniques and much hand labor to the use of modern "jet" dryers which use air heated between 300° and 350°F. Air movement within the dryers uses air jets that impinge air on both veneer surfaces at velocities of 2,000 to 10,000 feet per minute. Microprocessor controls have been utilized in some prototype models of European origin to increase the efficiency of the heating process. This is accomplished by monitoring the drying conditions within the dryer and then controlling the flow of heat to various locations within the dryer. The design of some jet dryers has been changed so that increased drying speeds can be achieved without increasing the physical length of the dryer (Fig. 9-5).

Several innovations have improved the productivity of the clipping process. The most significant is the change to a "European style" of clipping in which the operator of the clipper always stays on the feed side of the clipper and never has to rotate the packs of veneer. This change is made possible by the incorporation of a hydraulically-operated table on the non-feed side of the clipper which enables the operator to make a longitudinal cut in the middle of the pack of veneer and recover the portion on the non-feed side (Fig. 9-6).

Table 9-7

U. S. Exports of Domestic Non-Reinforced Hardwood Veneer, by Quantity and Value, 1978 - 1989

	MSF	$1,000s	MSF	$1,000s	MSF	$1,000s	MSF	$1,000s
	1978		1979		1980		1981	
Red Oak	68,899	4,324	70,223	4,991	120,060	10,041	107,403	12,755
White Oak	407,184	29,360	451,781	40,959	511,180	56,780	401,811	41,205
Black Walnut	128,923	10,264	80,590	8,503	117,842	11,640	107,869	8,546
Other (Total)	184,833	11,226	283,369	23,954	287,003	23,845	260,959	20,005
Birch	7,208	284	2,266	160	3,121	219	5,682	268
Maple	15,694	913	12,327	899	13,982	1,499	12,188	1,029
Other	161,931	10,029	268,776	22,895	269,900	22,127	243,664	18,708
Total	789,839	55,174	885,963	78,407	1,036,085	102,306	878,617	82,511
	1982		1983		1984		1985	
Red Oak	140,958	12,054	236,881	18,928	250,118	18,685	209,167	16,489
White Oak	371,939	31,075	387,484	42,660	385,915	37,318	272,377	25,056
Black Walnut	78,620	7,546	106,002	9,623	84,443	8,488	85,641	6,423
Other (Total)	180,407	13,412	251,697	19,210	250,982	15,417	203,127	13,248
Birch	6,980	359	7,001	325	14,101	542	4,936	310
Maple	10,259	846	19,533	1,139	21,523	1,019	12,825	901
Other	163,168	12,207	225,163	17,746	215,358	13,856	175,813	11,067
Total	771,924	70,801	982,064	90,421	971,458	79,908	703,857	58,897
	1986		1987		1988		1989	
Red Oak	236,318	29,352	296,506	22,403	360,720	27,528	(a)	28,204
White Oak	402,960	37,646	560,849	53,041	698,018	72,776	(a)	76,049
Black Walnut	72,643	6,300	90,688	8,961	69,888	8,490	(a)	13,429
Other (Total)	220,868	16,974	254,330	21,290	248,978	28,834	(a)	39,613
Birch	5,225	312	2,669	168	10,441	611	(a)	720
Maple	24,441	1,496	32,153	2,287	63,764	4,241	(a)	5,299
Other	191,202	15,166	219,508	18,835	174,773	23,982	(a)	33,594
Total	873,321	82,290	1,141,811	105,795	1,377,604	137,628	(a)	157,295

(a) Comparable export quantity data for 1989 are not available.

Source: USDA Foreign Agricultural Service, Forest Products Division.
Note: In 1988, exported veneer which was reinforced amounted to an additional 12.5 percent of the total hardwood veneer exported.

Table 9-8
Relative World-Wide Distribution of U.S. Hardwood Veneer Exports, by Species and Year

	Red Oak			White Oak			Walnut		
Country	1978	1984	1988	1978	1984	1988	1978	1984	1988
					— percent —				
Canada	27.3	31.7	15.5	1.1	6.4	1.4	21.9	22.9	5.7
W. Germany	39.2	35.1	41.7	61.2	57.1	45.9	35.7	36.3	50.5
Taiwan	4.2	16.2	19.4	0.1	0.7	5.7	0.0	0.7	1.3
Other	29.3	17.0	23.4	37.6	35.8	47.0	42.4	40.1	42.5
World	100.0	100.0	100.0	100.0	100.0	100.0	100.0	100.0	100.0

Source:
USDA Foreign Agricultural Service, Forest Products Division.

Table 9-9
U. S. Imports of Wood Furniture and Furniture Parts, 1977-1981

	Years				
Source	1977	1978*	1979*	1980	1981
		— thousands of dollars —			
Taiwan	46,352	74,792	13,569	108,944	134,139
Yugoslavia	47,904	53,699	60,067	70,084	82,597
Canada	23,973	30,710	46,269	58,620	68,885
Denmark	32,975	45,861	53,881	58,559	60,109
United Kingdom	24,420	31,931	36,306	35,189	35,354
Italy	17,395	24,148	30,157	30,755	29,701
Singapore	5,501	8,178	15,628	20,046	25,825
Romania	6,818	10,874	14,431	15,615	18,182
All others	80,732	112,527	129,790	128,540	146,808
Total	286,070	392,721	500,099	526,353	601,600

*Revised

Source:
Compiled from official statistics of the U. S. Department of Commerce.

Figure 9-5
A modern, high-performance veneer dryer at the Lake Jericho mill.

Figure 9-6
An efficient "clipping line" in which export veneer "books" are cut to eliminate ragged edges and defect. The veneer is subsequently measured and banded automatically as it proceeds down the line.

Although these technological improvements exist and capital has replaced labor at many points in the hardwood veneer manufacturing process, the industry still deals with identifiable "pieces" rather than homogeneous outputs generated by a flow process. As a result, the hardwood veneer industry remains labor-intensive when compared with modern dimension mills or the pulp and paper industry where the labor content per unit of output is minimal.

Productivity studies, with few if any exceptions, combine hardwood face and commercial veneer mills with hardwood plywood mills into the single Hardwood Veneer and Plywood Industry (SIC 2435) which consists of about 275 firms. The component representing hardwood face-veneer mills is submerged in the data and this makes it impossible to separate its 20-percent contribution from the other contributing components. An example of this type of research (Table 9-10) indicates only modest annual rates of change in productivity for the period 1972-1986, but the annual rate of change for 1982-1986 has been substantial.

Table 9-10
Indices of Total Output Per Production Employee-Hour, and Output Per Production Employee, for Hardwood Veneer and Plywood Industry, 1972-1986 (1977 = 100)

Year	Output per Employee-Hour	Output per Employee	Total Output
1972	83.5	86.6	103.0
1973	78.8	80.4	98.0
1974	81.7	81.1	88.7
1975	92.4	89.2	77.3
1976	88.7	88.7	83.7
1977	100.0	100.0	100.0
1978	99.8	98.6	104.1
1979	97.7	96.1	102.0
1980	104.9	100.7	94.7
1981	101.6	100.5	94.1
1982	104.6	97.2	79.2
1983	111.8	110.3	94.6
1984	104.8	103.3	94.8
1985	119.6	117.9	99.6
1986	123.4	123.1	107.2
	Average Annual Rates of Change (Percent)		
1972-1976	0.6	2.6	0.4
1982-1986	9.1	5.5	6.8

Source:
U. S. Department of Labor, Bureau of Labor Statistics, Office of Productivity and Technology.

An unpublished report by Professor Craig J. Davis of the New York State College of Environmental Science and Forestry contains an analysis of hardwood veneer and plywood collected by the U. S. Department of Commerce in its Annual Survey of Manufacturers. He concluded that technological change had had little positive effect on the industry. He also reported that the real monetary value of the industry's output, as measured by value added to the product as a result of manufacturing, had declined 23.9 percent between 1972 and 1981. If correct, his statements add strength to the argument that the SIC 2435 industry is plagued with over-capacity to produce and that its overall productivity is controlled by exogenous factors such as the business cycle, the international strength of the dollar, and domestic disposable income. This would seem to fit the hardwood face-veneer industry as a whole since it apparently has had no more than 20 really "good" years in the six decades since 1929, according to several long-time veneer producers.

Chapter **10**

THE BATTLE FOR CONTROL OF THE AMERICAN BLACK WALNUT RESOURCE: 1959 - 1970

"An intelligent trend today is for governments to discourage, limit, and prohibit exports of logs..." (Kelleher, 1979). Although written in 1979 about some Southeast Asian nations, the thrust of the statement could have been easily applied to other attempts by countries to protect valuable timber resources from destructive exploitation and to export semi-finished and finished products rather than unprocessed logs. The reasoning is sound and it is grounded in economic logic. "While log exports are useful in creating foreign exchange, the [country] may sacrifice...domestic employment when it exports logs rather than processed products" (Office of Technological Assessment, 1983). The additional "value added through manufacture" may also be significant for the veneer producer and for secondary manufacturers of furniture and panel products. These benefits should impact positively on local and regional economics.

The first request for control of black walnut log exports was made to the U. S. Department of Commerce on December 1, 1959, by a walnut veneer manufacturer in Indiana. This was undoubtedly a reaction to the numerous foreign buyers in the state who were out-competing his and other local companies for the best walnut veneer timber and cut logs. At the time of the request to control the export of black walnut logs, 17 of the 23 manufacturers of walnut veneer were located in Indiana. Between 1958 and 1959, the total export of walnut logs from U. S. ports increased nearly 65 percent, and the average value of the logs exported increased from \$454 to \$584 per thousand board feet. This export situation was to increase in overall intensity over the next decade, with the possible exception of 1964 (Table 10-1).

Although black walnut's botanical range extends into parts of 33 states, its prime commercial range is limited principally to the six midwestern states of Illinois, Indiana, Iowa, Kentucky, Missouri, and Ohio. This has been well known by American black walnut veneer and lumber manufacturers for more than a century.

As walnut log exports increased exponentially, primary manufacturers of black walnut products became acutely aware of a potential imbalance between growth and harvested volumes. This was particularly true for old-growth black walnut trees suitable for veneer (Fig. 10-1). Although rebuffed in 1959 by the U. S. Department of Commerce, manufacturers continued to petition the Department, requesting that the federal government control the export of walnut logs. On April 6, 1960, George C. Romeiser, secretary-manager of the American Walnut Manufacturers Association (AWMA), and a group of walnut manufacturers met with Commerce officials in Washington, D.C., to request the imposition of controls (Fig. 10-2). Verbal and documentary evidence of overcutting and inflationary increases in the prices of walnut timber and logs were presented. Romeiser was informed by mail, on April 27, 1960, that the impact on domestic supplies was not such as to warrant imposing restrictions.

Another meeting was held on November 29, 1961, by the AWMA members with Commerce officials specifically requesting the imposition

Table 10-1
United States Exports of Walnut Logs, 1954-1970

Year	Export Volume	Total Value	Average Value
	- MBF -	- M dollars -	- dollars/MBF -
1954	608	206	339
1955	1,199	508	424
1956	1,138	517	454
1957	1,395	631	452
1958	2,280	1,035	454
1959	3,734	2,182	584
1960	10,185	6,802	668
1961	7,171	5,908	824
1962	10,320	9,918	960
1963	14,392	13,525	940
1964	9,640	10,654	1,105
1965	17,990	25,023	1,391
1966	12,498	16,726	1,338
1967	16,129	22,936	1,422
1968	21,840	33,310	1,525
1969	20,829	31,646	1,519
1970	17,229	27,186	1,578

Source:
U. S. Department of Commerce, Business and Defense Services Administration, 1969. United States Exports of Logs and Related Products 1958-1968, unnumbered and unpaged December publication; and industry sources for 1956-1958.

Figure 10-1
Prior to 1960, domestic hardwood veneer companies generally had little difficulty in purchasing adequate supplies of quality walnut logs at competitive prices.

of an embargo or quota on exports of walnut veneer logs based on provisions of the Export Control Act of 1949 and on the evidentiary documents provided by the association. Supplemental data were supplied at this time and again in late May 1962, and on July 18 and 31, 1962. All requests for action were denied by the Department of Commerce.

Michael Joelson, a David R. Webb Company executive, returned in October 1962, after a five-week study trip of veneer mills in England, France, Switzerland, Germany, and Denmark. In a letter to Don Gott, he reported that the overall log supply conditions had definitely worsened for the Europeans since his last visit in 1960.

> As you know, the majority of their log supply comes from Africa and Asia. The emergence of new African countries complicated the problem because several of them required a government authorization to export logs; from others you can't get any logs unless you have a concession. All these difficulties make it hard for the European veneer industry to get good logs in sufficient quantity and the result is that only large firms can keep up with their production. In Asia the situation is no better. The most beautiful wood, somewhat similar to our walnut, used in Europe in quantities is teak and this wood comes, as you know, mostly from Burma. However, there is a timber control in Burma and European buyers have to deal directly with the government.

Figure 10-2
George C. Romeiser served as secretary-manager of the American Walnut Manufacturers Association between 1956 and 1961. He was actively involved in conservation programs which encouraged the culture and management of black walnut and other fine hardwood species on private forest lands.

> Among the native European woods, German oak is available, but is in limited use; and as far as French walnut is concerned, due to the embargo, no logs are allowed to be exported from France.
>
> The overall log situation in Europe makes our American walnut logs an object of great desire and intense demand. For, with American walnut, there is no embargo for the foreign buyer, no red tape to go through, no necessity to spend millions in concessions, no auction bidding against powerful competitors. Our borders are wide open and, considering the high price at which the American walnut veneers are sold, particularly in Germany, European log buyers have it easy when competing against the American buyers.
>
> Considering that American walnut is, for all practical purposes, the only beautiful wood in this country available in large quantities and lends itself to such a perfect finish, the disappearance of it would be a calamity...

A single page from an unreferenced 1962 publication (probably British), discovered in a veneer firm's file, reveals a collateral side to the drive by European veneer companies to acquire increasing amounts of American black walnut. It is excerpted as follows:

> Teak was always a special wood. It became sought after and fought after by architects and designers hungry to work in "luxury" materials after the war... it became the "ideal" medium for the interpretation of contemporary design. Of course, change is inevitable. Are there challenges to teak coming forward? There is one wood in the "class" field which stands alone. It has hit West Germany with a tremendous impact and the reverberations continue to ripple outwards with effect. The wood? American black walnut. The seekers from West Germany have stimulated air travel

> and freight traffic across the Atlantic. The Italians followed close on the heels of the Germans. And now the Swedes, the Danes, and the Dutch are looking westwards. There is even the rumor that England is wondering.

An indication of the building frustration of the American industry is evident in a letter written by Robert E. Hollowell, Jr., (1962). The letter describes an Indiana sawmill operator's sale of 220 veneer quality logs, scaling 25,700 board feet, which had taken place a few days earlier. This was and is a common practice of sawmill owners who can realize a higher return from selling high-quality veneer logs acquired in the normal course of securing sawlogs for lumber production. Seven domestic veneer mills and five foreign buyers were represented at the auction. Two of the domestic mills were also known to export walnut logs.

Hollowell estimated the value of logs to be $24,000 or $0.93 per board foot. The bidding began at $25,000 and moved quickly up to the $30,000 level. Ten bids later, the price of the logs reached $34,500 or $1.34 per board foot. At this point there was a break in the auction "...and it was announced by Mr. Bane that it would take a bid of at least $35,100 in order to purchase the logs..." since a buyer not present had made a flat offer of $35,000 for the logs. After some delay, the bidding proceeded as follows:

$35,100 Alfredo Tragni—Milan, Italy
$35,200 Mr. Davis—New Orleans exporter
$35,300 Tragni
$35,500 Davis
$36,000 Michael Joelson—David R. Webb Company
$36,250 Tragni

The final bid averaged $1.41 per board foot, "...a record for such a group of logs of average length and diameter." Hollowell's letter continued:

> When we asked the representatives of David R. Webb Company why they were bidding so high, they said that they were merely seeing how high the exporters would go. They admitted that they would lose money on these logs if they converted them into veneer for the domestic market. Since Mr. Joelson was quite familiar with the European veneer markets, he stated that he thought Mr. Tragni might make a profit on the sale of veneer cut from these logs, since his veneer customers would pay almost any price necessary to purchase quality Walnut Veneer.

Perhaps, this sale was a statistical accident, but the winning bid was 51 percent greater than Hollowell's estimate of the value to a domestic producer such as the Pierson-Hollowell Company. Since domestic producers were cutting veneer at a standard 1/28-inch thickness while Europeans cut veneer for their customers at 1/42 or 1/44 inch, or 50 percent thinner than the Americans, Europeans could cut about one-half again as much veneer from an equal volume of wood. Clearly, manufacturers intending to produce veneer for the domestic market were at a tremendous competitive disadvantage since veneer in both markets was sold by surface rather than volumetric measures. This inequity was not lost on domestic manufacturers or the U. S. Department of Commerce. The highest quality black walnut logs were destined for a foreign veneer firm unless some further action was taken.

On January 15, 1963, AWMA representatives presented the Department of Commerce staff with supplemental statements in support of their request for an embargo on the export of walnut veneer quality logs. The document contained the following:

> The scarcity and short supply of walnut veneer quality logs has been demonstrated. The alarming growth of exports from 1959 through 1962 has created this condition and the "snowball has started down the hill" and it must be stopped before more damage is done to this invaluable natural resource.
>
> It is respectfully submitted that the American walnut veneer manufacturers, through the American Walnut Manufacturers Association, have established that walnut logs are in short supply; that there is an excessive drain of such logs caused by abnormal export demand, and that such excessive drain has an inflationary impact.
>
> Congress has stated in the Export Control Act [of 1949] *that it is the policy of the United States to impose* export controls to the extent necessary to protect the domestic economy from the excessive drain of scarce materials and to reduce the inflationary impact of abnormal demand. Each of the pertinent elements has been shown to exist in the current log crisis (AWMA, 1963).

Meanwhile, exporters and others opposing export controls on walnut logs organized the Committee of American Log Exporters. This group had presented arguments and documents supporting their position at U. S. Department of Commerce Meetings on September 12, 1962, and January 15, 1963. These hearings provided a forum for both the proponents and opponents. Meetings were also held by the Department of Commerce with representatives of the furniture, plywood, and federal purchasing agencies involved in the procurement of walnut-veneered wood products for the government.

A resolution passed at a meeting of the Southern Furniture Manufacturer's Association (1963) in High Point, North Carolina, noted in the preamble the "...unprecedented export increases for the nine years since 1954 [and the] ...annual cutting total considerably in excess of the annual growth increment which threaten this valuable resource with extinction..." The resolutions called for a change in the Commercial Standard No. CS 35-61 so as to standardize hardwood veneer thickness at 1/36 inch instead of the then current 1/28 inch "...in the interest of resource conservation." The Association also urged "...the Secretary of Commerce of the United States to assist in this conservation effort by imposing at once an embargo on further export of American Black Walnut Logs."

To give some impetus to the resolution, a prominent group of Southern furniture manufacturers met with Secretary Hodges on August 26, 1963. The National Association of Furniture Manufacturers concurred with the resolutions and supported the 1/36-inch thickness in a letter to the Secretary on August 27 (Freeman, 1963). Other domestic user-groups soon followed in approving the change.

Twenty-one months elapsed between the formal request for a log embargo and an indication that an embargo was being seriously considered by the Department of Commerce. In a letter of October 1, 1963, to the Hardwood Plywood Institute, Secretary Luther Hodges (1963) wrote:

> We are concerned about walnut veneer log consumption, both foreign and domestic. We have found this to be a very complex problem which cannot be solved by any one single action. We are pleased that we now have complete industry acceptance of the conservation proposals made by industry leaders as a guide to the possible application of export controls.

Export controls on walnut logs were imposed on February 14, 1964. Exports were limited to 7.3 million board feet and a "...new program is designed to minimize depletion...by reducing total annual consumption... to 22.3 million board feet" (U. S. Dept. of Commerce, 1964a). This was intended to balance approximately the timber harvested with the estimated growth of the black walnut resource. In order to achieve this goal, the domestic industry was to limit its procurement to 15 million board feet through voluntary reductions in the total domestic cut of black walnut and through certain conservation measures. The most important of these was the manufacture of 1/36-inch veneer instead of the nominal 1/28-inch thickness which had become accepted as standard. Theoretically, this change in thickness would result in about a 20 percent increase in yield from a given quantity of wood suitable for veneering purposes.

The American walnut veneer industry responded quickly to these conditions and, on February 19, 1964, formed an advisory committee "...to serve in any capacity [Secretary Hodges] may wish in connection with the effectuation of such measures." The manufacturing change-over to thinner veneer was accomplished about April 1, 1964.

Peace and harmony were soon to be disrupted. On May 8, 1964, the Southern Furniture Manufacturers requested an opportunity to present to the Department of Commerce their problems in converting to 1/36-inch walnut veneer. This meeting was held on May 12, 1964, in which the furniture people recommended that a slightly thicker veneer, 1/32 inch, be adopted as the standard. Supporting letters from individual furniture firms were sent to Secretary Hodges following the meeting.

The restrained outrage of the Department of Commerce was expressed at the meeting on May 15 by a departmental official from the Business and Defense Services Administration.

> Gentlemen, we must face the facts! The foremost fact is that a change to 1/36-inch veneer, as almost universally agreed upon last year by the

industries involved, including many of you who are here today, was the principal basis on which the log export controls were justified to other government agencies, to Members of Congress, to Embassies of foreign countries, and to a great many other interested parties (U. S. Dept. of Commerce, 1964b).

The speaker went on to say that a change in commitment would reopen the entire question of walnut conservation which required two and one-half years of effort to implement and that the Department was disturbed by the sudden opposition to the Secretary's conservation program. After offering the technical services of governmental agencies to assist in providing answers to problems associated with abrasives, machinery, and finishes and suggesting the use of substitute species or non-wood materials, the Commerce official concluded with: "Again let me emphasize that unless we achieve the domestic objectives of the Secretary's program, of reducing log consumption to about 15 million board feet during the first year, export controls on walnut will have to be abandoned."

The veneer industry vigorously fought back at the proposal by the Southern Furniture Manufacturers Association to relax the veneer thickness standard. A 28-point brief was prepared for a hearing held in Washington, D. C. on June 9, 1964, before the Honorable George Donat, Administrator of the Business and Defense Services Administration. Prominent members of the veneer industry were present, including Bice A. Roth, Amos-Thompson Corporation; Don Bell, Wood-Mosaic Company; Ted Swain, National Veneer and Lumber Company; Robert E. Hollowell, Pierson-Hollowell Company; and Donald H. Gott, secretary of the American Walnut Manufacturers Association.

Their lawyer, J. Edward Day (1964), carefully reviewed the history of the walnut controls agreement, cited precedent court cases about the binding nature of pledges to abide by agreements, and implored the Department of Commerce to amend CS 35-61 without delay in order to promulgate 1/36-inch veneer as the legal standard. Attorney Day pointed out that the furniture manufacturers were holding back on purchases in an attempt to change the standard to 1/32 inch. "As of May 30, 1964, production of 1/36-inch veneer by the walnut veneer cutting industry was 100,104,963 square feet; sales: 35,598,748 square feet; shipments 18,781,686 square feet." The brief concluded "...that the problem will clear up rapidly once all appearance of uncertainty about the Department's support of the 1/36-inch thickness is officially dispelled."

An uneasy truce existed during the rest of 1964. Secretary Hodges recognized his limitations to enforce domestic controls on either the domestic furniture or veneer industries. The only legal response he could make to conserve the black walnut resource was in the export market arena. "If we feel that [we] do not have proper conservation of walnut timber, ...we could put an embargo on exports." Instead, "...to improve the effectiveness of export control on walnut logs, the United States

Department of Commerce announced..." restrictions on the export of walnut lumber, walnut furniture stock, and walnut small dimension stock when any of this material was two and one-half inches or more in thickness and two feet or more in length (U. S. Dept. of Commerce, 1964c).

Secretary Hodges left the Cabinet effective with the reelection of President Johnson. John T. Connor replaced Hodges. Following his Senate confirmation, Secretary Connor and his staff lost little time in reviewing the effectiveness of the entire walnut resource conservation program since a decision to extend or end controls was necessary by February 14, 1965. Senator Vance Hartke, in a letter dated January 19, 1965, reminded Connor of this:

> It is my understanding that consideration is being given to a long-term extension of this measure to preserve our dwindling supplies of black walnut... A change in policy would go counter to the considered conclusion and action arrived at within the Department and supported by Secretary Hodges. I trust you will find the facts warrant continuation rather than change, should the subject be brought to your personal attention (Congressional Record, 1965).

Responding to Hartke's letter 10 days later, Secretary Connor politely gave assurance that all factors would be considered and that he was pleased to know that Hartke was interested in the walnut conservation program. At this point, neither of these participants realized that this brief skirmish would lead to a bloody political battle.

On February 12, 1965, Secretary Connor announced that he would not extend export controls on walnut logs. In the text of the Secretary's message, Connor justified his decision in the following paragraphs:

> Under the reduced-consumption program, an export quota was fixed at 7.3 million board feet for the first year. At the same time a domestic consumption "target" of 15 million board feet was set.
>
> Reduced *domestic* consumption of logs, though not mandatory, was an essential element of the total program because it was needed to help achieve the goal of balancing consumption and growth and also to help comply with the requirements of GATT. It was provided for only after full consultation with all segments of the veneer-producing and veneer-using industries, and on the strength of an understanding that they would move promptly to the cutting and use of thinner veneers, so as to reduce the number of logs needed to meet their veneer requirements. The United States standard veneer thickness was then 1/28 inch. The U. S. industries committed themselves to adopt 1/36 inch as the new commercial standard. This was expected to make it possible to hold domestic log consumption down to the 15 million board feet target, and continuation of the export controls after February 13, 1965, was expressly made contingent on domestic consumption being reduced to approximately 15 million board feet.
>
> Notwithstanding the industries' commitment to adopt 1/36 inch as the commercial standard for veneer thickness, this was never done. After some

weeks, the veneer cutters did begin to slice a thinner veneer, which they reported as 1/36 inch but which field surveys have shown to average about 1/34 inch. Furniture manufacturers protested they could not use this thinner veneer, and for a long time failed to do so. We now find that domestic consumption of logs for the control year has exceeded the 15 million board foot target by over 4.5 million board feet. On this ground alone there is sufficient basis for not continuing the controls, since the condition for their continuance was not fulfilled.

In conservation terms, the results of the first year of controls are quite disappointing. The excess of consumption over growth, instead of being 6 million board feet as originally allowed for the first year of controls, is more than 10 million board feet, or approximately 2/3 more than the total amount of new growth.

The situation does not seem to involve the possible extinction of the walnut resource. On the contrary, walnut trees are constantly being planted and constantly maturing. Among the important factors which could help to reduce domestic log consumption are not only price shifts, but shifts in consumer preferences and increasing use of substitute materials.

Even if technical difficulties were to develop in making the transition to use of other woods, it would seem preferable to try to ease these transition difficulties rather than continue export controls. For, while the controls may benefit log cutters and users, they also work to the detriment of the log growers and log exporters by restricting the marketing opportunities for their products. Moreover, controls on the export of walnut logs are clearly detrimental to our balance of payments position (U. S. Dept. of Commerce, 1965).

An almost immediate response came from Vance Hartke on February 18, 1965, in a Senate speech which follows in abstracted form:

Mr. HARTKE. Mr. President, I invite the attention of the Senate to a situation of genuine gravity involving not only my own State of Indiana but an entire industry. In practical fact, it is a decision with serious consequences to the entire furniture and plywood industry, a billion dollar business.

Mr. President, I am convinced that the removal of export controls, imposed only after $2^1/_2$ years of study by the Department of Commerce, is a grave mistake. It may benefit a few who will reap the high prices which foreign buyers are eager to pay. But it will make a shambles of the orderly self-restraint which has been successfully practiced by the walnut veneer industry on a voluntary basis.

Mr. President, the order issued by Secretary Hodges on February 14, 1964, which has now been discontinued by Secretary Connor, was in harmony with the congressional declaration of policy as set forth in the Export Control act of 1949. That declaration, a part of the code, reads as follows:

"The Congress declares that it is the policy of the United States to use export controls to the extent necessary (a) to protect the domestic economy from the excessive drain of scarce materials and to reduce the inflationary impact of abnormal foreign demand; (b) to further the foreign policy of the United States and to aid in fulfilling its international

> responsibilities; and (c) to exercise the necessary vigilance over exports from the standpoint of their significance to the national security."
>
> Mr. President, I fail to see any logic in the process of justifying removal of the export controls, in effect inviting the rape of a natural resource, by blaming failure to achieve 100 percent of the conservation goal set forth originally. Yet that is precisely what the Secretary's statement does.
>
> This was based on the Department's own advance calculations that 1963 usage would be 20 million feet, so they undertook to cut it to 15 million. Only when the 1963 actual figures became available did it become apparent that a cut to the goal of 15 million would involve reducing domestic use not by 5 million but by 8.5 million feet. Had the 5 million reduction been based on the actual 1963 [usage of] 23.5 million, with a target of 18.5 million, it would have been exceeded by only the margin of 1 million board feet.
>
> The furniture manufacturers, Mr. President, are now cooperating with the veneer people in using the thinner veneer. Among the scores of telegrams which I have received in the past few days are dozens from furniture manufacturers, all pleading for cancellation of this rescinding order at least long enough for full hearings to be held.
>
> Is a shift to something else, perhaps plastics or imports of other varieties from the Philippines, desirable? Here is what Home Furnishings Daily said in its issue of Monday in discussing the matter:
>
> > "Walnut is a top wood in the furniture industry especially in modern and contemporary styles, and no one wants to give up a good thing. Go to a market and check the case goods houses, especially the promotional and medium price makers, and practically all you see is walnut. It's that overwhelming."
>
> Mr. President, I call upon the Secretary of Commerce to hold in abeyance, to revert to the status which existed before last Friday, the withdrawal of export controls, until a hearing can be held at which the industry views can be presented (Congressional Record, 1965).

Senator Hartke had entered in the Congressional Record more than 100 letters and telegrams from every region of the country calling for export controls and asking for a public hearing of the issues involved. At the same Senate session, Hartke was supported by Senators Javits and Bayh. Senator Bayh, in particular, attacked Secretary Connor's statement that: "Reduced domestic consumption...was an essential element...to help comply with the requirements of GATT" (General Agreement on Tariffs and Trade). Bayh's research found "...60-odd countries producing at least 20 million board feet of hardwood logs a year..." which embargoed or controlled the export of logs. He listed 34 of these countries "...of which 21 were contracting parties to GATT..." Among the three Senators, almost every justifying argument advanced by Secretary Connor had been seriously challenged on the basis of logic or interpretation or on the accuracy of information and data used in Connor's analysis.

Senator Hartke followed-up on the speeches given on the floor of the Senate by writing Secretary Connor on February 18, 1965, suggesting that export controls be temporarily reinstated, pending completion of hearings before the Senate Commerce Committee. Connor (1965) replied on February 26 that: "I have given this suggestion very careful consideration and regretfully conclude that a...reinstatement of export controls does not appear to be appropriate under the circumstances." His letter continued by stating that he was aware of all the significant facts and "...that whatever decision I might reach most certainly would not be pleasing to all interested groups involved." The new element Connor interjected was: "Our export control laws are today maintained primarily to regulate trade in strategic materials." At the conclusion of the letter, Connor reiterated that voluntary control of domestic consumption has not worked. He suggested that, for government to control domestic walnut consumption effectively, some agency comparable to the wartime Office of Price Administration (OPA) and War Production Board (WPB) would be required.

An angry Senator Hartke (1965) wrote Secretary Connor on March 5, 1965, that the Secretary's letter of February 26 had greatly disturbed him:

> [Your letter] confirms the feeling I had when you let controls expire that your real position is, not that the conditions required for application of the Export Control Act are not present in the case of walnut logs, but rather that you do not approve of the Export Control Act to the extent it applies to nonstrategic short supply materials.
>
> This represents a reversal—-without notice, without warning, without a hearing and without a chance of rebuttal—-of the views of your predecessors. The Act has been applied many times to nonstrategic materials.
>
> Nearly three years ago, on July 18, 1962, in a [Department of Commerce] conference on this subject..., this exact issue was discussed and cleared...I am informed that this meeting was considered to have settled this question...Now, unexpectedly, there is a complete reversal in interpretation and three and a half years of careful work on this subject are in danger of being thrown away.
>
> Your letter also says that if the public interest requires controls, this should include controls on domestic consumption which you say the Export Control Act cannot accomplish effectively. Section 2 of the Export Control Act says nothing about domestic controls. That aspect got into this picture purely because of Article XX of GATT. Article XX has been and will continue to be fully complied with by the significant reduction in thickness of veneer being cut by American producers.
>
> I must also question the relevance and accuracy of your statement that "the major demand for the item is domestic rather than foreign." In the period 1954-1960 alone, consumption of veneer quality walnut logs by domestic producers increased only 3 times but the volume of exports increased [20] times...[Table 10.2]. As a member of the Senate Commerce Committee, I am deeply concerned about the rigid and unsympathetic attitude shown by your department...

Senator Hartke's hearings on the "Export Controls on Black Walnut Logs" opened in Indianapolis eleven days later, on March 16, 1965. The walnut industry was well prepared, and extensive testimony was given by Donald H. Gott, executive director and secretary of the American Walnut Manufacturers Association. Gott built the case for continuing controls by citing statistical evidence and sections of the Export Control Act of 1949. He also cited GATT agreements and Section 2022 of the United States Code, which specifies the federal policy for imposing controls when the domestic economy is threatened by an excessive drain of scarce materials. He was followed by nine other supporting witnesses representing allied associations, furniture and panel manufacturers, and suppliers. Only three people came forward in Indianapolis to testify against export controls. None had prepared statements. The brevity of the Indianapolis hearing is indicated by the complete transcript printed in the final Senate report on the Hearing. Only 37 pages were required (U. S. Senate, 1965a).

Table 10-2
Estimated Consumption and Growth of Veneer-Quality Black Walnut Timber

Year	Consumption[1]			Growth[2]	Growth surplus (+) or deficit (-)
	Domestic	Export	Total		
	--- million board feet, Doyle log rule ---				
1954	5.5	0.5	6.0	20	+14
1955	7.3	1.2	8.5	20	+12
1956	9.9	1.1	11.0	20	+9
1957	11.0	1.4	12.4	20	+8
1958	12.3	2.3	14.6	20	+5
1959	17.0	3.7	20.7	20	-1
1960	17.5	10.2	27.7	20	-8
1961	17.8	7.2	25.0	20	-5
1962	22.4	10.3	32.7	20	-13
1963	23.5	15.0	38.5	20	-19
1964	19.8	9.9	29.7	19	-11

[1] U. S. Department of Commerce estimates. Includes timber cut from both forest land and other sources such as fencerows and farmsteads.
[2] Estimated average growth on commercial forest land only.

The hearing was resumed in Washington, D. C., on March 31, 1965. Besides Vance Hartke, Senators Monroney (Oklahoma), Daniel P. Brewster (Maryland), Robert Bass (Tennessee) and Thruston B. Morton (Kentucky) completed the Senate Commerce Committee panel present for the hearing. This was a day-long meeting at which 39 private citizens presented statements to the Commerce Committee. In addition to them, 13 Senators and Congressmen, representing nine states, presented statements. Without exception, all spoke favorably for the restoration of controls. Eighty-nine letters and telegrams were also submitted for the record from both proponents and opponents of exporting walnut logs with governmental controls. The oral statements and submitted materials required 174 pages in the printed proceedings. Undoubtedly, John T. Connor was the witness with "star" quality who held his ground while being aggressively challenged by Senator Hartke after delivering his prepared statements and oral arguments. Their interchange is abstracted in Appendix C-1.

The questioning of other Senators was equally incisive as they attempted to understand the reasoning used by Secretary Connor in removing export controls on walnut. After a lengthy dialogue between Senator Monroney, an authority on trade legislation and GATT, and the Secretary, the obviously disgusted Senator Hartke succinctly summed up his conclusion about the justification he had just heard as follows: "I can't buy this GATT idea in any way, shape or form, and I think it is as phony as a three-dollar bill."

The opposition to log export controls had organized the Committee of American Log Exporters in August 1962, after they had become sufficiently concerned about the actions of the American Walnut Manufacturers Association to secure log export controls. Their executive secretary, Harry J. Smith, Jr., testified that:

> ...it has been consistently felt by most exporters, loggers, and farmers interested in free and expanding world markets for log and timber products, that the purpose behind the Indiana-centered veneer manufacturers' petition was to protect the prices which they paid to farmers and log producers...prices which had increased due to a new foreign market for veneer quality American black walnut logs.

Smith made essentially four points in favor of abandoning controls:

1. There is no threat of exhausting the black walnut trees of veneer quality.
2. It is unreasonable to expect foreign and domestic demands for walnut to continue indefinitely.
3. By ending controls, trade in walnut logs will return to more normal and understandable patterns.
4. The Export Control Act 1949 cannot be employed as a price controlling apparatus.

Consultants and members of the Committee spoke to each of these points. Albert G. Hall of A. G. Hall and Associates, a respected consulting forester whose firm was managing private forest lands in Maryland, Virginia, West Virginia, and Pennsylvania, spoke knowingly of forestry conditions in that region and the landowner loss of an export market for lower quality Number 2 logs during the 1964 control period. In his concluding remarks, Hall said:

> The freedom of choice in the sale of the products of forest management, the freedom to sell in an unrestricted market, at a price which offers an attractive return, is one of the incentives to better forestry on private woodlands. If we are to maintain our forestry advances, and are to stimulate increasing interest in timber growing, we should consider very carefully any restriction we put on the marketplace.

Professor Harold Heck, Director of the Department of Trade and Transportation at Georgetown University, provided professional expertise for the Committee of American Log Exporters in the area of foreign trade.

> There seems to be some question as to the propriety of reimposing export quotas because of the General Agreement on Tariffs and Trade. Three articles of this instrument appear to touch on export quotas.
>
> Article XI provides for the general elimination of quantitative restrictions except for—
>
> (a) Export prohibitions or restrictions temporarily applied to prevent or relieve critical shortages of foodstuffs or other products essential to the exporting country, and
> (b) Export prohibitions necessary to the application of standards or regulations for the classification, grading, or marketing of commodities in international trade.
>
> Article XX specifies a number of exceptions to the agreement, stating that the agreement does not prevent the adoption of measures:
>
> (g) Relating to the conservation of exhaustible natural resources if such measures are made effective in conjunction with restrictions on domestic production or consumption;
> (h) Undertaken in pursuance of obligations under any intergovernmental commodity agreement conforming to GATT criteria; and
> (i) Involving restrictions on exports of domestic materials necessary to assure essential quantities of such materials to a domestic processing industry during periods when the domestic price is held below the world price as part of a governmental stabilization plan.
>
> Article XX (g) is perhaps nearest in applicability to the walnut log situation. However, one may question whether walnut logs are in this sense an "exhaustible" natural resource; they are certainly replaceable over time by good forestry practice.
>
> It may be that the contracting parties to the agreement intended this to apply to something that would not be reproduced, such as minerals.
>
> Moreover, we must bear in mind that the GATT is an intergovernmental agreement and the provisions of XX (g) appear in context to assume restriction imposed by a government on domestic production or consumption.
>
> As far as I know the United States does not have legislation that would permit this type of imposed action on the private domestic economy; and

> if this is so and we were to reimpose an export control of the type just terminated, it may become necessary to appeal to the contracting parties of GATT for an exception to the agreement to which we subscribe.
>
> Such a request for exception to the agreement requires a defense and, in the light of the record mentioned above, I believe it would be hard to sustain.

Interestingly, Professor Heck failed to make any reference to subparagraph (j) of GATT's Article XX which follows subparagraphs (g), (h), and (i), which he used to substantiate his arguments against the wisdom of reimposing walnut log export controls. J. Edward Day, lawyer for the American Walnut Manufacturers Association, pointed out in earlier testimony that subparagraph (j) permits export controls on short supply items without any restrictions of any type on domestic production or consumption. This subsection says:

> ...nothing in this agreement shall be construed to prevent the adoption or enforcement by any contracting party of measures...essential to the acquisition or distribution of products in general or local short supply: *Provided*, That any such measures shall be consistent with the principle that all contracting parties are entitled to an equitable share of the international supply of such products...

Lawyer Day then went on to state that subsection (j) had been reviewed and retained by the contracting GATT parties two weeks earlier, on March 15, 1965. He concluded that: "This sudden extreme reliance on GATT to defeat export controls in our case is but the leading example of many examples of the Department looking for excuses for not applying the Export Control Act short supply provisions to this clear-cut case."

Following all of the emotional and legalistic testimonies of the day, Paul R. MacLean, president of Wood-Mosaic Corporation, made the last statement prior to adjournment at 4:36 p.m. It was brief and prophetic:

> If the price exceeds market acceptance, then the users turn to other species. There is no reason to single out walnut for deification among woods. It is only since 1957 it has assumed worldwide popularity. No one knows when this popularity will wane in favor of other woods as has happened so many times in the history of the veneer business.
>
> Frankly, I feel we have a tempest in a teapot.

After the foreign and domestic mining of the veneer quality black walnut resource had been nearly completed, the rush to exploit white and red oak began. Although MacLean dismissed the hearing as a "tempest in a teapot," serious questions about natural resource policy in the United States were raised as well as questions concerning what are the best interests of the country when given a choice of either exporting unprocessed or manufactured products.

The black walnut log export control issue was back on Capitol Hill on June 15, 1965, at a Hearing before the Senate Committee on Banking and Currency. Being considered was the extension of and amendments to the Export Control Act of 1949 which was due to expire on June 30,

1965 (U. S. Senate, 1965b). The Committee was chaired by Senator A. Willis Robertson (Virginia). Other committee members present were: Paul H. Douglas (Illinois), William Proxmire (Wisconsin), Harrison A. Williams (New Jersey), Maureen B. Neuberger (Oregon), and Bourke B. Hickenlooper (Iowa).

Robertson set out the purpose of the hearing in his introductory remarks:

> The CHAIRMAN. What we really have to consider is not the discontinuance of previous controls but certain changes which are being recommended in the control program.
>
> We have a proposal for the benefit of manufacturers of walnut veneer to put quotas or an embargo on the export of walnut logs, which would eliminate the farmers' market for the sale of walnut logs and not give them the advantage of world prices.
>
> The administration is against putting an embargo on the shipment of walnut logs, but there are some who want the Congress to use its power to give preference to the manufacturers of lumber over the farmer who grows the walnut trees. That is a real difference in viewpoint.

A second proposal considered by the Banking and Currency Committee involved changing the provision that a certain proportion of domestic wheat, if exported, should be transported in American ships. A third issue, pertaining to Arab countries boycotting American firms who trade with Israel, was also on the agenda. The walnut log export opponents were not helped by this or by the rural orientation of the committee chairman who tended to put complex issues into a perspective of flue-cured tobacco farmers and walnut in terms of qualities for gunstocks.

Senator Hartke had introduced S. 1986 on May 4, 1965. It had been read twice and referred to the Committee on Banking and Currency. The bill consisted of just 14 lines and did not specifically refer to black walnut. The intent was to clarify the Export Control Act 1949 so that it included any domestic commodity in short supply whether it was or was not of strategic importance to national security. The President would make the determination of export control based on the guidelines contained in the bill. As Hartke explained:

> The reason for the introduction of this legislation is the apparent lack of clarity of interpretation of the present act as it relates to the export of black walnut, an exclusively American commodity in great world demand and in clear danger of becoming in short supply as veneer-size trees are being cut down and used more rapidly than they mature.
>
> My amendment would make [the] intent more explicit and make it clear that short supply controls are to be imposed when the conditions set out in the act are met, without any consideration of extraneous conditions which are not set forth therein. I base this judgment of the intent of Congress on the legislative history of this act, which includes this statement by your own committee in your report on the 1962 bill which continued the act. And here I quote again:

"The act is not limited to strategic materials or to critical material or to essential commodities. It will support a total embargo or the mildest of restrictions. The requirements of foreign policy, national security, and domestic shortages are the only test."

Hartke was questioned extensively by Senator Neuberger about the broad nature of the amendment and its relationship to the export of West Coast logs and canned salmon. She was particularly concerned about the potential for a reduction in the competition for walnut as it affected the price of walnut timber received by farmers. Although not antagonistic in her questioning, the Senator was obviously not sympathetic to Hartke's S. 1986 proposal. This may have set the tone for the rest of the hearing.

Secretary John T. Connor presented a prepared statement before Senator Robertson's committee pertaining to the walnut export issue. His statement was similar to the one he had prepared for the Senate Commerce Committee. The counsel for the American Walnut Manufacturers Association, Mr. Day, attempted to "set-the-record-straight" by briefly reviewing the history of the attempt to obtain log export controls and by responding to the questions from the panel of Senators. Understandably, some of the senators displayed a great lack of information and experience regarding hardwood veneer-quality trees or their specialized manufacture. Donald H. Gott, secretary of the AWMA; Burton F. Swain III, a veneer manufacturer; and J. O. Petrus, Jr., a walnut lumber manufacturer, strengthened the case for export controls from their own perspectives.

Harry J. Smith, secretary of the Committee of American Log Exporters, presented counter-point oral arguments against the imposition of log export controls. He also raised questions about the "...list of people associating themselves with the petitioner..." and asked "...if these organizations actually support [S. 1986]... or are they in fact aligning themselves with a previous position which the AWMA has taken?" Smith submitted statements for the record which, for the most part, had been prepared for the Senate Commerce Committee Hearing. He also noted that about 125 telegrams and letters from log producers, coastal port authorities, and exporting companies in opposition to the reinstatement of log export controls had been sent to the Senate Banking and Currency Committee. If the number of messages sent to the Banking and Currency Committee was a decision-making criterion, the walnut veneer and furniture manufacturers had clearly lost their case. The pro-export control constituency sent barely 25 percent as many letters and telegrams.

The American walnut industry had some reason to be optimistic, however. On June 15, 1965, the Senate Committee on Commerce sent a strongly worded letter to Secretary Connor. The committee urged Connor to reconsider his February 12 decision which allowed controls to expire. Chairman Warren G. Magnuson and other committee members who signed the letter declared they "...are concerned over the serious situation caused by removal of the export controls on walnut logs." Donald H. Gott believed that the walnut industry had won Round 1 and that "...Round 2 is coming up soon and we feel confident" (Anon., 1965).

An apparent knock-out blow was landed by the Commerce Department in Round 2, on September 22, 1965, when Secretary Connor replied to Senator Magnuson's letter of June 15 that:

> I have considered the record of the hearings held by Senator Hartke last spring and I have considered additional information and comments received by the Department from interested persons and groups since it became known that I was reconsidering my original decision [and]...after examining once again the basic facts and arguments, I [have] reaffirmed my decision of February 12, 1965.

The Secretary's lengthy letter presented a background summarizing the events preceding his review of the question of exports of black walnut logs (U. S. Senate, 1965c). The review did not reflect any of the substantial legal and technical arguments presented by industry at the public hearings. Detailed contents of Connor's letter can be found in Appendix C-2.

Stunned by Secretary Connor's refusal to reimpose export controls on black walnut logs, the American Walnut Manufacturers Association "...vowed not to give up the fight." The Association's public reply to Connor's decision was to the point:

> We are shocked and disappointed because his argument is faulty. It reveals a complete lack of understanding of the crisis nature of short supply of walnut. Moreover, the Secretary delayed his answer... for more than three months. As a result, his decision requested in June was not received until two weeks before adjournment of the Congress. This precluded any possible action before the adjournment of the 89th Congress. The issue is bigger than walnut alone. A similar problem is developing rapidly in other major hardwoods. It is our hope that Secretary Connor will once more review the facts and figures we will present between now and January in an effort to save this American heritage wood, its respective processing manufacturers, and other walnut using industries (AWMA, 1965).

To provide the domestic walnut manufacturers with a public forum and, possibly, to placate his constituency back in Indiana, Senator Hartke reopened the Commerce Committee's hearing on walnut export controls on October 15, 1965. It was a short meeting, beginning at 10:12 a.m. and adjourning at 11:20 a.m., "subject to the call of the chairman." As far as the Senator and the walnut industry were concerned, the hearing was not satisfactory. If this was Round 3, their principal opponent did not even show up. Instead, Secretary Connor sent an Acting Assistant Secretary without a prepared statement. The hapless Secretary, James F. Collins, was then subjected to a pummeling by an infuriated Hartke in the form of biting statements and questions about the quality of the facts used by Secretary Connor in reaching his decision. These extend over nine pages of the printed proceedings. In summing up, Hartke said: "I think the attitude of the Department [of Commerce] has been typical of its cavalier attitude in this entire proceeding. It seems to indicate that there is no intention to be concerned about this industry."

The next witness was David E. Herrick, Chief of the Forest Products Marketing Section of the U. S. Forest Service, who had also not prepared a statement for the hearing. He was curtly dealt with by the Chairman.

> Mr. HERRICK. Good morning, sir. I'm here to answer any questions you might have.
>
> Senator HARTKE. You have no statement?
>
> Mr. HERRICK. No, sir.
>
> Senator HARTKE. I have no questions, sir. And I am asking you and directing you forthwith to submit a statement...you are certainly aware of what these hearings are about. You were permitted to come as you were an expert in this field to state the situation.
>
> If you people can't do any work down there, maybe we had better find somebody who can.
>
> You're excused (U. S. Senate, 1965c).

The Forest Service statement which was later added to the hearing record briefly described research on black walnut culture and utilization, studies that were under way to determine the extent of walnut on "non-forested" areas, and technical assistance available to landowners. A more relevant document prepared by Dwight Hair, a staff member of the Economics and Marketing Research Division in the Washington office of the Forest Service, was presented as a speech to members of the Hardwood Plywood Manufacturers Association at their annual meeting two weeks earlier, on September 30, 1965. Titled "Effective Action Must Be Taken To Meet Future Hardwood Log Needs," the address was entered into the hearing document by Donald Gott. Although Hair's speech broadly included all of the hardwoods in the eastern United States, the basic contentions which had been made by the walnut industry in the earlier hearings were clearly substantiated with statistics and illustrative charts.

Three representatives from the veneer industry were the only other witnesses at the hearing to present oral testimony. All three had testified at earlier hearings. They were: Burton F. Swain III, a veneer manufacturer and president of the AWMA; Donald H. Gott, and J. Edward Day. Swain strongly refuted many of the statements made in Secretary Connor's letter of September 22 and concluded:

> I consider this letter to be a hoax. It is my opinion that Mr. Connor made a hasty and ill-advised decision last February, and since then has tried to make the facts justify that decision. Of course, they don't justify it.
>
> This is a serious problem, and no amount of ignoring will make it go away. Like a disease, it started very small, but it is continually getting worse and is beginning to spread. And until the Department of Commerce is willing to face its responsibility, it can only continue to plague us.

Gott also attacked the letter by stating that it "...seems to be a remarkable combination of apparent sincerity and some very strange logic, if not actual misinformation." The AWMA's lawyer, Mr. Day, focused on just two legal aspects relating to Secretary Connor's letter of September 22:

1. The Department's invalid requirement for reduction in domestic consumption, and
2. The Secretary's statements about the price effects of controls, ignoring the clear language of the Export Control Act.

Mr. Day's discussion of these points may be found in Appendix C-3.

After the new Congress convened in 1966, the ferment continued among those legislators who had been witness to the testimony or were continuing to be appraised of the effect that the accelerating walnut log export and price situation was having on affected constituents. The competitive edge remained with foreign buyers for the increasingly scarce best walnut. Domestic buyers bought what was left over at inflated prices and were, of necessity, forced to cut greater amounts of veneers from other species.

Reacting, undoubtedly, to this persistent political headache, the Executive Departments of Agriculture, Commerce, and Interior established the Hardwood Timber Conservation Committee. Its purpose was to study the current domestic supply of high quality hardwoods and related factors so that it could produce recommended measures for the conservation and improved utilization of hardwood forest resources. Six months later, in April 1967, the Committee issued a public report making various recommendations.

The first recommendation suggested the means for long-term improvements in supply. These were already well known to field foresters who provided landowner services and to forestry researchers in private and public institutions. The last two recommendations showed the strong input of the Department of Commerce because they used language that could have been taken directly from one of Secretary Connor's statements two years earlier.

In a brief concluding statement, the Committee wrote that it had "...determined the following facts."

(1) The current rate of removal of high-quality fine hardwoods exceeds growth.
(2) Forest management programs supported by the Committee may achieve a balance between growth and removal over the long-term, but will not solve the problem during the intervening years.
(3) Export controls without concurrent reduction in domestic consumption would not achieve the desired balance with growth.
(4) Under the authority of the Export Control Act, no mandatory reductions in domestic production or consumption may be imposed [and] "...that expansion of certain Federal and State forest conservation programs could materially improve the hardwood timber situation in the long run by increasing supplies and improving utilization...(U. S. Dept. of Agriculture, 1967).

The 19-page Committee report appended 20 pages of tabular information on general and specific data relating to the fine hardwoods, including black walnut stocking, growth, harvested volumes, and import/export statistics. Most of the data were extracted from *Timber Trends in the United States,* Forest Resources Report No. 17, published in 1965 by the U. S. Forest Service. The remaining data were generated by the Forest Products Division, Business and Defense Services Administration, or from Bureau of the Census statistics.

Most knowledgeable observers would agree that the document illustrated the need for the three federal Departments (Forest Service, Commerce, and Interior) "to circle their wagons" by producing a bland document rationalizing a previous administration's decision. Very little new information was offered in addition to the standard, off-the-shelf, remedies and a non-feasible solution requiring legislation to control, at the federal level, the domestic consumption of "fine" hardwood species.

The walnut log export issue remained officially dormant once again until February 26, 1969. The AWMA and the Fine Hardwoods Association (FHA) then petitioned the Department of Commerce that the Export Control Act of 1949, as extended and amended, be imposed on the export of black walnut logs and veneer (AWMA/FHA, 1969).

After providing an historical review of the issue, the petition pointed out that the "crisis" had continued to worsen.

> In the time since the Departments considered the problem in October of 1967, the situation on excessive drain and inflationary impact of abnormal foreign demand has continued to worsen with an enormous increase in exports of both logs and veneer in 1968 over 1967.
>
> The following factors have brought about a markedly changed situation since the time of consideration of the issue by the Hardwood Timber Conservation Committee:
>
> 1. The shortage has become worse.
> 2. The volume of exports of both logs and veneer has increased at an abnormal pace.
> 3. Prices have continued to spiral at an inflationary rate.
> 4. Logs available at walnut veneer and lumber mills in the United States are of increasingly poor quality and the yield of usable veneer and lumber from such logs is deteriorating.
> 5. The conservation efforts of the black walnut manufacturers have increased and become even more significant and effective.

In some detail, the petition emphasized that the necessary conditions of the Export Control Act had been met fully, viz., that the material was in short supply, that an "abnormal demand" had caused an inflationary impact to become apparent, and that there was an urgent need to protect the domestic economy.

The petition reiterated that the veneer industry had made 1/36-inch veneer a standard thickness as requested and had doubled its financial contributions and had continued to cooperate with the U. S. Forest Service on research and on walnut forest conservation and management programs. Additionally, the two hardwood associations had jointly employed Larry R. Frye, a graduate forester, to carry out "...the program exactly as recommended by the Hardwood Timber Conservation Committee." The final evidence presented consisted of government statistics showing the downward trend in domestic consumption and a rising trend in walnut exports (Table 10-3).

In the petition, the following rhetorical questions appeared:

> If the Forest Service Branch of the U. S. Department of Agriculture substantiated the shortage of quality black walnut by testimony, both written and oral, during 1963, why then should not this fact be more pronounced and self-evident now?
>
> If the current importing nations of black walnut logs from the United States have either total embargoes or export restrictions on their own logs, why, then, cannot the United States impose similar restrictions to overcome shortages and inflationary spiraling prices in the interest of conservation...a restoring of the growth vs. cut balance?

When no action had been taken by Secretary of Commerce Maurice Stans during the 13 months after the petition was filed, the American Walnut Manufacturers Association began to rally its members in a memo dated March 23, 1970. They were exhorted to contact the top executives of the firms they serviced and to convince them that they should contact their Congressional delegation or "...at the very least, write them a strong letter requesting that they demand a favorable decision on this vital matter" (Gatewood, 1970). They were also challenged not to ask their customers to do more than they had done themselves. Immediate

Table 10-3
Domestic and Export Consumption of Walnut in the United States, 1964-1969

Year	Domestic Consumption	Export Consumption
	— thousand board feet —	
1964*	17,571	9,640
1965	19,334	18,077
1966	20,413	12,763
1967	14,389	16,586
1968	14,308	21,896
1969	12,455	20,829

* Both values for 1964 are less than the established quota.

action was requested as there was reason to believe that the Department of Commerce would shortly reach some kind of decision on the matter.

Whether or not the blitz of letters, telegrams, and contacts with Congressmen and Senators triggered the next public hearing is questionable. The 1967 report of the Intergovernmental Hardwood Timber Conservation Committee had recommended such a meeting and the Department of Commerce was undoubtedly being pressed to respond to the petition of the walnut associations. The hearing, billed as the "Interdepartmental Public Hearing on Supply of Veneer Quality and Veneer," was convened in the Department of Commerce Auditorium on May 25, 1970. Statements prepared by the two cooperating industry associations closely paralleled the materials included in the petition they had submitted to the Secretary of Commerce on February 26, 1969.

Their position was supported by 22 U. S. Senators, 14 Congressmen, and 25 walnut industry leaders, as well as by Dr. F. Bryan Clark, a top U. S. Forest Service research administrator. The only new element proposed by the walnut associations was the recommendation "...that a maximum level of total consumption of 23.2 million board feet would be feasible and would realistically achieve meaningful conservation of walnut. This would be approximately 15 million board feet for domestic consumption and 8.2 million board feet for export, including exported veneer converted to a board-foot basis" (Day, 1970).

More than four months passed before the Department of Commerce broke its self-imposed silence on the walnut log and veneer export determination. On October 6, 1970, a terse news release was issued from the Office of the Secretary announcing that the U. S. Department of Commerce "...would not impose quotas on the export of walnut logs, veneer, or timber at this time." Assistant Secretary Robert McLellan "...noted that the Department had received many requests from individuals and firms engaged in walnut production and processing to announce what, if any, action was planned regarding the possibility of export controls on walnut logs and veneer." He also pointed out that "...on the basis of data now available, the Department believes that the Export Administration Act's prerequisites to the imposition of quotas have not been met." No details or explanations of the decision were given (U. S. Dept. of Commerce, 1970).

Legal counsel for the American Walnut Manufacturers Association immediately responded by letter to the Department of Commerce's general counsel, James T. Lynn, declaring:

> If these few sentences constitute a "decision" in any proper sense of the word on a matter on which there was formal application, a lengthy public hearing, and a hearing record, then there is urgent need for some change in the notion of the Department of Commerce about due process and administrative procedure.
>
> At a minimum in a matter of this kind, the applicants, which we represent, are entitled to findings on the various issues that were raised and the

> various pertinent provisions of the law, including what the Department found as to the allegation of short supply, what it found as to the allegation of abnormal export demand, and what it found as to the allegation of inflationary impact.
>
> The fact is that it is clear that the specific provisions of the Export Administration Act are being ignored simply because certain people in the Department do not like the idea of having the Act effective in the walnut situation.
>
> To dismiss an important piece of legislation as being inapplicable with a brief unexplained statement that the "Act's prerequisites to the imposition of quotas have not been met" is an example of government by men rather than government by law. The applicants are entitled to an explanation in a proper and valid decision of the Department as to what findings were made on the evidence produced which resulted in the conclusion that "prerequisites" have not been met (AWMA, 1970).

At this point it is reasonable to ask at least two questions:

1. How does one begin to explain the actions of the Department of Commerce during both Democratic and Republican administrations, several Secretaries of Commerce, and numerous Under and Assistant Secretaries, given circumstances where the federal law and the evidence for export controls were so clear?, and
2. What were the consequences of a decade-old struggle of the walnut industry to staunch the flow and loss of the best black walnut timber to foreign markets?

With respect to the first question, it is highly likely that internal consensus within the Department was not easily achieved if political pragmatists were pitted against those schooled in international trade theory. Such theory is based on free trade and minimum governmental controls. This is particularly true in situations where raw materials are involved since it is assumed that economic efficiency will be improved through specialization as determined by markets for the factors of production. Theoretically, "free trade" helps everyone a little and protection from it helps a much smaller group a great deal. The scale of the log export problem of the walnut industry was small in dollar terms. It may also have been unique enough so that there was little in the way of precedence for establishing a uniform policy for this commodity as well as for others.

During the decade of 1960-1970, all Secretaries of Commerce were under pressure to reduce balance-of-trade deficits. These were also Viet Nam years when the concerns and priorities of the federal government lay elsewhere. Other areas of federal policy which may have influenced the decisions of the Commerce Department included the requirements of GATT, the goal of full employment, and the plight of the rural poor. Secretary Connor specifically referred to GATT and to Appalachian woods workers and landowners who might be worse off under a log control situation. Senator Robertson of Virginia, who chaired the 1965 Hearing on the Export Control Act, was also most sympathetic with respect to poor rural farmers in his state when questioning witnesses. Commerce officials were also quick to indicate that black walnut was not a strategic resource.

The delicate policy question of what to do about the Canadians if export controls were instituted may have been an important factor as well. Veneer mills in the Northeast and in some areas of the Lake States were dependent upon Canadian yellow birch logs. This material crossed the border into the United States with a minimum of intervention by the federal government. If controls were instituted, retaliation from Canada was a near certainty. Finally, all of the firms which processed walnut in the United States were not fully in accord with export controls, particularly if they were involved in exporting walnut logs and veneer. The multinational timber companies based in the United States were known to be "very cool," if non-receptive, to the idea of export controls on forest products. Whatever the real reasons were, they may be lost in time or in memos gathering dust in a warehouse. They will remain unsatisfactory and unsatisfying reasons for the destruction of a resource essential to the health of many American manufacturing entrepreneurs and their employees.

The consequences of this protracted effort to limit exports had at least two perverse effects. Besides increasing the necessity for the veneer industry to exploit other "fine" hardwood resources as substitutes for veneer-grade walnut, the threat of an embargo or strict export controls caused foreign buyers to develop strategies to overcome either situation. One foreign buyer agreed to discuss the strategies developed by his company on the condition of anonymity:

> My business partner called me and said we are going to have a problem because the Nixon administration's Under Secretary of Commerce Peterson is trying to impose a log embargo, prohibiting the exportation of walnut logs and, later on, possibly other species too. Walnut was always considered in America not only a very valuable species, but a very sentimental species. We, the big European operation which depended so much on handling walnut veneer, would have been without a primary source. Therefore, consideration was given to building a new veneer plant in America so that the product could be produced in America. The exportation of walnut veneer would be permitted and even promoted because the raw material would be utilized together with American labor to have an upgraded dollar-value going out instead of just being an exporter of raw materials. This makes a lot of sense. So then, the two of us worked together and at first tried to get some leniency with regards to the export quota system. We went to all kinds of hearings in Washington in order to try and establish for Canada a different treatment than for the rest of the world [so we could transship to Europe], saying we are neighbors and we have many privileges among each other. You should treat Canada really as North America, and then look at us as being your neighbor country. While the rest of the world can have a log embargo, we don't want you to do it to us because we're going to ship you birch logs which you want for your New York State mills.

> As another Canadian said, "Well, if they want to stop our veneer mills, we may just stop their automobile plants in Windsor, Ontario, and not let them ship any more automobile parts from Detroit to Windsor. We want to have a most-favored status with the United Status." So, we went to all these hearings but could see the handwriting on the wall and believed that one day there would be a log embargo. The two of us then discussed the real possibility of building a veneer plant in America.

The seeds for the virtual takeover of the American hardwood face-veneer industry had been sown. Domestic firms would soon be fighting to remain competitive in their own domestic markets. The bloody competitive battles caused a number of casualties among the traditional and family-owned and operated veneer mills, the mills which had been operating as they had in previous decades using average machines and technology.

Chapter 11

THE VENEER TRADE ASSOCIATIONS

Trade associations are formed by firms attempting to realize certain explicit and implicit objectives. The formation may also be prompted by governmental intervention to secure public benefits. The long-term viability of these associations is dependent upon the continuing relevancy of the objectives sought. They are weakened and threatened by changing patterns of consumer tastes and preferences for the particular industry's product, by economic depressions, by poor association leadership, or by a change in the political climate which may affect access to raw materials or other factors of production. One survival defense for an association is to remain flexible by changing its objectives and absorbing membership from related organizations through mergers.

It is apparent from the literature that the several segments of the forest products industry began to "mature" into national organizations about 1900. For example, the "...National Hardwood Lumber Association saw the first light of day in the modest, dimly lit office of the Chicago Hardwood Lumber Exchange, replete with a roll-top desk, bare maple floor, well-worn hardwood chairs and the highly utilitarian cuspidors" (Pendleton, 1973). This occurred on April 8, 1898, and was attended by 37 men representing firms located in a variety of hardwood regions in the eastern one-half of the country. The formation of this association was predicated on the development of national markets for hardwood products. Local and regional markets were supplanted as urban population centers expanded, communication systems improved, and as consumer-product industries became centralized to provide better service to urban markets. Distance meant that producers no longer personally knew their customers. Brokers, wholesalers, and manufacturers demanded common standards if products were to be uniformly priced when pooled from several sources. Strenuous negotiation and a great deal of compromise were eventually required to iron out sectional prejudices and differences in lumber inspection grading rules.

The National Lumber Manufacturers Association (present National Forest Products Association) was organized in 1902 and was a federation of regional, essentially softwood, associations. It was prompted by the need to provide a unified voice in influencing national resource management policy and to mediate "turf" and lumber grading disputes among powerful regional associations and companies. The American Walnut Manufacturers Association (AWMA), which was organized in 1914, and the Mahogany Association, organized in 1918, are more clearly identified with a uniform desire to maintain or increase the market share of their specialized member firms.

The early regional hardwood veneer associations are less well known in the literature. The Commercial Rotary Cut Gum Manufacturers Association and the Quartered Oak Veneer Manufacturers Association are examples of regional firms which amalgamated into the National Veneer and Panel Manufacturers Association. This association later merged with the Plywood Manufacturers Association in 1922. The treasurer of the merged association, H. E. Kline of the Louisville Veneer Mills, provided a clear explanation as to why a trade association was so important. Similar messages have, without doubt, been uttered by frustrated industry association executives in the more than six decades since he spoke:

> Gentlemen, we are standing today at a crisis in the industry. We are being sorely tried and assailed from all sides. We are bowed down with taxes, and we are racked with excessive freights both on inbound and outbound materials. We are handicapped by overheads that cannot possibly be applied to the decreased business we are doing today, but by standing together, pooling our interests, making our fight where necessary, a fight of an association instead of individuals, we may not hope to gain any ground but we may at least hope that we can hold our own and maintain an even keel" (Anon., 1922b).

H. Brook Sale, of Hoffman Brothers in Fort Wayne, Indiana, had the dubious honor of being the last president of the National Veneer and Panel Association and the first president of the resulting organization. In his first address to the combined membership, in Chicago on June 20, 1922, he recalled his experience over the previous five or six years during which he had been strongly persuaded by associates to assume the presidency. "The strongest point which [my associates] made was that the position, having been tendered to me, would make it very difficult to find anyone who would accept the office, and which would very likely result in the disintegration of the association." Mr. Sale was leading up to the salient point that good leadership is an essential ingredient for a successful association. He put it this way:

> When an office is refused, it is primarily because the association is not worth a great deal to the members, in that they have not put in sufficient funds to make it possible for any one person to draw a salary that would compensate an experienced person to spend the time necessary to properly direct the functions of the association (Anon., 1922a).

However, one can infer that, without a sufficient incentive for firms to maintain an actively functioning national association, the best leadership will fail to sustain an organization.

This association, which retained the identity of the Plywood Manufacturers Association, merged in 1953 with the Hardwood Plywood Institute. In 1964, it merged again, with the Southern Plywood Manufacturers Association, to form the Hardwood Plywood Manufacturers Association (HPMA).

FINE HARDWOOD VENEER ASSOCIATION

The National Industrial Recovery Act of 1933 (NIRA), a part of the National Recovery Act (NRA), provided the "inspiring urge" for hardwood veneer firms to band together in 1933 as a "single" product trade organization. NIRA was designed to encourage economic recovery, to combat serious employment problems of the Great Depression, and to institute fair market competition by imposing minimum price and wage standards. To accomplish these objectives, codes were drafted which would permit self-regulation of given industries. Industrial groups were charged with the responsibility for determining the length of the work week, rates of pay, production quotas, and prices. For the veneer industry, plagued by excessive competition, the prospects of an association potentially capable of curing their individual "bottom line" problems was a flame in the dark which these free-enterprise capitalists could not resist.

On June 8, 1933, a select group of face-veneer manufacturers met in Indianapolis to organize the Veneer Association (Fig. 11-1). Preliminary meetings of essentially the same group of veneer manufacturers had taken place in Indianapolis in 1931 and 1932. Consequently, the Veneer Association was, in reality, an outgrowth of these earlier meetings and was an organization poised to take immediate action under the auspices of the NIRA. Those at the June 8 meeting recognized, however, that the association would be truly national only if its membership were broadened. It would need to go beyond the sliced, face-veneer firms and include companies producing rotary-cut furniture and panel grades of veneer as well as producers of commercial grades of hardwood veneer. Original membership totaled 39 firms. The association quickly established its headquarters at 616 South Michigan Avenue in Chicago and a blue-ribbon panel of officers was elected to lead the organization:

President: J. C. Rodahaffer—Penrod, Jurden & Clark

Vice President: Harry Hartke—Acme Veneer & Lumber Company

Secretary/Treasurer: Burdett Green

By July, the Veneer Association had made substantial progress in formulating its recommendations. It had a fully functioning committee which was collecting and tabulating statistics on hours and wages, labor conditions, and production information. The organization was fully legitimized by the Emergency National Committee of the National Lumber Manufacturers Association on July 6 when it announced that the Veneer Association had been "...officially delegated as representative of the

Figure 11-1
J. C. Rodahaffer, elected in 1933, was the first president of the Veneer Association. The organization eventually evolved into the Fine Hardwood Veneer Association in 1986.

veneer industry...and provision was made for adequate representation through the Veneer Association before the Emergency National Committee on all matters pertaining to Veneers" (Anon., 1933).

Many executive committee meetings were consumed with a myriad of details, including proposals which needed to be acted upon by the membership at its second meeting in Chicago, on November 15, 1933. The Veneer Association had been incorporated as a Delaware corporation two weeks earlier; therefore, one of the items of business was the affirmation of members switching allegiance from the Veneer Association to the Veneer Association, Inc. The rest of the all-day meeting was devoted to the more serious matters of quotas, violations of rules, and the various provisions of the production code. The meeting also authorized the board of directors to implement a working agreement with the Mahogany Association on matters of joint interest. Five weeks earlier, on September 8, a close relationship between the Veneer Association and the American Walnut Manufacturers Association had been established. That relationship eventually led to merger of the two associations.

The first official meeting of the Veneer Association, Inc., was held in Chicago on February 21, 1934. Burdett Green reported that, in its nine months of existence, the Veneer Association had become "...a comparatively well organized industry...from what were formerly extremely individualistic (or groups of) individualistic manufacturers...This is true even though, in respect [to] certain activities, especially production control, greater misunderstanding and confusion exist today than at any time during the first eight or nine months."

Minutes of later executive committee meetings, preserved in files of the Fine Hardwood Veneer Association (FHVA) are filled with information about cases which were acted upon by a single individual, the chairman, who functioned as a veneer industry czar. Difficulties of making quota decisions fairly were clearly apparent. The matter was partially resolved, quantitatively, by determining the length of knife edge on slicers and rotary lathes (plus diameter of veneer saws, if present) and then multiplying that total by a constant. This mathematical product became the square footage of permissible veneer production for a given veneer mill. Certain other matters were clearly defined. For example, "face-veneer" was defined as:

> All rotary cut, half round, sliced and sawn veneer made from American Walnut, Butternut, Cherry, and Red Cedar logs or flitches, all foreign woods, Birds Eye Maple, Blister Maple, Curly or Blistered Poplar, Holly, Figured Myrtle, and all half round, sawn and sliced veneers of whatever species.

On July 1, 1934, the Commercial Veneer Subdivision was separated from the Veneer Association to form its own NIRA group. This left the hardwood face-veneer "interests" united by a single national association for the first time.

In January 1935, the Veneer Association was notified that its secretary-manager, Burdett Green, could no longer perform NIRA code activities for the association because of his promotional activities with the American Walnut Manufacturers Association. Green's resignation was accepted "...fully mindful and appreciative of his efficient service during the twenty months of the association's existence." L. W. Anderson, an assistant secretary, was promoted to fill the vacancy (Anon., 1935).

In July 1935, the Veneer Association asked to be freed from the onerous task of attempting to control production. Claiming certain difficulties peculiar only to the face-veneer industry, the association affirmed that a balance could be maintained between production and shipments by disseminating relevant control statistics to its members. Shortly thereafter, the U. S. Supreme Court declared the entire National Recovery Act unconstitutional and that resolved the issue of controlling production.

The transition to a voluntary association was welcomed by the membership which had become increasingly disenchanted with the NRA prior to the court decision. At its meeting on May 24, 1935, the Association membership voted to resign as Administrative Agency of the Veneer Division of the Lumber and Timber Products Code. Objectives of the "new," non-NIRA, association were expanded to include "...promotional, statistical, and such other work... as may be deemed advisable...and to include the members of the Commercial Veneer Group." The Association continued to endorse the "...continuance of the present arrangement of Hours and Wages." The impossible task of regulating prices was abandoned. Socialism in the veneer industry had been tested and there emerged few prospects for its eventual resurrection.

Minutes of business and committee meetings for the next 50 years reveal periods of both optimism and despair, depending upon the economic health of the industry and the outside threats of substitutes. Membership numbers ebbed and flowed with the business tides. An unwavering, solid core of firms, some dating back to creation of the Veneer Association in 1933, provided the organizational support required to see the association through the worst of times. The Minutes also provide a perspective on how the association leadership developed strategies to meet the realities of changing market and political forces.

Following the end of the brief NIRA period, the association continued in an attempt to regularize inventories and the terms of sale. "Rules of Fair Trade Practice" was a booklet distributed to member and nonmember firms. It outlined specific contract language and credit terms which would not unduly penalize the veneer producer when dealing with a distant broker or buyer. Market surveys and credit reports furnished by the veneer manufacturers were processed and analyzed by the association, and the resulting statistical reports were returned to members for their edification. The Veneer Association also worked diligently for veneer firms to adopt uniform cost accounting systems. The purpose was to achieve fairer market prices if and when firms were able to know actual production costs of the veneer they produced. Market expansion was enhanced by trade promotion projects, research directed toward new uses for veneer, a film ("Forest Treasure"), and by educational efforts directed toward architects and furniture designers.

Although 1932 is thought of as the year the Great Depression bottomed out, the veneer industry continued to be sorely pressed until 1939. Prewar infusions of public money then began to activate that part of the lagging economy dependent upon hardwood veneers. For example, the value of veneer sales in 1938 was less than one-half what it had been in 1937. Membership remained relatively constant at about 60 firms despite the liquidation of several businesses because of the lingering business depression. The association estimated that member firms continued to cut about 90 percent of the face-veneer cut in the United States. Efforts to increase the number of commercial hardwood veneer manufacturers in the association were not successful because most of their markets were local or "captive" and no membership benefits were apparent to them.

By 1940, the fortunes of the hardwood veneer group decidedly began to improve. Business was better as new markets were discovered. Dues, which were based on veneer sales, increased the financial resources of the association so that new projects could be proposed which would benefit members in the longer term. These proposals took the form of coordinated activities with the American Walnut Manufacturers Association, the Mahogany Association, and the Plywood Manufacturers Institute, and in the form of a "scheme" to buy up surplus and idle veneering machines. The object was to scrap them or resell the better machines to member firms as they expanded to meet the increasing demands for war-related veneer production. With the industry operating at full capacity, more than 385 million square feet of hardwood face-veneer were produced in 1944, the peak year for war-time production.

The problems of reconversion to peace-time production began to draw the attention of the Veneer Association as it became increasingly clear, prior to 1945, that the end of the war was in sight. The association envisioned both domestic and tropical log shortages. It was, for example, anticipated that the British and French would embargo the export of tropical timber from their African colonies in order to use it exclusively in rebuilding their own devastated economies. British quotas and embargoes did occur but were lifted late in 1947. The Veneer Association countered a potential domestic log shortage threat in 1948 by instituting a joint forestry program with the American Walnut Manufacturers Association. W. C. Finley and Burdett Green were made responsible for encouraging private forest owners to plant walnut seedlings, for promoting "good" woodland management, and for improving cutting and utilization practices. Printed materials were prepared and the association's film, "Faces and Figures," was updated to better inform the public of the virtues of veneered furniture and panels. As the decade of the 1940s ended, an association member was quoted as saying, "The plush days are over, temporarily at least."

Chicago was still the hub for hardwood associations in 1950. Of the 12 hardwood trade organizations then existing, six were near each other in Chicago. They were:

Veneer Association
American Walnut Manufacturers Association
Mahogany Association
Hardwood Plywood Institute
National Plywood Distributors Association
National Hardwood Lumber Association

Regional hardwood associations were also located in Cincinnati, Louisville, New York City, Oshkosh, Los Angeles, and Memphis.

The decade of the 1950s was about to bring about substantial changes to the overall hardwood industry. Decorative Masonite, grained metal, and photo-copied plastic laminates such as Formica began to make significant inroads on the traditional markets for veneer and panel products. Veneers and plywood imported from Japan and Southeast Asia (later) wiped out scores of domestic commercial veneer producers and a significant number of domestic face-veneer manufacturers. A substantial portion of the market for veneer virtually collapsed in 1951 as wood substitutes took the place of hardwood veneer in the manufacture of TV cabinets, except for the more expensive units. Furniture manufacturers made similar substitutions using plastic laminates instead of veneer in surface applications.

The Veneer Association fought back in several ways. Research was funded to discover more durable, protective finishes which would do for veneer what the plastic laminates accomplished in resisting defacing marks caused by water, alcohol, and heat. Educational efforts were increased through the use of films and slide/sound programs which were distributed widely. The association also provided money and leadership in establishing a noteworthy hardwood exhibit at the Museum of Science and Industry in Chicago. In a continuing struggle to fight the substitutes, a Deceptive Trade Practices Committee was formed in 1953. Its purpose was to eliminate the misrepresentation of consumer goods which were purported to be genuine hardwood products but were actually made from some other material. Hearings with the Federal Trade Commission and involvement with consumer groups brought the issue to the attention of the general public.

Charles E. Close, who had been the chief executive officer of the Veneer Association since 1936, resigned his position to accept a similar post with the Hardwood Plywood Association. Burdett Green was chosen to succeed Close as a permanent secretary and executive vice president. Near the end of 1953 (October 21), the name of the Veneer Association was changed to the Fine Hardwoods Association to permit a broader inclusion of members. The association's headquarters was also moved to the American Furniture Mart at 666 North Lake Shore Drive in Chicago.

The rest of the decade was troublesome. Members were lost by reason of business failures or when they switched into some other form of business, including the importation of tropical veneers and plywood. Robert P. Thompson, president of the Mahogany Association, proposed in 1956 that his association, particularly hard hit by "...recent serious changes due to business and economic conditions...," merge with the Fine Hardwoods Association to protect the mutual interests of both groups. This useful proposal was eventually turned down by its membership. The Mahogany Association, left to its own devises, "slowly twisted in the wind" before it disappeared from the scene in 1964 with hardly a whimper.

The plight of the veneer and plywood people in the middle of the 1950s is evidenced by increases in the value of plywood imported from Japan. It increased from $7 million in 1953 to $16.5 million in 1954. As W. F. Ashby reported at the 1956 annual meeting of the Fine Hardwoods Association, the flow of imports had neither been staunched by the association's promotional efforts nor by the federal government through its regulatory agencies. He suggested the need for more promotion and stronger linkages to other related associations. More money was needed because 80 percent of the association's budget was now allocated to campaigns against substitutes and foreign imports. The forestry program and other "non-essential" projects were eliminated in this fight for survival.

The next major action taken was the hiring of E. Howard Gatewood in 1957 as a replacement for Burdett Green. Gatewood was familiar with the Washington, D. C., political arena and had a business background with IBM and with other trade associations. He was well aware of dwindling association membership, the declining sales volume of the industry, and the relationship between the two. At the 1958 annual meeting, he quantified the problem by pointing out that face-veneer sales had decreased nine percent between 1951 and 1958, during a period when the GNP had increased 32 percent. At the same time, household furniture sales had increased 24 percent and hardwood plywood consumption had increased 89 percent. The figures for domestic commercial veneer sales had decreased 237 percent.

Among the antidotes Gatewood recommended were the following: trade legislation, product improvement, promotion, better trained sales people, better customer service, and the required labeling of "genuine" hardwood products. Individual firms were challenged to be more responsible for several of these items. He also strongly recommended that veneer manufacturers employ their own salesmen rather than trusting unsupervised broker-sales agents who got their "cut" regardless of the price they negotiated.

It was also noted that a number of large veneer manufacturers were not going to join the Fine Hardwoods Association until the association could show some positive results by stimulating sales. The last major effort of the association in the final year of the 1950s was to seek relief from deceptive marketing practices. Labeling bills were introduced in both houses of the Congress. As Gatewood said, "You can call it anything it is, but you can't call it anything it ain't without the word 'imitation'." In various guises, the labeling bill went through the grist-mills of Congress for eight years without being enacted into law. The retail furniture associations flooded Congress with personal letters and visits to key Congressional leaders to negate the legitimacy of this legislative attempt to protect the unwary buyer. On top of that, the Federal Trade Commission provided Congress with a terse and adverse report that there was "...no consumer demand for either bill."

The decade of the 1960s started with improved situations for both the industry and the Fine Hardwoods Association. Face veneer sales were up 21 percent in the first half of 1960 but they slowed in the second half of the year. They then continued at a good pace until 1968. Membership in the association stood at 35 in 1960 and it slowly tailed off to 29 in 1969. The association had a budget of $130,514 in 1960. It peaked in 1968 at $180,700 before slipping to $163,700 at the end of the decade.

To counter the lack of consumer "demand" for the labeling bill, Gatewood began writing articles targeted for popular magazines and by contacting consumer interest groups. These groups, which had been pressing Congress on "truth-in-advertising" legislation, "discovered" that the "hardwood veneer case" fit their legislative model perfectly.

The following year, a consumer education specialist was hired by Gatewood to work with action groups, widely-circulated magazines, and furniture dealers. In one sense, the program was notably successful. Many potential retail customers became aware of the quality attributes of genuine hardwood products. This knowledge was relayed by furniture salesmen to quality furniture manufacturers. As a result, more than 300 furniture manufacturers voluntarily adopted the "Genuine Hardwoods Seal." This gold seal certified that their products contained only fine hardwoods, either in the form of plywood or lumber, on all exposed surfaces.

Aware of the shift away from fine hardwoods to furniture which incorporated glass, marble, metals, cane, and similar materials, the Fine Hardwoods Association assembled an exhibit at the 1964 World's Fair in New York. The exhibit provided an opportunity for manufacturers of hardwood veneers and furniture to meet the competition with its stylish displays. Educational activities were also expanded with architects and furniture designers.

The mid-years of the decade found the association involved with the American Walnut Manufacturers Association in requesting that the Department of Commerce reduce the export of walnut logs. Excessive exporting of walnut, the argument went, was threatening the supply of walnut timber available for U. S. manufacturers. To strengthen their credibility with the Department of Commerce, the two associations hired Larry R. Frye, a professional forester, on January 1, 1967, to conduct a forestry education program with woodland owners and the general public. The purposes of the "United Hardwood Forestry Program" were to improve the culture of hardwood stands through better management practices and tree planting and to inform the public about the economic and social values associated with a properly managed hardwood resource. Frye, in the next several years, was to make hundreds of mass media television and radio presentations. He also worked with dozens of interested woodland owner groups by providing them with printed materials and films he had prepared for these purposes.

The Fine Hardwoods Association continued to spend about one-half of its budget on promotional activities as it had in the recent past. These activities took the form of publicity releases, exhibits, magazine articles, slide/tape lecture kits, and motion picture productions. In 1968, $94,000 of the $181,000 budget was spent in these ways.

The poor sales year of 1969 carried over into the first years of the 1970s. In fact, the 1969-1970 period was referred to as "these troubled times." Veneer shipments had declined 21.5 percent. E. A. Heidt, a prominent member of the association, reported that since "...50 percent of the hardwood veneer market has disappeared, all other association activities should be terminated except those directly aimed at curtailing imitations." He then called for a reevaluation of the problems facing the industry and a new set of priorities for the association.

Association president, George T. Kinley, suggested that, instead of fighting plastics, the association should limit itself to promoting hardwood strongly and fighting vigorously for the high-value furniture market. Merging with the American Walnut Manufacturers Association was another option which came out of the 1970 annual meeting. By 1971, there was an obviously "desperate need" to keep both the veneer and walnut associations from "going under." Polls of members in both associations provided the thrust for the merger to occur.

On March 9, 1971, the two merger committees met in Chicago to work out the details of the merger. Because "...the Illinois Not-For-Profit Corporation Act was more favorable than comparable legislation in Delaware, it was agreed to merge the Fine Hardwoods Association into the American Walnut Manufacturers Association rather than the other way around." E. Howard Gatewood was assigned to a new sales office to be established in High Point, North Carolina, where he would be responsible for sales promotion and publicity in the East, as well as for government activities in Washington. Donald H. Gott was assigned administrative and statistical activities at the headquarters office, to be located in the Midwest, with additional responsibilities for regional sales promotion and publicity. The Agreement and Plan of Merger were ratified by both boards of directors on March 25, 1971; and the agreed-upon name retained the identities of both associations. Objectives of the Fine Hardwoods/American Walnut Association were stated to be the following:

> To perform all functions of a trade association, to promote the development and expansion of the walnut and other fine hardwood industries, and to promote the welfare of growers, manufacturers, distributors, dealers, users and consumers of the products of such industries.

This relationship lasted until 1985 when the two hardwood groups again sought separate identities. The actual split occurred in 1986 when the American Walnut Manufacturers Association and the Fine Hardwood Veneer Association emerged as separate organizations. As had been the situation with similar joint office arrangements in Chicago, the two associations chose to remain headquartered in Indianapolis and Larry R. Frye continued as the executive head for both associations.

PROFILES OF ASSOCIATION EXECUTIVES

Burdett Green

The Veneer Association, in all its succeeding organizational names, has had excellent executive-level leadership from men truly dedicated to the hardwood industry. Burdett Green was its first executive officer (Fig. 11-2). He was born on October 15, 1898, in Independence, Missouri. His forestry degrees were obtained at the University of Missouri and Yale University, where he was granted a Master's degree from the School of

Figure 11-2
Executives of related fine hardwood associations (left to right): Charles E. Close, Veneer Association (1936-1953); Charles H. White, American Walnut Manufacturers Association (1953-1956); Burdett L. Green, Veneer Association (1933-1935), American Walnut Manufacturers Association (1929-1953), and Fine Hardwoods Association (1953-1957); and George N. Lamb, American Walnut Manufacturers Association (1918-1930) and Mahogany Association (1934-1964).

Forestry. After a short stint at the Forest Products Laboratory in Madison, Wisconsin, he set up the headquarters of the National Lumber Manufacturers Association in 1927. Two years later, he was appointed assistant to George Lamb, secretary-manager of the Mahogany Association. When Lamb retired in 1930, Green was appointed secretary-treasurer and he retained that position for 25 years.

When the Veneer Association was organized in 1933, it was Green who provided the executive-type functions for both associations for a period of 20 months. In 1953, he returned to the Veneer Association to become its executive vice president until 1957. In the intervening years, between 1935 and 1953, the Veneer Association employed L. W. Anderson and Charles E. Close as executives. Anderson served one year and Close for about 17 turbulent, "boom and bust" years. Association minutes reveal that Close was effective in his position. Unfortunately, little was left of his personal "paper trail" in the association's files or in the trade literature. A number of rare stories about Charlie Close, perhaps somewhat apocryphal, still exist in the memories of veneer people who knew him. The anecdotes clearly identify him as a man-of-the-times who enjoyed a lusty good time when not rendering outstanding service to the Veneer Association.

Burdett Green is, perhaps, best known for his efforts to create, finance, and build a permanent hardwood exhibit in Chicago's Museum of Science and Industry. He was involved in the association's 1953 move to the Furniture Mart and the name change to the Fine Hardwoods Association in the same year. Outstanding programs which flourished under his leadership were the development of burn- and stain-resistant finishes for veneered surfaces, the preparation of authoritative hardwood reference sources, and the creation of a library of more than 200 fine hardwoods (Anon., 1957). Green spent his retirement years in California where he died on February 9, 1972.

E. Howard Gatewood

E. Howard Gatewood was born in Virginia in 1914 (Fig. 11-3). He graduated from the University of North Carolina and obtained additional academic training at King's Business College and George Washington University. His early business experience was with IBM, in the Washington, D. C., area where he was employed for 12 years. Gatewood then served as executive secretary of the Wood Office Furniture Institute for eight years and as vice president for Taylor Chair Company for two years after that. At a turning point in his career, in 1937, he joined the staff of the Fine Hardwoods Association as executive vice president. There, he was strongly identified with publicity efforts, educational meetings for members, and the classic traveling exhibit assembled from the association's New York World's Fair display. Gatewood was an aggressive lobbyist in Washington. He testified before Congressional committees on walnut exports and on behalf of the Wood Labeling Act which he wrote. Gatewood followed Burdett Green's lead in developing

Figure 11-3
E. Howard Gatewood provided executive leadership to the Fine Hardwoods Association between 1957 and 1971.

new finishes for hardwood case goods and in producing an award-winning promotional film. In 1968, the National Association of Furniture Manufacturers presented an award to the Fine Hardwoods Association, recognizing it as the organization contributing most significantly to the furniture industry in the previous two years. As secretary of the Hardwoods Exhibit at Chicago's Museum of Science and Industry while he was with the association, Gatewood renovated all displays and secured new sponsors to support the cost of the exhibit. In 1971, he moved on to the larger National Hardwood Lumber Association as its secretary-manager where he remained until he retired to the Outer Banks of North Carolina in 1978 (Anon., 1976). Gatewood died on April 9, 1986.

Donald H. Gott

When the American Walnut Manufacturers Association merged with the Fine Hardwoods Association in 1971, Donald H. Gott became executive director of the merged association (Fig. 11-4). Long associated with the American Walnut Manufacturers Association, he provided a continuum of leadership during some difficult years. Gott was born in Brooklyn, New York, in 1913 and grew up in Scarsdale. In a published profile, he was described as having "...a resonant voice, easy manner, a winning smile...[and] a sense of humor" which remained a constant throughout his life (Anon., 1971). He was also characterized as "...a brash sort of a fellow [and] hard on the outside but a softie inside...[and] totally familiar with the hardwood story" (Anon., 1983). This all added up to a hard-driving inner force and provided Gott with an unusual capacity for sizing up problems and taking appropriate action.

Figure 11-4
Donald H. Gott had a long and distinguished career in the hardwood industry, serving in executive positions with several associations. He was especially active in his work to promote the culture of black walnut and to stem the excessive flow of walnut out of the United States.

During his early years of involvement with the hardwood industry, Gott was general sales manager of Roddis Plywood Division of Weyerhaeuser, a major producer of hardwood veneer and plywood. In this capacity, he also served as president of the Hardwood Plywood Institute which eventually evolved into the present Hardwood Plywood Manufacturers Association.

Gott joined the American Walnut Manufacturers Association in 1961 as secretary-manager and was later named executive director of the same organization. He was co-founder in 1963 of the Hardwood Action Council and was its first president. This organization became the Hardwood Institute which represented the interests and provided a united front for the entire range of hardwood industries.

Gott is, perhaps, best remembered for his work to promote the culture of black walnut trees and his efforts to obtain relief from the excessive flow of walnut logs to foreign ports. He reactivated the forestry education program in 1967 and helped organize the first Walnut Workshop in 1966. He lobbied successfully in Washington to create the Walnut Research Center at Southern Illinois University. Cooperative studies involving the U. S. Forest Service, industry, landowners, and academic institutions followed. These efforts were designed to increase the volume and quality of walnut through a better understanding of walnut ecology and appropriate cultural techniques. Larry R. Frye, current executive director of the Fine Hardwood Veneer Association, believes that Gott's proudest moments were associated with his work to bring about export controls of walnut logs. Although these were strenuous and tenacious efforts which culminated in a nine-month embargo of walnut logs in 1962, the countermanding economic and political Goliaths eventually proved to be too strong for Gott and his hardwood allies.

Ill health eventually forced Don Gott to resign from the Fine Hardwoods/American Walnut Association on September 1, 1978. He retired to Kitty Hawk, North Carolina, where he lived until October 6, 1982, as a much-respected and trusted man who had contributed greatly to the hardwood industry.

Larry R. Frye

Larry Frye was the hand-picked successor to Don Gott (Fig. 11-5). In Gott's letter to A. B. (Dude) Cox, president of the Fine Hardwoods/American Walnut Association, in which he announced his decision to retire, he wrote: "I do not have the time to extol his other virtues after eleven years of devoted and exemplary service to us...in forestry, promotion, publishing, radio and television...you name it. If Larry seems not to measure up to your desires as a successor, I can suggest other names who are qualified, but not of his caliber, ability, know-how, loyalty and experience."

Frye was born on August 15, 1938, in Logansport, Indiana. A Purdue University forestry graduate, Frye worked as a State of Indiana Service

Figure 11-5
Larry R. Frye's professional experience as a forestry school graduate, timber buyer, and extension specialist for Purdue University provided an excellent training background for a veneer trade association executive. Hired by the Fine Hardwoods/American Walnut Association in 1967 to develop a United Hardwood Forestry Program, he became executive director of the association in 1978.

Forester, as a timber buyer for Pike Lumber Company, and as a Purdue University Extension Forester prior to accepting employment as director of forestry programs for the United Hardwood Forestry Program in 1967. Between then and late 1978, Frye traveled in 20 states promoting hardwoods and hardwood management. He directed much of his attention to radio and television appearances and the creation of booklets, slide sets, newsletters, and video segments to reach a broad-based audience. During this time, he also managed the Hardwood Institute, the trade promotion group for the hardwood industry.

On September 1, 1978, Larry Frye took on the responsibilities of executive director of the Fine Hardwoods/American Walnut Association. Between 1978 and 1983, he was assisted by James H. Lee who created publications and conducted meetings of interest to woodland owners and the industry.

It was not an easy task for Frye to take over responsibilities for an association that had become severely weakened. Membership was perilously small and internal stresses became increasingly apparent as foreign veneer mills and conglomerates became dominant producers of veneer in the United States. This had led to conflicts among members. One can easily speculate that the splitting apart of the Fine Hardwoods and the American Walnut Association, in 1986, was a manifestation of an "us-and-them" conflict. Somehow, through it all, Frye and both associations survived the turmoil. He has remained executive director of both groups, operating from his Indianapolis office.

MAHOGANY ASSOCIATION

In January 1918, the federal government called the "mahogany people" together and told them to organize "...and get busy to furnish the government with over one hundred million feet of mahogany annually." A reporter for the Hardwood Record publication noted that:

> It is very pleasing to note the good fellowship existing between these men who never worked together before, but seemed to think the other fellows wore horns. However, now there is a happy feeling and the best kind of cooperation can be anticipated not only in government needs, but with the consuming trade in mahogany (Anon., 1918).

The newly formed organization was composed only of domestic mahogany importers, manufacturers, and dealers. They were charged with the responsibility of working together cooperatively to expedite the delivery of mahogany products for airplane and ship construction purposes. The only constraint placed upon them was that the government insisted that all mahogany logs should come from Central America in order to reduce transportation costs during this wartime period.

Membership consisted primarily of the elite Atlantic and Gulf Coastal mahogany mills. The Midwest was represented by the C. C. Mengel & Bros. Company in Louisville and, probably, the Willey veneer mill in Chicago. Importers and dealers were New York-based. The list of first elected officers clearly indicates that the association was dominated by East Coast manufacturers:

Thomas Williams—Ichabod T. Williams & Sons, New York
R. S. Huddleston—Huddleston-Marsh Mahogany Co., New York
Charles H. Thompson—Lewis Thompson & Co., Philadelphia

The committee formed to cooperate with the government's request consisted of Charles H. Thompson; Alex Williams, of Long Island's Astoria Veneer Mills & Dock Company; Frank Otis, representing the Otis Manufacturing Company of New Orleans; J. C. Wickliff, of C. C. Mengel & Bros. Company of Louisville; and F. C. Leary, of I. T. Williams & Sons.

Incorporated in 1922, the Mahogany Association supported a major advertising and trade promotion campaign for the next three years. At that point, it became essentially dormant, being reduced to the bare essentials of corporate existence. By 1934, the mahogany industry had reached a 50-year low, having been battered by the recession following World War I, the Depression of the 1932 era, and substitutes which misused the mahogany name.

The Mahogany Association began to revive when it hired George N. Lamb in 1934 as secretary to formulate and direct a trade promotion campaign. His selection was due in part to his experience in conducting a similar and successful campaign in the 1920s when he was secretary-manager of the American Walnut Manufacturers Association. The Mahogany Association's trade extension committee met on July 17, 1934, and justified its actions on the basis of the "...increased interest in mahogany" and concluded that "...the industry felt that the time was opportune for an intensive trade extension campaign" (Anon., 1934). Recognizing the urgent need for revitalization, association members agreed reluctantly to move the organization's headquarters from New York City to East Wacker Drive in Chicago. Lamb had insisted on that as a condition of employment.

The obstacles encountered in reviving the association and the market for mahogany veneer and lumber were numerous and difficult to overcome. Mahogany was out of style and other hardwood species were well entrenched. Traditional mahogany finishes were so dark that the aesthetic qualities of the wood were obscured. The tactics Lamb used in pursuing his goal of expanding the use of mahogany were direct and, probably, ingenious. He set out to promote the revival of classical 18th Century furniture designs by working with trend-setters, cultivating latent tastes and demands of the consuming public, and by working with furniture designers from the major quality furniture factories. He also encouraged the development of new finishes and new finishing techniques.

After World War II, competition with other wood species was a continuing challenge and threat because many were less expensive and could be finished to look like mahogany. This dilemma was solved, in part, by the issuance of official Mahogany Association labels and tags utilized extensively by quality manufacturers. As Lamb wrote in an unpublished document: "From the low ebb of the depression years, imports rose to over ten-fold in 1948. The curve has been steadily upward except for import restrictions due to the war or unfavorable logging conditions in the tropics." Membership in the association at this time consisted of eight mahogany veneer or lumber mills and five wholesalers or other market intermediaries.

The 30th annual meeting was held in Chicago in 1952. Essentially the same East and Gulf Coast firms were in control of the organization: Ichabod T. Williams & Sons of New York, Freiberg Mahogany Company of New Orleans, and the Thompson Mahogany Company of Philadelphia. The officers and directors of these firms approved the largest budget for promotion ever passed by the association. Their special target was the architectural field where the demand had been increasing for structures and room paneling. New mahogany finishes were developed for use on traditional and contemporary designs. There was general consensus that sales conditions for mahogany had been particularly good in 1950 and 1951, but the prospects for 1952 were probably not as good (Anon., 1952).

Recognition that mahogany was a satisfactory material for contemporary designs was a clear indication that 18th Century furniture designs were in trouble. It was necessary for them to compete "head-to-head" with the other fine hardwood species for the consumer's preference. Another potential trouble spot was the TV cabinet business. In 1950, it was speculated that the TV industry would produce six million sets, with as many as 85 to 90 percent of them being installed in a mahogany cabinet. This situation was bound to change as less expensive, mass-marketed models became available at discounted prices.

The mahogany industry suffered through 1954, with pessimists predicting that there would be a further "slide-off in business." Lamb presented an optimistic viewpoint, writing that "...we are brazenly predicting at least six months of good business for mahogany in 1955 [while admitting] this prediction is not altogether based on the past record" (Lamb, 1955). Association membership stood at 14 firms, only five of which were mahogany manufacturers (three coastal mills and two in the Midwest). By 1962, membership had dwindled to 10 firms and included only four mahogany producers along with six importers and wholesalers. The final death throes of the Mahogany Association took place early in February of 1964 when the office was disbanded and the assets were dispersed.

MAHOGANY ASSOCIATION LEADERSHIP

George N. Lamb

A legitimate question arises as to the nature of the man who led the Mahogany Association as his single business passion for nearly three decades. Lamb was born in Stanton, Nebraska, on August 12, 1887, where his father was a high school superintendent. He was an honor student at the University of Nebraska and earned degrees in both forestry and botany. Upon graduation, Lamb became a Forest Examiner for the U. S. Forest Service and conducted technical tree and wood utilization studies throughout the United States. A Hardwood Record publication testimonial sums up his contributions to the hardwood industry:

> When the United States entered the war [World War I] the industry needed a direct representative in Washington, and George Lamb was engaged. He soon demonstrated his ability to aid the government as well as the walnut manufacturers. The smooth and satisfactory handling of the government's walnut needs is still a matter of record at Washington and a monument to George Lamb's genius.
>
> The real test came at the end of the war when large supplies of walnut were left in the hands of manufacturers with an uncertain consuming demand. Mr. Lamb demonstrated his ability to create new markets in such an efficient manner that for the past ten years walnut has been the leading cabinet wood in America, and is used in quantity for many other special purposes as well. His knowledge of the industry was intimate. He has never exposed a confidence. All association members have been treated alike. His code of ethics has created a unanimous, consistent confidence in his leadership which, combined with his knowledge of the business and his untiring energy, has sustained one of the most successful of trade associations (Anon., 1930).

Between 1918 and 1930, Lamb was secretary-manager of the American Walnut Manufacturers Association as well as managing director of the Walnut Export Sales Company (WESCO). WESCO consisted of a group of (mainly) Indiana manufacturers who organized in 1919, under the Webb-Pomerene law for export/import purposes, in order to protect themselves from anti-trust prosecution. Lamb resigned from these positions to become secretary-treasurer and sales manager of the Flexwood Company between 1930 and 1934. It was at this point that he brought his noteworthy business and technical experience to the struggling Mahogany Association.

His extensive writings appeared as government publications, popular articles, book chapters, and at least seven editions of *The Mahogany Book*, a small volume of less than 100 pages. The book, first published in 1935, contained a succinct history of furniture styles, technical and illustrative information on mahogany "figure" types, and other reference material. It was refined and revised with each succeeding edition. Current publications of the Fine Hardwood Veneer Association are patterned after Lamb's mahogany books. They include: *Fine Hardwoods Selectorama*, written during Green's tenure and revised by John Vollette, Richard Stem, and Larry R. Frye; and *Hardwoods of the USA*, compiled by Frye. This latter publication had been translated by 1989 into seven foreign languages for world-wide distribution. Among his other accomplishments, Lamb produced three major and award-winning mahogany films, collected nearly 2,000 wood specimens of domestic and foreign species, and assembled an extensive library of books relating to mahogany and the fine hardwoods. Above all, he was the model of ethical conduct and had the essential skills required by association executives to manage and promote the fine hardwoods effectively.

Chapter 12

FUTURE EXPECTATIONS

One of the several reasons for studying and attempting to understand the past is that history has a bearing on the living present and a potential for revealing a glimpse of the future. Saint Augustine expressed this thought somewhat differently when he wrote: "Time is a three-fold present: The present as we experience it, the past as a pleasant memory, and the future as a present expectation..." Certainly, the folly of slavishly following trend lines and computer-generated projections to distant points of time is well documented. Nevertheless, the history of the hardwood face-veneer industry does present a clear picture of an industrial life cycle which has frequently occurred.

Typically, companies have been created by an entrepreneurial burst of energy and inspiration. Following the physical beginnings of these companies, a period of rapid growth ensues for firms that initially survive. Diversification and plant expansion may eventually follow. The businesses mature with their owners. Succeeding generations of owners and partners may begin to take on important management responsibilities during this process. The declining years are marked with a reluctance to invest in new tools and technology and a disdain for seeking out new and evolving domestic and foreign markets. The "status quo" becomes and remains the company goal. Weakened as they increasingly fail to compete effectively with more aggressive members of the industry, companies die with the owners or are eliminated by one of the periodic traumas which beset the industry. This cycle is of indeterminate length and may extend for more than 100 years depending upon the quality of interim management, economic conditions, and other factors.

The sequence is similar to the familiar biological cycle of germination or birth, early rapid growth, maturity, senescence, and death. However, conditional exceptions to this fatalistic viewpoint are evident in, at least, the business world. For example, there is strong evidence in the hardwood face-veneer industry that decadent firms can be revitalized through change of ownership or management, new capital investments, or a refocus on products and services offered to customers. One must, however, conclude that the industrial life-death cycle will remain a potential eventuality and a continuing challenge to entrepreneurs faced with changing production and market environments.

Another, more academic, way to look at the future of an industry is through its structure. Structure refers to the number and size of the firms, their location, ownership, and market practices. Changes in the structure provide perspectives of the industry's vitality, competitive status, and market performance. Periodic "snapshots" taken of the industry permit one to chart the progression of an industry's growth or decline and to provide evidence of any "discontinuity" which may have impinged on the industry during a given time interval.

As determined in this study, the structure of the hardwood face-veneer industry has been greatly modified and reindustrialized by foreign investors competing in domestic and foreign veneer markets. Their state-of-the-art equipment (either new or modernized), in conjunction with well-trained management and aggressive marketing behavior, has influenced new cadres of domestic producers to adopt similar management characteristics. Together, they form a new wave of veneer manufacturers. The shock wave of this discontinuity began between 1960 and 1965 when three mills owned by foreign nationals were established in the United States. That wave has continued to build in strength as the number of mills foreign-owned or partially financed by foreign investors has doubled every 10 years since that time. The estimated production potential of these mills increased more than 10-fold between 1965 and 1990 (Tables 12-1 and 12-2).

Table 12-1
Numbers of Hardwood Face-Veneer Firms in the United States with Foreign Ownership or Financing, Numbers of Veneer-Producing Machines, and Estimated Production Potentials, 1960-1990

Year	Number of Firms	Equipment		Estimated Production Potential
		Slicers	Staylog lathes	— million sq ft —
1960	0	0	0	0
1965	3	6	1	200
1970	3	7	1	280
1975	6	17	4	680
1980	11	36	10	1,650
1985	13	44	11	1,975
1990	14	48	12	2,260

Sources:
Industry interviews and records of the Capital Machine Company, Indianapolis.

Table 12-2
U. S. Hardwood Face-Veneer Mills Owned Entirely or Partly by Foreign Nationals, 1990

Nationalities of Owners	Number of Mills Owned
German	8
Canadian	2
English	1
Italian/American	2
Asian	1
Total	14

Since the die-off of domestic mills in the Depression era of the 1930s, the whole industry has been remarkably stable as to the number of firms. As companies disappear, they are replaced either by new ownerships or new high-tech mills. The trend is toward high-capacity mills and this is evidenced by the nearly 60-percent increase, from 60 to 95, in the number of slicers in domestic mills. During the 38 years between 1950 and 1988, no net increase occurred in the number of firms within the industry. Staylog lathes remained at an average of one per firm. Total production potential, however, increased about 250 percent in this same period. All regions within the eastern one-half of the United States have shared in this increase since 1930. Companies in Indiana and other midwestern states remain the predominant leaders in terms of number of mills and productive capacity. They are vulnerable, however, to competitive dangers as their older plants are pitted against the newer mills being built in the East and, in particular, the South (Table 12-3).

So, what can we expect in the next decade or two, barring another discontinuity? The whole industry can be expected to be "leaner" with respect to labor inputs and more cost-conscious by utilizing appropriate technology as it becomes available. World markets and forward vertical integration will take on greater importance in view of the potential excess capacity for producing veneer in the United States.

The trend toward concentration of production in fewer firms can be expected to continue as some companies increase their market shares (Table 12-4). However, the extent of concentration is and has been relatively limited in comparison with that in other domestic industries. This provides a theoretical basis for certain statements made by face-veneer manufacturers. They have said that the industry has been traditionally plagued by excessive competition. In their view, this has put them at a distinct pricing disadvantage when dealing with furniture firms and other more concentrated industries. The importance of foreign firms in the

Table 12-3
Number of Firms, Equipment Owned, and Estimated Hardwood Face-Veneer Production in the United States, 1930-1988

Year	Location[a]	Firms	Veneer Machines Owned			Estimated Production Potential	Average Annual Production per Veneer Machine
			Slicers	Staylog lathes	Total		
						— million square feet —	
1930	Indiana	17	23	17	40	500	
	Other Midwestern	19	17	24	41	475	
	East	5	11	9	20	255	
	South	7	7	8	15	200	
	Totals	48	58	58	116	1,430	12.3
1950	Indiana	16	25	16	41	710	
	Other Midwestern	10	16	12	28	455	
	East	8	14	7	21	470	
	South	4	5	3	8	180	
	Totals	38	60	38	98	1,815	18.5
1970	Indiana	15	27	15	42	1,090	
	Other Midwestern	10	19	11	30	600	
	East	6	11	8	19	455	
	South	6	12	5	17	490	
	Totals	37	69	39	108	2,635	24.4
1988	Indiana	14	30	14	44	1,415	
	Other Midwestern	11	28	10	38	1,225	
	East	4	8	7	15	530	
	South	8	29	6	35	1,280	
	Totals	37	95	37	132	4,450	33.7

[a]Geographical definitions: Other Midwestern—includes Kentucky, Ohio, Missouri, Iowa, Wisconsin, and Michigan; East—states east of Ohio and north of Virginia; South—Virginia and Tennessee southward, including Louisiana.

Sources:
Industry interviews and records of the Capital Machine Company, Indianapolis.

concentration of hardwood face-veneer production in the United States cannot be understated. Many of the largest firms in the United States today are foreign-owned (Table 12-5).

Further shake-outs of domestic firms are anticipated as firms with "tired machines and personnel" fall victims to price and other competitive factors. During the past decade, their casualty rate has averaged one firm per year. Some smaller firms will survive because of their particular expertise in producing superior veneer. Others will survive by defending their specialized market niches or by continuing to cultivate personal relationships with secondary manufacturers and architectural clients. It can be conjectured that it will be the non-integrated, mid-sized domestic veneer producer with two to three machines who will be hardest pressed in the future by the arrayed competitive forces when buying timber or selling veneer. This could be particularly true if a domestic producer does not have a global market strategy and if domestic veneer markets become more attractive to foreign-owned mills located in the United States.

The future for the hardwood face-veneer industry does have some downside elements. Future productivity of veneer cutting machines will have less to do with the machines, themselves, and more to do with limiting the intervals of time currently required to install and disengage the flitches from the machines. The machines presently perform at rates approaching a reasonable maximum rate. Direct drying speeds may control slicing rate in the future.

Uncertainties will continue to exist with respect to consumer preferences for particular species of wood or styles of interior furnishings. The threat of additional off-shore furniture production is more imminent and carries uncertain ramifications for the veneer industry based in the United States.

Perhaps most serious is the threat of an uncertain future supply of quality domestic hardwood resources. The drain of veneer timber has increased since the 1960s as domestic use has more than doubled and as European and Asiatic countries have had unchecked access to American

Table 12-4
Estimated Proportion of U. S. Hardwood Face-Veneer Produced by the Industry's Largest Firms, 1930-1988

Year	Share of Total U. S. Hardwood Face-Veneer Produced By			
	4 largest Companies	8 largest Companies	20 largest Companies	Total Industry (37)
		— percent —		
1930	12	23	46	100
1950	20	35	71	100
1970	18	33	69	100
1988	26	42	75	100

Table 12-5
Estimated Proportion of U. S. Hardwood Face-Veneer Produced by Companies with Foreign Financing or Ownership, 1930-1988

Year	Foreign Share of Total U. S. Hardwood Face-Veneer Produced By			
	4 largest Companies	8 largest Companies	20 largest Companies	Total for Foreign Firms
		— percent —		
1930	0	0	0	0
1950	0	0	0	0
1970	56	35	16	11
1988	83	50	62	55

timber for export to their own veneer mills. Remedial action by the federal government appears unlikely in the foreseeable future.

The other basic conflict is between the long-term supply objectives of the veneer industry and the short-term objectives of the private forest landowners who seek to maximize either cash or non-cash benefits. This is further complicated by the role of environmentalists who may continue pursuing legislative solutions to forest management problems associated with some small forest holdings. Their challenge to the timber objective in multi-use management on state and federal forest lands may, likewise, continue to complicate the supply issue.

The fact remains that the successful companies of the future will be those which have managers with a preference for speedy, thoughtful action when problems arise, which stay close to their domestic and world-wide customers, which encourage the productive potential of every employee, which stay with the business the company knows best, which minimize the administrative structure and the barriers of status and hierarchy, and which create an environment in which employees are inspired to defend company values as loyal associates in the business. This is an industry dependent upon the strength of the construction and furniture industries. Those veneer firms—large or small—which possess the traits just listed are those which will prosper during the "booms." And they are the firms that will be best able to survive the "busts" when the "gales of creative destruction," as Joseph Schumpeter described them, sweep away the obsolete and the inefficient.

BIBLIOGRAPHY

Chapter 1

EVOLUTION OF THE VENEERING PROCESS

Anonymous. 1903. The manufacture of fine veneer. American Lumberman, December 19 issue, p. 52.
Anonymous. 1907. A new veneer slicer. Wood Craft, December issue, p. 89.
Anonymous. 1908. Historical: John Strong, pioneer veneer manufacturer. Veneers, July issue, p. 3.
Anonymous. 1919. How the ancients cut veneers. Hardwood Record 47(3):26.
Anonymous. 1924. History and romance of plywood and veneer. Hardwood Record 58(5):36-37, 40.
Anonymous. 1951. 100 years of progress. Southern Lumberman, December 15 issue, p. 326.
Anonymous. 1957. Veneer shavers. Veneers and Plywood 51(1):15.
Anonymous. 1971. Need a hot shave? Plywood and Panel Magazine, May issue, p. 27.
Bardin, P. C. 1922a. Some veneer and panel history. Veneers 16(6):22.
Bardin, P. C. 1922b. Some veneer and panel history. Veneers 16(7):22.
Coe Manufacturing Company. 1977. 125 years young and still growing. (Company publicity publication.) Painesville, Ohio. 8 p.
Esarey, Logan, ed. 1924. *History of Indiana: From its exploration to 1922*. Vol. 4. Dayton Historical Publishing Co., Dayton, Ohio. p. 556.
Hyman, Max R. 1902. *Journal Handbook of Indianapolis*. Indianapolis, Indiana. 380 p.
Miller, Paul, Jr. 1985. Cummings Lumber and Oak Hill Veneer—A winning pair. National Hardwood Magazine, August issue, p. 41.
Taylor, J. Crow. 1917. Historical and status data on veneering. Veneers 11(9).
Wilkinson, Sir John Gardner. 1883. *Ancient Egypt. Vol. II*. p. 199. (See also: Newbury, Percy E. 1900. *The Life of Rekmara*. Constable & Co. London.)

Chapter 2

THE PIONEERING EAST COAST VENEER MILLS: MASSACHUSETTS AND NEW YORK

Anonymous. 1917. The Astoria Veneer Mills and Dock Company. Veneers 11(11):14.
Anonymous. 1920a. Astoria mahogany lumber and veneers. Hardwood Record 49(10):44.

Anonymous. 1920b. New Otis mahogany mill is operating at New Orleans. Hardwood Record 49(10):36.

Anonymous. 1930. Ichabod T. Williams & Sons, 1930. After 88 years: A history of the firm. (Privately published for publicity and sales purposes.) 50 p.

Anonymous. 1933. One hundred years of fine wood production. Hardwood Record 71(2):15-16.

Anonymous. 1935. One hundred and two years of successful business: 1833-1935. Hardwood Record 73(3):10.

Anonymous. 1938. A century of leadership in fine woods: Ichabod T. Williams & Sons. Hardwood Record 76(5). (See also: Booth, Frank L. 1938. A modern veneer manufacturing plant. Veneers 25(12).)

Anonymous. 1939a. First full cargo of African mahogany. Hardwood Record 77(2):25.

Anonymous. 1939b. Genuine mahogany. Hardwood Record 77(7):27.

Anonymous. 1955. Nation's oldest mahogany mill receives largest mahogany shipment in U. S. history. Veneers and Plywood 49(10):13.

Anonymous, c.1956. Untitled typescript on file at Palmer & Parker. Tewksbury, Massachusetts. 2 p.

Anonymous. 1967. Palmer & Parker increase storage and kiln space. The Commercial Bulletin newspaper, (The North of Boston Supplement), Sept. 23, 1967. p. 45, 55.

Brady, Philip. 1956. America's oldest mahogany mill. Worcester Sunday Telegram, Oct. 7, 1956. p. 14-15.

Burrell, Jim. 1979. Historical highlights department. Plywood and Panel 20(2):6, 8.

Lamb, George N. 1948. *The Mahogany Book.* 7th ed. The Mahogany Association, Chicago. p. 34.

Payson, William F. 1926. *Mahogany: Antique and Modern.* E. P. Dutton & Company, New York. p. 111.

Sullivan, James (ed.). 1928. *History of New York State: 1523-1927.* Lewis Historical Publishing Company, Inc., New York and Chicago. 412 p.

Volmer, Jamie Robert. 1982. *Ninety-Nine Year Lease on Cloud Nine: An Oral Biography of James Stone Stenersen.* American Memoirs Publishing, Stuart, Florida. 239 p.

Williams, Ichabod T. 1963. Peruvian mahogany. National Hardwood Magazine 37(8):38-39, 44-45, 47, 51.

Williams, Ichabod T. 1983. A history of Ichabod T. Williams & Sons and affiliates. Williams family archives, Hillsdale, New York. 16-page typescript revised from earlier 1972 draft.

Wisdom, Harold W, and Carmen Del Coro. 1983. Wood use in the American furniture industry. Journal of Forest History 27(3):122-125.

Chapter 3

COASTAL MAHOGANY MILLS: PENNSYLVANIA TO LOUISIANA

Anonymous. 1883. Manufacturing and mercantile resources of Indianapolis, 1883. Part IV in: *Resources and Industries of Indiana.* p. 520.

Anonymous. 1908. Plant of Williamson Veneer Co. Veneers 2(2):3.
Anonymous. 1910. The Freiberg Lumber Company. Hardwood Record 38(4):37.
Anonymous. 1912a. Going abroad to buy logs. Veneers 7(9):19.
Anonymous. 1912b. Penrod Walnut and Veneer Co. Veneers 6(3):16.
Anonymous. 1918a. Modern hardwood operation. Hardwood Record 45(2):31.
Anonymous. 1918b. A southern mahogany mill. Veneers 12(4):16.
Anonymous. 1919. The modern manufacture of walnut. Hardwood Record 47(12):25, 29.
Anonymous. 1920a. New plans for old concern. Hardwood Record 50(1):36, 40.
Anonymous. 1920b. Stories of the hardwoods: Making mahogany lumber. Hardwood Record 48(6):25-28, 30.
Anonymous. 1921. Williamson plant one of the best and largest. Hardwood Record 51(9):36.
Anonymous. 1925a. John N. Penrod dies. Hardwood Record 60(1):36.
Anonymous. 1925b. R. L. Jurden heads Penrod Walnut & Veneer Company. Hardwood Record 60(1):32.
Anonymous. 1928. Three prominent walnut producing firms merged into $5,000,000 corporation. Southern Lumberman, June 9 issue, p. 30.
Anonymous. 1936. Familiar facts, faces and figures. Hardwood Record 74(9):23.
Anonymous. 1938. Williamson veneer lines. Veneers and Plywood 32(3):1.
Anonymous. 1949. Romance in mahogany. Cincinnati Enquirer, April 11 issue, p. 2.
Anonymous. 1954. Veneer mill consumes 6000 tons of mahogany logs. Veneers and Plywood 8(10):31-32.
Anonymous. 1957. Dixie Veneer and Thompson Mahogany share plant. Veneers and Plywood 5(12):46.
Anonymous. 1969. Harry A. Freiberg dies, civic, charitable leader. Cincinnati Enquirer, May 17 issue, p. 25.
Anonymous. 1977. A century of achievement, a public relations release. Mallinson-Denny Group, London.
Anonymous. 1986. Thompson Mahogany Company: Quality imported hardwood supplier. Import/Export Wood Purchasing News, August/Sept. issue, p. 7, 16.
Burrell, Jim. 1964. Williamson red oak: from new state, new city, new plant, new machinery. Plywood Magazine 5(1):20-24.
Green, Joe. 1968. Harry Freiberg has accomplished much in a lifetime. Cincinnati Enquirer, June 30 issue.
Pickrel, Ray E. 1946. Untitled three-page narrative, dated June 1946, on file at Fine Hardwood Veneer Association, Indianapolis.
Ross, L. F. 1972. Jake's Corner. The Plywood and Panel Magazine 13(11):12.
Vollmer, Jamie R. 1982. *Ninety-Nine Year Lease on Cloud Nine: An Oral Biography of James Stone Stenersen.* American Memoirs Publishing, Stuart, Florida. p. 108.
Williamson, D. W. 1908. Is the price of veneer high enough? Wood Craft IX(4):104-105.
Williamson, Dwight W. 1939. Untitled, unpublished typescript memoir in possession of James S. Stenersen, Jr., High Point, North Carolina. 55 p.

Chapter 4
INDIANA'S EARLY VENEER MILLS: 1850–1920

Adams, Frank M. 1963. A history of the woodworking industry in New Albany, Indiana. Typescript on file in the Floyd County Library, New Albany, Indiana. 6 p.

Anonymous. 1903. Indiana's candidate. Hardwood Record, February 21 issue.

Anonymous. 1910. New plant of Central Veneer Co. Veneers 4(6):9.

Anonymous. 1913. A new veneer plant. Veneers 7(12):22-23.

Anonymous. 1915. Men you should know. Veneers 9(10):19.

Anonymous. 1920. A modern sawed-oak veneer mill. Veneers 16(4):17.

Anonymous. 1925a. An exceptionally convenient warehouse. Veneers 19(8):27.

Anonymous. 1925b. Evansville Veneer Company burns. Hardwood Record 58(9):48.

Anonymous. 1929a. Harry E. Daugherty. In: *A Book of Indiana.* The Indiana Biographical Association (Compiled under the direction of the James O. Jones Co.). 694 p.

Anonymous. 1929b. Announcing the formation of the Amos-Thompson Corporation. Hardwood Record 67(12):62-63.

Anonymous. 1930. Fire destroys plant of Floyd County veneer company. Hardwood Record 69(12):42.

Anonymous. 1931. An Indiana hardwood pioneer. Southern Lumberman 144(1817).

Anonymous. 1932. Evansville veneer sale reopened. Hardwood Record 70(9):25.

Anonymous. 1937. Familiar facts, faces and figures. Hardwood Record 75(9):29.

Anonymous. 1950-1960. Shots, Chips, 'N Slices. Bound copies of Amos-Thompson in-house newsletter on file with Bice A. Roth, Edinburgh, Indiana.

Anonymous. 1954. John N. Roberts, a leader. New Albany Tribune, August 27 issue.

Anonymous. 1960. New ownership for Central Veneer. Plywood Magazine, September issue, p. 12.

Anonymous. 1965. Four ways to modernize a face veneer operation. Wood and Wood Products 70(4):52-55.

Anonymous. 1973a. Amos-Thompson head retires after 50 years. Plywood and Panel 14(6):27.

Anonymous. 1973b. Cross goes to Indiana...and Amos-Thompson. Plywood and Panel 14(8):18-19.

Anonymous. 1973c. Jake Ross learned by doing...and helping others learn. Plywood and Panel 14(Oct.):22-23.

Anonymous. 1978. Merger leads to expansion of Amos-Hill's veneer and lumber works. Plywood and Panel 19(9):51-53.

Anonymous. 1981a. The Soundboard, News and Views of Kimball International, Inc. 16(8).

Anonymous. 1981b. Thiesing endures hardships with top quality cutting. Plywood and Panel 22(8):34.

Anonymous. 1982. Buyers adding 40-70 jobs at Edinburgh veneer plant. Indianapolis Star, October 13 issue, p. 17.
Anonymous. 1987. G. R. Wood...producing and exporting quality veneer. Import/Export Wood Purchasing News, August/Sept. issue, p. 4.
Balika, Mary Beth. 1984. Home towners revive the "Veneer Capital." Indianapolis Star, March 25 issue, p. 11, 19.
Burrell, Jim. 1974. New equipment, methods are Evansville Veneer's answer to rising costs. Plywood and Panel 15(7):14-15.
Burrell, Jim. 1979-1980. The Daugherty Manuscript - A Series: Insight into the Search for Wood. Plywood and Panel: Sept., Oct., Nov., Dec. 1979; Feb. 1980.
Burrell, Jim. 1979. Note in "Log Book" column. Plywood and Panel 20(6):8.
Graham, Paul H. 1956. Quality veneers. Veneers and Plywood 50(6):12-13.
Hill, Harry D. 1946. Lathe apparatus for cutting comb grain veneer. U. S. Patent 2,397,110.
Lee, James. 1981. National Veneer: A history of innovation and success. Plywood and Panel 22(3):20-21.
Millett, B. Z. (Zach). 1963. Letter from Millett to Chester B. Stem, dated July 17, 1963. On file at Fine Hardwood Veneer Assn., Indianapolis.
Otto, Sarah B. 1972. *A History of Edinburgh [Indiana]: 1822-1972.* (1977 printing.) Edinburgh Sesquicentennial Commission. 74 p.
Rohr, W. H. 1921. Efficiency retained in expanding. Veneers 15(6):38-39.
Visitor. 1927. Veneer mill under new management. Veneers 21(6):34-35.

Chapter 5

INDIANA VENEER FIRMS ESTABLISHED AFTER 1920

Adams, Frank. 1963(?) undated. A history of the woodworking industry in New Albany, Indiana. Typescript on file in Floyd County Library, New Albany, Indiana. 6 p.
Anonymous. 1908. Veneer plant destroyed by fire. Veneers, Vol. 1, November issue, p. 22.
Anonymous. 1910. An advertisement in Wood Craft Magazine 12(3).
Anonymous. 1919a. The plant of Hoffman Brothers. Veneers 13(2):20.
Anonymous. 1919b. An electrically-driven saw and veneer mill. Veneers 13(11):26.
Anonymous. 1920. Walnut veneer mill changes hands. Veneers 14(3):19.
Anonymous. 1921a. An advertisement in the Hardwood Record 51(12):37.
Anonymous. 1921b. Veneer company plans new plant. Hardwood Record 52(5):44.
Anonymous. 1925a. American Veneer and Walnut Company incorporates. Hardwood Record 58(7):76.
Anonymous. 1925b. Untitled. Hardwood Record 59(5):52.
Anonymous. 1925c. Veneer plant burns. Hardwood Record 59(5):56.
Anonymous. 1930. Fire seriously damages veneer mill. Hardwood Record 68(3):60.

Anonymous. 1937. Lumber firm is ordered sold. Miami (Ohio) Valley (newspaper), April 7 issue, p. 1.

Anonymous. 1944. History: Pierson-Hollowell Company, Inc. Unpublished ms. on file at corporate office of Pierson-Hollowell, Indianapolis. 3 p.

Anonymous. 1950. Hoosierland is veneer "Capital." Indianapolis News, October 24 issue, p. 41.

Anonymous. 1965a. Lawrenceburg plant of Pierson-Hollowell Company, Inc. News release on file in corporate office of Pierson-Hollowell, Indianapolis. 2 p.

Anonymous. 1965b. Veneer firm buys Byron Jackson plant. The Lawrenceburg Press, September 30 issue, p. 1.

Anonymous. 1966a. American Face Veneer Corporation..new Indiana producer. Plywood 6(May):36-38.

Anonymous. 1966b. Untitled. Plywood 6(12):36.

Anonymous. 1972. History of Pierson-Hollowell Co., Inc. Unpublished ms. on file at corporate office of Pierson-Hollowell, Indianapolis. 4 p.

Anonymous. 1974. Indiana veneer maker is NHLA head. Plywood and Panel 15(12):24-25.

Brown, Paul Donald (ed.). 1923. *Indianapolis Men of Affairs*. The American Biographical Society, Indianapolis. p. 597.

Burrell, James F. 1967. Pierson-Hollowell celebrates 50 years by expanding, modernizing. Plywood and Panel 8(1):42-45.

Burrell, James F. 1972. Where they do things nicely. Plywood and Panel 13(5):22-23.

Dunn, Jacob Piatt. 1910. *History of Greater Indianapolis. Vol. II*. The Lewis Publishing Co., Chicago. p. 1002-1003.

Graham, Paul H. 1955. A trip through Pierson-Hollowell. Veneers and Plywood 59(3):13.

Loehr, Rodney C. 1959. Saving the kerf: The introduction of the band saw mill. Agricultural History 23(3).

Miller, Paul, Jr. 1984. Pierson-Hollowell Company moves sales office and showroom...modernizes facilities. National Hardwood 58(9):35.

Pierson-Hollowell Company, Inc. 1946. Advertisement in the Gate City and Constitution Democrat (newspaper). Keokuk, Iowa. May 29 issue, p. 12.

Pollock, D. 1931. Invention and development of the sewing machine. Unpaged historical document in files of Singer Company, Stamford, Connecticut.

R. H. W. 1920. The latest veneer and flitch mill. Veneers 14(12):32.

Roamer. 1922. Some veneer plant innovations. Veneers 16(6):38.

Ross, L.F. 1972. "Jakes's Corner..." column. Plywood and Panel 13(12):14.

Saul, James D. 1973. Innovative veneer mill sets new record. Wood and Wood Products 78(12):36-38.

Stem, Chester B. 1964. Correspondence with Vernon L. Clark, on file in office of Fine Hardwood Veneer Assn., Indianapolis.

Williams, Gene. 1975. Once-homely warehouse contains teak, not eggroll. Louisville Times, May 13 issue.

Williamson, Dwight W. 1939. Untitled, unpublished typescript memoir in possession of James S. Stenersen, Jr., High Point, North Carolina. 55 p.

Chapter 6
MIDWESTERN VENEER MILLS

Adams, Frank M. 1963(?). A history of the woodworking industry in New Albany. Typescript on file in the Floyd County Library, New Albany, Indiana. 6 p.
Anonymous. 1895. Veneers. Hardwood 8(11):453-454.
Anonymous. 1904. Advertisement in The Hardwood Record, August 25 issue, p. 39.
Anonymous. 1905a. President Kline and his veneer mill. Veneers, January issue, p. 14.
Anonymous. 1907b. Willey veneer plant at Memphis. Veneers 1(1):22.
Anonymous. 1908. Veneer mill of the Manley, Thompson & Moffett Company. Wood Craft 9(1):34.
Anonymous. 1909a. C. L. Willey. Hardwood Record 28(4):28.
Anonymous. 1909b. Ohio Veneer Company. Veneers 3(3):17-18.
Anonymous. 1909c. Fred W. Black Lumber Company. Hardwood Record 28(4):28.
Anonymous. 1911a. New plant of Southern Veneer and Manufacturing Co. Veneers 5(12):18.
Anonymous. 1911b. Veneer manufacturers visit Kentucky plant. Veneers 5(3):15.
Anonymous. 1913. The Ohio Veneer Co. Veneers 7(9):19.
Anonymous. 1916a. A model rotary veneer mill. Veneers 10(3):11.
Anonymous. 1916b. An electrically driven plant. Veneers 10(4):18.
Anonymous. 1917a. In three states. Veneers 11(5):10.
Anonymous. 1917b. The walnut mills in Piqua. Veneers 11(12):14.
Anonymous. 1917c. Picture story of a modern veneer plant. Hardwood Record 43(8):26.
Anonymous. 1918b. A modern veneer mill. Veneers 12(11):21.
Anonymous. 1918c. Untitled trade journal story. Hardwood Record 45(5):28.
Anonymous. 1919. A modern veneer mill of the North. Veneers 13(9):20.
Anonymous. 1920a. A mahogany lumber and veneer mill. Veneers 14(2):14.
Anonymous. 1920b. Announcing our new million dollar plant. Veneers 14(11).
Anonymous. 1920c. Clarence H. Mengel, Jr., dies in Africa. Hardwood Record 49(11).
Anonymous. 1920d. Veneer plant makes improvements. Veneers 14(1):24.
Anonymous. 1921a. A bit of veneer history. Veneers 15(10):32.
Anonymous. 1921b. New lumber and veneer concern starts auspiciously. Hardwood Record 52(4):36.
Anonymous. 1921c. One of veneer industry's leaders dies. Hardwood Record 51(12):34.
Anonymous. 1922a. Dean-Spicker Company makes changes. Hardwood Record 52(4):44.
Anonymous. 1922b. President of Birds Eye Veneer Company dies. Hardwood Record 54(4):54.
Anonymous. 1922c. Statement regarding C. L. Willey Company. Hardwood Record 52(8):40.

Anonymous. 1923. Made sales manager of a great veneer company. Hardwood Record 54(6):49.

Anonymous. 1925a. By way of announcing completion of our new mill. Hardwood Record 59(7):51.

Anonymous. 1925b. Charles J. Byrns, the man. Hardwood Record 58(8):35.

Anonymous. 1927. A story of the Dayton Veneer Company. Hardwood Record 62(6):34.

Anonymous. 1928a. A pictorial story of Dixie Veneer Co. Hardwood Record 64(11):29-30.

Anonymous. 1928b. Frank H. Scott dies suddenly. Hardwood Record 66(2):28.

Anonymous. 1928c. Julius T. Spicker dies. Hardwood Record 66(11):42.

Anonymous. 1929a. C. J. Byrnes elected president of Birds Eye Veneer Company. Hardwood Record 66(7):36.

Anonymous. 1929b. Daniel E. Kline, leader in the veneer industry passes. Hardwood Record 66(5):30.

Anonymous. 1929c. Henry Walsh becomes R. S. Bacon executive. Hardwood Record 66(10):41.

Anonymous. 1930a. A new Cincinnati veneer mill. Veneers 24(7):29.

Anonymous. 1930b. An interview with C. J. Byrnes, pioneer in the development of birds eye maple. Hardwood Record 68(9):46.

Anonymous. 1930c. Bird's Eye completing $75,000 expansion program. Hardwood Record 68(1):60.

Anonymous. 1930d. Herbert W. Mengel. Hardwood Record 68(2):74.

Anonymous. 1930e. Wood-Mosaic combines New Albany and Louisville. Hardwood Record 68(8):52.

Anonymous. 1931. Louisville—no longer a sawmill center, but still highly important. Southern Lumberman 144(1817).

Anonymous. 1932. C. J. Byrnes dies. Hardwood Record 70(8):17-18.

Anonymous. 1933. Charles T. Tuxford. Hardwood Record 71(12):20.

Anonymous. 1934a. Mengel mahogany. Veneers and Plywood 28(9):1.

Anonymous. 1934b. W. A. MacLean dies of heart attack. Hardwood Record 72(2):16.

Anonymous. 1935a. Birds Eye ships 60,000 feet for monster liner. Hardwood Record 73(3):17.

Anonymous. 1935b. Thirty-year Club dinner an inspiring event. Hardwood Record 73(10):12.

Anonymous. 1937a. Familiar facts, faces, and figures. Hardwood Record 75(5):21.

Anonymous. 1937b. Two modern plants to serve you. Veneers and Plywood 31(7). (Cover.)

Anonymous. 1941. A large parcel of genuine Honduras mahogany logs. Veneers and Plywood 35(8):21.

Anonymous. 1943a. Aircraft veneer plant plans production increase. The Timberman 44(7).

Anonymous. 1943b. Special supplement: March of the Year. Louisville Courier Journal (January).

Anonymous. 1952. Veneer Association re-elects Mertes. Veneers and Plywood 46(12):43.

Anonymous. 1953. Wood Salutes: T. J. Connelly. Wood and Wood Products 58(8):74.

Anonymous. 1954. Birds Eye production unaffected by March fire. Veneers and Plywood 48(7):47.

Anonymous. 1955. Southern veneer mill covers two acres. Veneers and Plywood 49(1):14.

Anonymous. 1956. Egan dies at Louisville. Veneers and Plywood 50(3):38.

Anonymous. 1957. How to get 20% more salable veneer per log. Wood and Wood Products 62(5):23, 27, 53.

Anonymous. 1958. In-house historical account of the firm (untitled). On file at Wood Mosaic, Louisville, Kentucky. 4 p.

Anonymous. 1959. Foreign and Domestic Veneer Company's Knoxville mill. National Hardwood 59(11):177.

Anonymous. 1965a. Dean Company plans new veneer plant. Plywood and Panel 6(5):34.

Anonymous. 1965b. Veneer output boosted 35% by mill consolidation. Wood and Wood Products 70(2):41.

Anonymous. 1968. How much does veneer mill downtime cost? Wood and Wood Products 73(2):49-50.

Anonymous. 1969a. Robert N. Hartzell, Piqua, dies at age of 72. Plywood and Panel 10(2):38.

Anonymous. 1969b. Weber buys Panawall line from R. S. Bacon. Plywood and Panel 10(5):33.

Anonymous. 1969c. Clipping: Wood-Mosaic Planning Merger with Olinkraft. Louisville Courier-Journal (March).

Anonymous. 1973. The Dean Company board elects new officers. Plywood and Panel 14(9):28.

Anonymous. 1975. Hartzell: A century of progress. Hartzell Industries publication celebrating 100 years of the company (1875-1975). 19 p.

Anonymous. 1976. Wood Mosaic broadens concepts in new marketing strategy. Plywood and Panel 17(8):16.

Anonymous. 1977. Freeman rebuilds and upgrades after November fire. Plywood and Panel 18(9):9-10.

Anonymous. 1984. Wood Mosaic celebrates 100th anniversary. Import/Export Wood Purchasing News, June/July issue, p. 8-9.

Anonymous. 1985. Wood Mosaic ends veneer production. Southern Lumberman, November issue.

Anonymous. Undated. Veneer information. A Hartzell information mimeograph on file in Company office, Piqua, Ohio. 3 p.

Bardin, P. C. 1922a. Some veneer and panel history. Veneers 16(7):22.

Booth, Frank L. 1928. The manufacture of walnut veneers. Veneers 22(3).

Bowman, Denise R. 1982. Risk inherent in hardwood trade. Virginia-Pilot (newspaper), Section E, page E1-5, August 22 issue.

Burrell, J. F. 1962. Wood-Mosaic has strong market with plywood product: flooring. Plywood Magazine, Mid-January.

Burrell, Jim. 1965. Bacon streamlines operations for improved veneer service. Plywood magazine, September issue, p. 32.

Burrell, Jim. 1968. Dean's Dixie mill at Princeton is designed for efficiency. Plywood and Panel 9(9):22-27.

Burrell, Jim. 1972. Arkansas face veneer looks ahead, anticipating a sure future for wood. Plywood and Panel 13(3):16-18.
Burrell, Jim. 1979. Log book... Plywood and Panel 20(3):6.
Donnell, Rich. 1985. Dean Company modernizes veneer complex. Plywood and Panel World, February-March issue, p. 6-8.
Graham, Paul H. 1956. The story of the Birds Eye Veneer Company. Veneers and Plywood 50(8):17.
Hamann, Ding. 1975. "Cutting it fine" column. Plywood and Panel 16(1):8.
H. W. M.. 1927. A pioneer plywood concern. Veneers 21(4):32, 34.
M. F. J. 1922. As it arose from the fire. Veneers 16(11):34.
Moore, Jim. 1975. Foreign and Domestic's Knoxville mill continues to produce fine veneers. Plywood and Panel 16(5):15-16.
Rohr, W. H. 1921. Enlarged veneer and panel mill. Veneers 15(4):34-35.
Stem, Richard K. 1965a. The Stemwinder. Plywood, March issue, p. 14.
Stem, Richard K. 1965b. The Stemwinder. Plywood, May issue, p. 10.

Chapter 7

VENEER FIRMS ON THE FRONTIERS OF THE HARDWOOD HEARTLAND

Anonymous. 1911. Untitled. Hardwood Record 33(2):30.
Anonymous. 1916. A live wire in the veneer business. Hardwood Record 41(2):30.
Anonymous. 1925. Veneer plant increases equipment. Hardwood Record 58(9):40.
Anonymous. 1937. Familiar facts, faces, and figures. Hardwood Record 75(4):21.
Anonymous. 1952. Gogoll and Cummings combine. Veneers and Plywood 46(12):43.
Anonymous. 1958. Veneer of Value: A feature article about the Ashby Veneer and Lumber Company of Jackson, Tennessee. Southern Lumber Journal, January issue, 4 p. (reprint).
Anonymous. 1961. The Lane company...featuring cedar chests from quality hardwoods. National Hardwood, March issue, p. 26-27.
Anonymous. 1965. New hardwood veneer plant geared for 162MM sq. ft. capacity. Wood and Wood Products 70(10):42-43.
Anonymous. 1967. Veneer line modernization pays off with 33% more output. Wood and Wood Products 72(8):46.
Anonymous. 1968. New owners of eastern veneer mill hold open house, expand production. Plywood and Panel 9(5):23-25.
Anonymous. 1973. Nickey-Bellante International, Inc., to import Banak from South America. Plywood and Panel 14(3):23.
Anonymous. 1977. Is this the oldest working machine in the furniture industry? Wood & Wood Products 82(6):51.
Anonymous. c.1985. Walnut Products, Inc., of St. Joseph: The products...the progress...the people. In-house publication of company.
Ashby, William F. 1957. Untitled typescript of Ashby Veneer and Lumber Company, dated December 3, 1957, in files of William F. Ashby, Jackson, Tennessee. 10 p.

Cobb, Charles P. Undated. Untitled oral transcript, in Nickey Bros. files, Memphis, Tennessee. 8 p.

Graham, Paul H. 1956. The story of Pannill Veneer. Veneers and Plywood 50(1):10.

Lane, Edward Hudson. 1962. *Lane: The Company's First 50 Years*. The Newcoman Society in North America, New York. p. 9-10.

Lane, Helen H. 1963. *A History of the Lane Company: The First Fifty Years.* The Stone Printing and Manufacturing Co., Roanoke, Virginia. p. 26, 36.

Lee, Jim. 1981. Cutting fine veneers as a family tradition. Plywood & Panel 22(12):21.

McCarthy, Ethalind. 1953. Sam Nickey [Sr.]: The pace setter. The Southern Lumber Journal, November issue.

Moore, Jim. 1975. Erath Veneer Corporation completes agreement with Brazilian producers. Plywood and Panel 16(1):12-15.

Nickey, Sam M. 1953. Untitled oral transcript, in Nickey Bros. files, Memphis, Tennessee. 3 p.

Quigley. 1915. A new veneer mill. Veneers 9(2):14.

Traveler. 1927. A new Virginia veneer plant. Veneers 21(2):28.

Chapter 8

NORTH AMERICAN VENEER FIRMS WITH FOREIGN INVESTORS OR OWNERS

Anonymous. 1913. Veneer plant burns; loss put at $100,000. The Indianapolis Star, December 23, 1913, p. 1.

Anonymous. 1921. Amos K. Hollowell dies at ripe age. Hardwood Record 51(3):46.

Anonymous. 1922. A rebuilt veneer factory. Veneers 16(7):42.

Anonymous. 1946. Questionnaire completed by Swords-Morton Veneer and Lumber Co., on file in office of Fine Hardwood Veneer Assn., Indianapolis. 2 p.

Anonymous. 1965. Modern mill at Beaufort makes veneer for export. Plywood 6(5):42-43.

Anonymous. 1968. Canadian plant continues growth and production of fine veneers. Plywood and Panel 8(10):44-46.

Anonymous. 1971. Indiana Veneers, Inc., under new ownership; Pinsonneault, Edwards, new corporate head. Plywood and Panel 12(9):18.

Anonymous. 1975. Expansion program follows Webb Company's "deconglomeration." Plywood and Panel 16(5):14-15.

Anonymous. 1976a. General Woods and Veneers develops electronic veneer measurement device. Plywood and Panel 17(1):48.

Anonymous. 1976b. Moehring's big Beaufort face veneer plant completes newest expansion. Plywood and Panel 17(12).

Anonymous. 1977. Atlantic Veneer obtains rare walnut tree. Plywood and Panel 18(4):16.

Anonymous. 1979. A visit to the David R. Webb Company. National Woodlands, May issue, p. 10-12.

Anonymous. 1982a. Indiana Veneers Corporation, 90 Years. Publicity pamphlet, on file in office of Indiana Veneers Corporation, Indianapolis. 4 p.

Anonymous. 1982b. Pleasant Hill Veneer begins work. Pleasant Hill Times (newspaper), September 16 issue.

Anonymous. 1985. International Veneer Company now serving North American markets. Import/Export Wood Purchasing News, Dec. 1984/Jan. 1985 issue, p. 8.

Burrell, Jim. 1969. Montreal producer opens new offices. Plywood and Panel 10(7):24-25.

Burrell, Jim. 1973. Just stick it on and slice it. Plywood and Panel 14(6):14-16.

Dixon, Robert, and Howard Flock. 1980. Iseli now "universal" part of U. S. veneer production. Plywood and Panel 21(8).

Dunn, Jacob Piatt. 1910. In *History of Greater Indianapolis.* The Lewis Publishing Company, Chicago. p. 936-938.

Flock, Howard R. 1979. Flitching is an art (column). Plywood and Panel 20(3).

Gunnason, Jon. 1955. Swords-Morton: modern integrated veneer and lumber plant operation. Veneers and Plywood 49(7):12, 20-21.

Hamann, Ding [Ronald E.]. 1975. Cutting it fine (column). Plywood and Panel 16(12):10.

Hyman, Max R. (ed.).1897. In *Hyman's Handbook of Indianapolis.* M. R. Hyman Publishing Co., Indianapolis. p. 349.

Hyman, Max R. (ed.). 1902. *Journal Handbook of Indianapolis.* The Indianapolis Journal Newspaper Co., Indianapolis. 380 p.

Knowles, Doug. 1984. International Veneer expands to the domestic market. National Hardwood 58(10):55.

Knowles, Doug. 1986. International Veneer Company continues growth in export and domestic markets. Inport/Export Wood Purchasing News, October/November issue, p. 5.

Moore, Jim. 1976. Ultramodern Swiss plant produces wide variety of sliced and rotary veneers. Plywood and Panel 17(1).

Moore, Jim. 1977. Moehring sees Brazil as key to world's future timber supply. Plywood and Panel 18(3).

Otto, Sarah B. 1972. *A History of Edinburg [Indiana]: 1822-1972.* (1977 printing.) Edinburg Sesquicentennial Commission. 74 p.

Perry, Patrick. 1978. German partnership buys Indiana Veneer. Plywood and Panel 19(8):31.

Ross, L. F. 1970. Jake's Corner (column). Plywood and Panel 11(1):15-16.

Chapter 9

FINE HARDWOOD VENEER MARKETS AND MARKET FUNCTIONS

Anonymous. 1908. Historical John Strong, pioneer veneer manufacturer. Veneers, July issue, p. 3.

Anonymous. 1929. What is the matter with veneer sales. Hardwood Record 67(12):16.

Anonymous. 1976. Moehring's big Beaufort face veneer plant completes newest expansion. Plywood and Panel 17(12).

Lydens, Z. Z. 1966. *The Story of Grand Rapids.* Kregel Publications, Grand Rapids, Michigan.
Miller, Paul, Jr. 1984. Pierson-Hollowell Company moves sales office and showroom...modernizes facilities. National Hardwood 58(9):34-37, 49-52.
Montgomery, Vera. 1977. High Point—nation's furniture center. Wood and Wood Products 82(2):24.
T. C. 1927. The veneer dealer. Hardwood Record 63(2):42.
Thompson, Ed. 1988 (?). Number One and still stretching: Atlantic Veneer streaks on. Atlantic Veneer publicity release, on file in company office, Beaufort, North Carolina. 4 p.
Walscheid, Henry P. 1913. The jobber in the veneer business. Veneers 7(1):8.

Chapter 10

THE BATTLE FOR CONTROL OF THE AMERICAN BLACK WALNUT RESOURCE: 1959–1970

American Walnut Manufacturers Association. 1963. Second supplemental statement, on file at Fine Hardwood Veneer Assn., Indianapolis.
American Walnut Manufacturers Association. 1965. Walnut group to continue export fight. News release (October 5), on file in the office of Fine Hardwood Veneer Assn., Indianapolis. 3 p.
American Walnut Manufacturers Association and Fine Hardwoods Association. 1969. Request for imposition of export controls on black walnut logs and veneer. Dated February 26, 1969. Mimeo on file in office of Fine Hardwood Veneer Assn., Indianapolis.
American Walnut Manufacturers Association. 1970. Walnut Association bristles at Department of Commerce refusal on export controls. News release for The Commercial Bulletin's issue of November 7, on file in office of Fine Hardwood Veneer Assn., Indianapolis.
Anonymous. 1965. Commerce Department agrees to review walnut industry's export problems. Panel 6(9):10.
Connor, John T., Secretary. 1965. Letter to Vance Hartke (Feb. 26), on file in office of Fine Hardwood Veneer Assn., Indianapolis.
Day, J. Edward. 1964. Statement presented before George Donat, Administrator, Business and Defense Services Administration, U. S. Department of Commerce, Washington, D. C. (June 9), on file in office of Fine Hardwood Veneer Assn, Indianapolis. 11 p.
Gatewood, E. Howard. 1970. Memo to All Member's Sale Representatives, from executive vice president of Fine Hardwood Assn., March 23. Mimeo on file in office of Fine Hardwood Veneer Assn., Indianapolis.
Freeman, Sam W. 1963. Letter to Commerce Secretary Luther Hodges (April 15), on file at Fine Hardwood Veneer Assn., Indianapolis.
Hartke, Vance. 1965. Letter to Secretary Connor (March 5), on file at Fine Hardwood Veneer Assn., Indianapolis.
Hodges, Luther H. 1963. Letter to Clark McDonald (Oct. 1), on file at Fine Hardwood Veneer Assn., Indianapolis.

Hollowell, Robert E., Jr. 1962. Letter to Donald H. Gott (Dec. 12), on file at Fine Hardwood Veneer Assn., Indianapolis.
Kelleher, Jack. 1979. Importing woods. Wood and Wood Products 84(12):21.
Office of Technological Assessment. 1983. Growth of world demand for forest products expected to give U. S. producers new export opportunities. News release, August 25.
Roth, Bice A. 1964. Letter to Commerce Secretary Luther H. Hodges, on file in office of Fine Hardwood Veneer Assn., Indianapolis.
Southern Furniture Manufacturers Association. 1963. Resolution adopted at special meeting at High Point, North Carolina, on August 14, 1963, on file in the office of Fine Hardwood Veneer Assn., Indianapolis.
U. S. Dept. of Agriculture. 1967. Report of Hardwood Timber Conservation Committee by Departments of Agriculture, Commerce and the Interior. Memo on file in office of Fine Hardwood Veneer Assn., Indianapolis. p. 5-6.
U. S. Dept. of Commerce. 1963. Black walnut log export control proposal. Working paper (apparently), on file in office of Fine Hardwood Veneer Assn., Indianapolis. 53 p.
U. S. Dept. of Commerce. 1964a. Controls over exports of walnut logs, bolts, and hewn timber. Current Export Bulletin No. 888. Washington, D. C. 8 p.
U. S. Dept. of Commerce. 1964b. "Statement on Walnut Conservation Program." Undated speech (probably May 15, 1964) by unidentified speaker. 8 p. On file in the office of Fine Hardwood Veneer Assn., Indianapolis.
U. S. Dept. of Commerce. 1964c. Untitled press release, Bureau of International Commerce, FC 64-61, July 6, 1964, Washington D. C.
U. S. Dept. of Commerce. 1965. Decision of the Secretary of Commerce regarding walnut log export controls, G 65-20, Feb. 12, 1965, Washington, D. C.
U. S. Dept. of Commerce. 1970. News, G70-120, Decision on walnut log exports announced, Washington, D. C.
U. S. Senate. 1965a. Hearing before the Committee on Commerce, Eighty-ninth Congress, on Export Controls on Black Walnut Logs, March 16 and 31, 1965. Serial No. 89-9. U. S. Govt. Printing Office, Washington, D. C.
U. S. Senate. 1965b. Hearing before the U. S. Senate Committee on Banking and Currency on H.R. 7105 and S. 1986, Bills to provide for the continuation of authority for regulation of exports and for other purposes, and to amend Section 3 of the Export Control Act of 1949, June 16, 1965. U. S. Govt. Printing Office, Washington, D. C.
U. S. Senate. 1965c. Hearings before the U. S. Senate Committee on Commerce, on Export Controls on Black Walnut Logs, Part 2, October 15, Washington, D. C.

Chapter 11

THE VENEER TRADE ASSOCIATIONS

Anonymous. 1918. The mahogany operators get together: Close co-operation will follow years of friction. Hardwood Record 44(8):29.

Anonymous. 1922a. Sale talks on value of association services. Hardwood Record 53(5):47.

Anonymous. 1922b. Veneer and plywood merger effected. Hardwood Record 53(3):36.

Anonymous. 1930. A testimonial to Lamb from a "Charter" Member. Hardwood Record 68(10):52.

Anonymous. 1933. Veneer industry organizes. Hardwood Record 71(7):11.

Anonymous. 1934. George Lamb to run mahogany association at Chicago. Hardwood Record 72(8):18.

Anonymous. 1935. Petition to drop veneer production control. Hardwood Record 73(3):7.

Anonymous. 1952. Mahogany Association elects Williams. Veneers and Plywood 46(4):44.

Anonymous. 1957. Wood Salutes: Burdett Green. Wood and Wood Products 62(10):94.

Anonymous. 1971. Wood Salutes: Donald H. Gott. Wood and Wood Products 76(5):58.

Anonymous. 1976. Wood Salutes: E. Howard Gatewood. Wood and Wood Products 81(1):100.

Anonymous. 1983. Personal profile: Donald H. Gott. Wood and Wood Products 88(1):20.

Lamb, George N. 1955. How about mahogany in 1955? The Wood-Worker 73(11):56.

Pendleton, Maurice B. 1973. People–policies–progress; N.H.L.A.'s proud record of achievement, 1898-1972. Unpublished booklet on file with National Hardwood Lumber Assn., Memphis.

PHOTO CREDITS

Photographs which appear in this publication were provided by the individuals, firms, and institutions shown below. For ease of reference, credits are listed according to the pages on which photos appear.

Page	Source of Photo
5	Coe Manufacturing Company
7	Capital Machine Company
9	Capital Machine Company
11	Capital Machine Company
13	Capital Machine Company
16	Coe Manufacturing Company
17	Coe Manufacturing Company
18	Coe Manufacturing Company
26	William Farquhar Payson
27	William Farquhar Payson
30	Ichabod T. Williams
31	Ichabod T. Williams
33	Ichabod T. Williams
34	Ichabod T. Williams
35	Ichabod T. Williams
36	Ichabod T. Williams
37	Ichabod T. Williams
38	Ichabod T. Williams
39	Ichabod T. Williams
40	Ichabod T. Williams
46	Ichabod T. Williams
51	John A. Williamson
65	John S. Stenersen, Sr.
67	Fine Hardwood Veneer Assn.
69	Fine Hardwood Veneer Assn.
70	Fine Hardwood Veneer Assn.
72	Fine Hardwood Veneer Assn.
85	Fine Hardwood Veneer Assn.
86	Fine Hardwood Veneer Assn.
89	Fine Hardwood Veneer Assn.

Page	Source of Photo
91	Percy Simmons
92	Fine Hardwood Veneer Assn.
96	Fine Hardwood Veneer Assn.
99	Fine Hardwood Veneer Assn.
106	Fine Hardwood Veneer Assn.
113	Fine Hardwood Veneer Assn.
115	Fine Hardwood Veneer Assn.
119	Fine Hardwood Veneer Assn.
123	Fine Hardwood Veneer Assn.
125	Fine Hardwood Veneer Assn.
132	The Singer Corporation
135	Fine Hardwood Veneer Assn.
137	DeHavilland Aircraft of Canada
152	Ichabod T. Williams
153	Ross Kreamer
159	Fine Hardwood Veneer Assn.
167	United States Naval Institute
181	Fine Hardwood Veneer Assn.
224	John C. Callahan
225	John C. Callahan
231	Capital Machine Company
239	Fine Hardwood Veneer Assn.
251	Capital Machine Company
256	Fine Hardwood Veneer Assn.
257	Fine Hardwood Veneer Assn.
284	Fine Hardwood Veneer Assn.
292	Fine Hardwood Veneer Assn.
293	Fine Hardwood Veneer Assn.
294	Fine Hardwood Veneer Assn.
296	Fine Hardwood Veneer Assn.

Appendix **A**

ALPHABETICAL LISTING OF INDIVIDUALS INTERVIEWED

These are the people who gave time and support to the project by agreeing to share their knowledge and experience. Our conversations and taped interviews provided the important oral histories which were used extensively in the manuscript. Names of these individuals are listed below in alphabetical order. The organization or business to which they were affiliated is given followed by place and date of the interview.

Frank M. Adams
    Adams Custom Veneers, Inc.; New Albany, Indiana; November 11, 1981.
Robert C. Ashby
    Ashby Veneer and Lumber Company; Jackson, Tennessee; March 31, 1982.
William F. Ashby
    Ashby Veneer and Lumber Company; Jackson, Tennessee; March 31, 1982.
Michael Boel
    Lake Jericho Veneer, Inc.; Lake Jericho, Kentucky; November 10, 1984.
G.Karl Bordner
    Penrod, Jurden & Clark Company; Norfolk,Virginia; June 2,1981.
Harry T. Breece
    Breece Veneer Company; Kenova, West Virginia; May 14, 1982.
William Breece
    Breece Veneer Company; Kenova, West Virginia; May 14, 1982.
James Burrell
    Plywoods & Panel (magazine); Greenwood, Indiana; June 25, 1980.
Kenneth H. Christgen
    Missouri Valley Veneers, Inc.; St. Joseph, Missouri; November 9, 1985.
Ted J. Connelly
    The Dean Company; Princeton, West Virginia; June 11, 1982.
Jack Curry
    B. L. Curry & Sons; New Albany, Indiana; November 16, 1981.
Jeff Curry
    Erath Veneer Corporation; High Point, North Carolina; May 7, 1982.
Mary (Mrs. Carl) Daugherty
    Hoosier Veneer Company, Inc.; Indianapolis, Indiana; October 23, 1982.
Howard C. Davis
    Central Veneer Company; Indianapolis, Indiana; March 10, 1981.
Thomas A. Dean
    The Dean Company; Chicago, Illinois; June 29, 1981.

O. C. (Bo) Edwards
International Veneer Company; Coldwater, Michigan; October 29, 1984.
George S. Erath
Erath Veneer Corporation of Virginia; High Point, North Carolina; May 7, 1982.
Howard Flock
Veneer industry consultant; Louisville, Kentucky; May 15, 1984.
Eugene E. Freeman, Jr.
The Freeman Corporation; Winchester, Kentucky; March 29, 1982.
Harry Freiberg, Jr.
Freiberg Mahogany Company; New Orleans, Louisiana; December 17, 1982.
Larry R. Frye
Fine Hardwood Veneer Association; Indianapolis, Indiana; numerous times during the period 1980-1990.
Robert S. Garrish
Palmer & Parker Company, Inc.; Tewksbury, Massachusetts; October 13, 1983.
Leon Groscott
French Veneer Company; Paris, France; May 30, 1983.
John A. Grunwald
David R. Webb Company; Edinburgh, Indiana; November 5, 1981.
Arnold F. Habig
Kimball International, Inc.; Jasper, Indiana; October 24, 1981.
Tom Hale
Hale Engineering & Sales; New Albany, Indiana; March 28, 1982.
G. William Hartzell
Hartzell Industries; Piqua, Ohio; May 2, 1982.
Edward A. Heidt
Penrod, Jurden & Clark Company; Norfolk, Virginia; June 2, 1981.
Robert E. Hollowell, Jr.
Pierson-Hollowell Company; Indianapolis, Indiana; March 11, 1981 and March 29, 1983.
Edward V. Horton
The Dean Company; Princeton, West Virginia; June 11, 1982.
Charles Hudson
Fairfield Veneer Mill Company; Fairfield, Ohio; April 30, 1982.
Rainer Ingman
Fairfield Veneer Mill Company; Fairfield, Ohio; April 30, 1982.
Carlo Iseli
Universal Veneer Mill, Inc.; Newark, Ohio; May 3, 1982 and March 4, 1985.
Dieter Johnson
Erath Veneer Corporation; Martinsburg, West Virginia; May 4, 1982.
Max Joyce
Palmer & Parker Company, Inc.; Tewksbury, Massachusetts; October 24, 1983.
Fred Kirkfield
Central States Veneers, Inc.; Paducah, Kentucky; April 1, 1982.

L. Jack Koss
Capital Machine Company; Indianapolis, Indiana; numerous times during the period 1980-1989.
Ross Kreamer
Kreamer Veneers, Inc.; Memphis, Tennessee; December 16, 1982 and March 3, 1983.
J. Russell Lee
Birds Eye Veneer Company, Inc.; Escanaba, Michigan; October 25, 1985.
Michael J. Lefever
Erath Veneer Corporation of Virginia; Rocky Mount, Virginia; June 3, 1981.
William C. Linn
Hoosier Veneer Company; Indianapolis, Indiana; October 20, 1982.
Werner Lorenz
Indiana Veneers, Inc.; Indianapolis, Indiana; March 26, 1981.
E. K. Lovelace
Roberts & Strack Veneer Company; Clarksville, Indiana; November 17, 1981.
Smedley W. Lynn
Erath Veneer Corporation; Martinsburg, West Virginia; May 4, 1982.
Robert Mathers
Thiesing Veneer Company; Mooresville, Indiana; March 26, 1981 and March 27, 1984.
Sam Mengel, Jr.
Foreign & Domestic Veneer Company; Louisville, Kentucky; December 8, 1982.
Hermann F. Meyer
Atlantic Veneer Corporation; Beaufort, North Carolina; May 10, 1981.
Russell T. Miller
Curry-Miller Veneers, Inc.; Indianapolis, Indiana; May 6, 1981.
William R. Miller
Curry-Miller Veneers, Inc.; Indianapolis, Indiana; October 31, 1984.
Sam Nickey III
Nickey Brothers, Inc.; Memphis, Tennessee; December 16, 1982.
Roman B. Oser
Kimball International, Inc.; Jasper, Indiana; October 24, 1981.
Bud Price
Indiana Veneers, Inc.; Indianapolis, Indiana; March 26, 1981.
Peter Prizevoits
Amos-Hill Corporation; Edinburgh, Indiana; November 6, 1981.
John R. Roberts
Roberts and Strack Veneer Company; New Albany, Indiana; March 28, 1982 and October 20, 1982.
Bice A. Roth
Amos-Thompson Corporation; Edinburgh, Indiana; October 21, 1982.
Alfred Silvan
Hardwood Centre; Sommers, New York; October 4, 1986.
Percy Simmons
Log and veneer merchant; Indianapolis, Indiana; May 8, 1981, March 27 and April 12, 1984, and May 13, 1986.

Robert Solo
Genwove U. S. Ltd.; Indian Trail, North Carolina; May 6, 1982.
Russell Stadelman, Sr.
Russell Stadelman Company; Memphis, Tennessee; March 4, 1983.
Gus Stetter
Swords Veneer & Lumber Company; Rock Island, Illinois; March 13, 1985.
Richard K. Stem
Chester B. Stem, Inc.; New Albany, Indiana; March 2, 1981.
James S. Stenersen, Jr.
Stenersen Sales Corporation; Cockeysville, Maryland; May 13, 1981; and High Point, North Carolina, May 7, 1982.
James S. Stenersen, Sr.
Stenerson Mahogany Company; Stuart, Florida; January 4, 1982.
Roger W. Stenersen (by phone)
Norfolk Veneer Company; Portsmouth, Virginia; November 7, 1988.
Burton F. Swain III
National Veneer, Inc.; Seymour, Indiana; November 8, 1981.
D. Billings Swain
National Veneer, Inc.; Seymour, Indiana; November 8, 1981.
Carl Wheeler
Plywood and veneer entrepreneur; Savannah, Georgia; January 6, 1983.
Paul Widener
Swords Veneer & Lumber Company; Rock Island, Illinois; March 13, 1985.
George Wilhelm
R. S. Bacon Veneer Company; Hillside, Illinois; May 21, 1982.
Ichabod T. Williams
Ichabod T. Williams & Sons; Hillsdale, New York; October 25, 1983.
John S. Williamson, Jr. (by phone)
Williamson Veneer Company; Delray Beach, Florida; January 15, 1982.

Appendix **B**

THE QUEST FOR MAHOGANY LOGS

Direct importation of mahogany and other tropical timber by U. S. firms began about 1880. Ichabod T. Williams & Sons of New York and George D. Emery Company of Boston were among the first to do so. Their early exploits in West Africa apparently went unrecorded. However, a somewhat later account written in the 1920s probably captures much of the strenuous effort which was required to furnish mahogany logs to American veneer and lumber mills. Based on his own experiences, William F. Payson (1926) wrote:

> Mahogany hunting is almost as much of an adventure as big game hunting. The work of finding and felling the trees in their native jungles and of shipping the logs to the world's markets is still an enterprise scarcely less primitive than in the days when the mariners of Cortez and Raleigh first came upon this sovereign timber three hundred years ago in the wilds of tropical America.
>
> Whether in Mexico, Honduras, Nicaragua, or along the Gold or Ivory Coasts of West Africa, the work of logging still demands individual initiative and hardihood. There is still the difficult penetration of the jungle and navigation of torrential tropical streams; still the need of adroit negotiation with local concessionaires, and on their part the necessity of shrewd barter with landowners, government agents and tribal chiefs and the experienced employment and management of native labour. All these elements in the exportation of mahogany change little from generation to generation.
>
> In Africa the natives still haul the giant logs for miles through the bush, the prevalence of the tsetse fly making it impossible to use cattle. Attempts have been made to haul by tractor but the tangle of the bush is so thick and the terrain in the African mahogany districts so irregular that repair costs have thus far made the expense prohibitive.
>
> In certain parts of Central America the tractor has recently been introduced more successfully for log haulage, but the greater part of this operation is still carried on by cattle as it had been for centuries.
>
> In British Colonies, as for example the African Gold Coast in Ashanti, the traditionally liberal British policy as to native rights and privileges leaves the timber in that territory largely in the hands of the tribal chiefs. Protective ordinances are, however, enforced by the government as, for example, a restriction against the cutting of mahogany trees under nine feet in circumference and other regulations of the sort framed to conserve the timber. On the Ivory Coast the French Colonial government controls the timber-lands, deals directly with concessionaires and regulates all logging operations. The system is detailed and strict. As evidence of French methods of conservation it is interesting to note that every available mahogany tree on the Ivory Coast is numbered by government agents, and that no tree except those so numbered can be cut.

In Central America the different governments likewise control mahogany concessions and restrict the cutting of the timber, some of these having even passed laws requiring that for every mahogany tree felled two or more young mahogany trees or seedlings shall be planted.

One of the main reasons why the colour and romance of mahogany logging still survive wherever it is undertaken and why the adventure is still primitive and frequently even dangerous, lies in the fact that what might be termed the mahogany frontier has steadily receded, ever necessitating a deeper penetration into the bush on the part of the mahogany hunter. As the timber near the coasts and navigable rivers has gradually been cut away new trails have had to be blazed further and further into the tropical forests. The work thus constantly requires pioneer effort.

Many factors...have to be considered by those prospecting for mahogany before they select a tract of land on which to operate. Let us first consider typical methods of procedure in Central America.

After a preliminary survey has been made to determine if there is a sufficient stand of timber to warrant the establishment of a camp, it is necessary to study carefully numerous details of the timber's environment. Does it grow within hauling distance of a navigable river? Does the character of the terrain permit of economical hauling? Are the banks of the river high enough to contain its waters during periods of flood? Is the distance down the river to the sea short enough for the completion of the log drive before the dry season empties the streams? These and questions of a like nature present themselves to the prospector who contracts to supply mahogany at tidewater to European or American importers.

In Central America the contractor takes with him on his prospecting trip three or four "hunters" whose assistance is invaluable to him. Having established a small camp the hunters begin their work of locating and counting the mahogany trees. The task is arduous and taxes the endurance and ingenuity even of these wiry and practiced natives. Cutting their way through the dense tropical tangle of the bush they go in zigzag courses over the tract, counting the mahogany trees and marking each with a machete to prevent a duplication of the count by the other hunters.

Two or three weeks of this work provide an adequate estimate of available mahogany and enable the contractor to decide whether or not he can profitably conduct logging operations in the territory just cruised. If he decides in the affirmative he must next estimate the exact extent of the area to be worked, this decision being necessary because in Central American countries one person or company can secure a concession to operate [5,000 acres] in one year.

These concessions are obtained by [proclamation]. Having satisfied himself that the tract he has investigated will produce sufficient timber to enable him to fulfill his contracts, the contractor returns to the nearest city in which he can draw up [the proper] legal papers.

In the meantime the hunters thoroughly cover the [concession] area. This is done by cutting one or more trails straight through it. Additional trails, cut at right angles from these, block out the tract and systematize the work. There are also trails from tree to tree and others connecting these with the base or right angle trail of each section.

During these trailing operations the contractor sends to the concession ten or fifteen additional men to establish permanent camps for the loggers. Heavy timbers are cut for the framework of the houses and fastened together with vines. The sides of the houses are made of bamboo, the roofs thatched. Each house holds about thirty men. They sleep in hammocks or in beds constructed of bamboo cut in the bush.

After the contractor has taken title to the property, he takes steps to employ the necessary labour to operate the concession. The men are hired for minimum periods of six months and usually receive advance payments before being sent to the logging camp.

Preparatory work, which is done before the main operations are undertaken, includes the planting of corn and clearing of pastures for the cattle on whose fit condition for haulage a large part of the success of the venture depends. The rest of the preliminaries, including the cutting of the main track passes and smaller feeding wings, are completed in time to make everything ready for hauling operations by December or January before the ground is deluged by tropical rains and grows miry and impassable.

The tree is cut. One after another the lords of the forest are brought low, felled with axes swung by hand. As mahogany trees have stanchion-like buttress roots as supporting bases for their great height which sometimes exceeds a hundred feet, the axes attack the trunk ten or fifteen feet above the ground. To accomplish this a scaffold is built around the tree just above the buttress and the axemen or "fallers," taking their places on the scaffold, hew into the trunk at that level.

After the tree is felled it is cut into logs by hand with crosscut saws. There are several ways of hauling these mammoth logs out of the forest. If the land slopes toward the river bank the logs are often rolled at least part of the distance; otherwise oxen or tractors are used. Mahogany is economically hauled with cattle as far as five miles and with tractors and wagon trailers, if traction conditions are fairly favourable, perhaps three times that distance. From four to six yoke of oxen are used for each logging wagon, the wagons having four or eight wheels in accordance with local conditions and loggers' preferences. Two-wheeled "skidders" are also used for short hauls.

The effort involved in this work of hauling through the bush enormous logs each weighing tons and measuring as much as five feet in thickness and from twenty to thirty feet long is prodigious. Men and animals struggling along the track pass, dragging their kill with them—the great carcasses of slain trees their trophies in this strange big game hunt, they present a picture that strikingly suggests the days of the pioneer adventurers. When the logs arrive at the river they are scaled and branded with a number and kept on skids on the banks under a covering of leaves to await the coming of the floods which are to float them to the seaboard.

In Central America the rivers are in flood twice a year, first between June and August, then to a lesser degree during November and December. When the flood comes the drive instantly begins and from that moment all other logging activities are subordinated to the spirited task of floating the logs down to tidewater. Every minute is precious in carrying the drive forward to its goal. Not a moment must be lost in dumping the logs into the rush of the river before it subsides. The drive ends at the mouth of the river where at all seasons there is enough water to keep the logs afloat.

On the arrival of the steamer which usually drops anchor at least a mile off shore these ocean rafts, each as a rule composed of not more than one hundred and twenty-five logs, are towed out over the bar by tugs to the steamer's side. This is often the most perilous part of the entire logging operation.

There the Central American logging adventure ends. The logs are hoisted up the steamer's side and stowed in the hold for transportation to the purchaser's yards in some distant country.

The work of mahogany logging in British and French West Africa so closely resembles the same undertaking in Central America that a separate description of the African enterprise would be superfluous. The Gold Coast and Ivory Coast have policies of control, each differing from the other, and the method of work in both of those districts differs to a certain extent from the Central American method, but this is only because natural and racial conditions are different and the logger adapts himself as best he may.

The African scene with its West Coast fevers and reptiles, its native blacks, half-breeds and white settlers and traders seems more rankly tropical in many of its aspects than similar scenes in Central America. The sight of the shining ebony bodies of fifty or a hundred natives hauling a gigantic squarecut mahogany log for miles through the dense African bush, urged by the rhythm of leaders' chanteys and the strident commands of local loggers, black and white, presents a picture that cannot be duplicated in any other country.

A portion of an unpublished oral history interview conducted in 1961 between Thomas R. Williams and Elwood R. Maunder (1961) provides additional personal insights into the problems and opportunities of direct involvement in the procurement of tropical timber. Williams spoke from his work experiences with the firm of Ichabod T. Williams & Sons.

As a rule, we made it a practice not to engage in a timbering operation but to buy from the local people who got the timber out of the forest. As a young man I spent three months in the mountains back of Santiago, Cuba, where at that time we were getting quite a large amount of beautiful mahogany. I would ride for days back in the mountains selecting trees which were then cut and the logs were to be sent up to this country.

Of course, I didn't select every tree, but I indicated the kind that we wanted so that the people, after I left, would cut the right kind of a tree. Then later on the African mahogany that we used all came through the foreign ports of Liverpool and London where it was sold at public auction. The world would go to these auctions and bid on the timber which had been piled and stored in the lumber yard and lotted and would be valued by the buyers who would then go into the auction and bid and purchase the timber that they wanted. That was a very interesting part of the business, because the bidding would be very great on specially figured logs. I can remember one time when I bid $10,000 for one particular [mahogany] log and finally bought it, and that was the highest priced log they had ever sold in the history of the Liverpool market. I can remember the log very distinctly. It was 24 feet long, and it was 48 inches square. It had been shipped from the port of Sekondi on the coast of Ghana.

The custom of hewing logs square originated from the belief that squared logs gave better stowage in the hold of ships and as a result could move at a lower freight cost and since the cost of the labor to hew the logs square was almost negligible, the bulk of African mahogany logs, shipped to the great timber markets in England, were square logs and it was not until some time later that the custom was abandoned.

In Liverpool there were three auctioneers: Farnsworth and Jardine, Edward Challner and Son, and Alfred Nobel and Company. In London there was Churchill and Sim, Floyd Morgan, and C. Leary and Company. These firms were purely brokers. They sometimes might advance money to their clients who had shipped them logs on consignment to be sold at public auction, the seller realizing that in that way he'd have the world appraisal value for his timber and he'd get the highest possible price he could get. Generally, these brokers would only help their clients by financing them after they had a bill of lading. There were all kinds of log suppliers for these auctions.

Native African entrepreneurs and also British operators would go on down to the coast and organize timbering operations. For most of the very fine figured wood, a great deal of it came from a Negro named J. W. Morrison, who seemed to have the knowledge of what figured logs were. The bulk of the wood that he shipped into Liverpool was always selected figured wood. Then in Nigeria there were firms such as McNeil Scott and Miller Brothers who were producing mahogany and shipping it up to these auctions.

After my first training, say after six years, I went to Europe every year for the next eight or ten years to buy timber at the auctions in England and also in the markets of France and Germany. Later on the development of the African mahogany trade changed, and the timber was shipped direct from Africa to this country. The auctions were seriously injured [after World War I] and the volume went down because all the American demand was supplied from Africa and didn't go through the European or English markets.

There are considerable risks in these foreign operations. For instance, we bought at a public auction from the government of Guatemala 10,000 mahogany trees in 1945. We were given a contract by the government limiting us to five years to take the trees out and pay for them as we took them out. Unfortunately, after we set up our operation with an American logger in charge with a great deal of mechanical equipment—some twelve trucks, twelve or fifteen caterpillar tractors, had built our roads and our camp and had started to cut timber and drag it out, the government of Guatemala closed the border on account of an incident between them and England, and the border remained closed for three to four years. So, it completely disrupted our operation and we were about to get out only a small quantity of the timber that we bought and not even all the trees that we paid for as we had to pay for them in batches as we went along under the contract. We were able to get our equipment out, but the entire transaction was a losing venture and most unfortunate.

The perceptions of an "on-the-ground" expediter of African mahogany logs being exported to the United States were obtained in an interview with Alfred Silvan who retired in 1981 as president of the Hardwood Centre, a division of Balfour, McClain International, Ltd. Silvan was born in Budapest in 1906 where his family had been in the veneer business for several generations working with logs from the famous Slovanian oak forest. After working for the family firm and an English company, Silvan made his way to the United States in 1936 and joined the U. S. Army in 1942. After the war, he went to West Africa where he organized the production and shipment of mahogany logs for Fordom Forest Products, a New York City company.

> I had my headquarters in what was at that time the Gold Coast in Takoradi which was a major port. We controlled quite a bit of the mahogany export from Ghana. Every state there had different rules and regulations. My major business was in Ghana, which was at that time a British colony. They were very careful how they handled the situation there and everything had to be government approved. I had several problems but I got along with them quite well, because I had lived in London for four years, and I got accustomed to their ways. I had the freedom to go any place I wanted, only they warned me to be careful with native suppliers because the suppliers would have preferred that we work through the English companies. The representatives of these London companies lived there for many years, even during the war, and supervised the export to England and France which were the main mahogany users before World War II.
>
> To get the timber out we made a contract with the chief, which included all the labor needed, including the loading of the logs on rail cars and taking them down to the port. These chiefs were pretty astute businessmen. Most of them were very smart people. There were only five or six who were important because they had huge areas of forests. The British could do very little there. There were all kind of regulations of this or that, but the chief was the main man. He picked out the trees, as a matter of fact, often I've been called by the chief, "I have twenty mahogany logs uncut on the ground. Do you want to see them." Well, sometimes you did see them and sometimes you didn't.
>
> The trees were cut with 10- to 12-foot long saws with two men on each side pulling, pushing, pulling, pushing. And that's how they cut—some of those trees were absolutely huge, but they managed. I got four large chain saws over there in 1947 or 1948. They lasted about four months and they were finished. No matter what the supervision, what the story or how you talked to them, they didn't oil them and left them laying outside in the dew. In the morning, they were all rusty. You had to work about two hours before they were back in operation. So it didn't work. I don't know if they ever used chain saws later on.
>
> After the rail cars arrived at the port, the logs were our responsibility. Usually, we had arranged shipment with either Elder Dempster, a British Company, or Barber Lines which was an American company.
>
> You had to be very careful because you had a lot of competition there for cargo space and boats, not only from Takoradi, but from other ports. Other things were exported too from Africa to the U. S. so you had to be friendly and understand shipping company problems and they would help

you out as much as they could. Sometimes we took entire boats and loaded up the boat completely with our merchandise. The Liberty ships were rather small and would only carry about five thousand tons of logs. If we had enough logs for a boat load, I would say: "Now look here, I need the boat on June 1, I'll be ready for it." Bringing all the logs down to Takoradi, jamming up the port completely, was also a problem. Nobody else could come in as the port wasn't so big. After you loaded for about two weeks, the boat sailed and in about three weeks it was in Norfolk.

Norfolk was a very good port for us. First of all, the rail connections were excellent to all of the veneer mills. Unloading was very efficient and quick and railroad cars were always available which was not the case in New York. We had trouble in New York. Sometimes, you had the logs lying around for a week or more in the sunshine and incurring defects and other problems. So, we actually preferred Norfolk. You could put the logs in the water five or six days before the ship was expected. And then you towed the logs out to the ship and loaded them with the ship's equipment. The logs were floated out in rafts of 10 to 12 logs pulled by a motor launch. Each log had a metal spike in it and they were then chained together for the trip of a mile or so to the ship.

When the ship arrived in Norfolk, we had a couple hundred railroad cars waiting along with five or six engines simply standing and waiting. I used to fly over in time for the arrival of the boat and my boss and I would notify all the American veneer mills that we are going to have five thousand tons of mahogany logs and you can buy them by the car load. If you want this car load or that car load, it's up to you. As we unloaded them, the buyers were standing there saying: "I'll take this, I'll take that." Usually, we could have sold more than we had. We never had any problems.

The logs were loaded on the railroad cars very carefully before they were sold. As they came out of the hold we determined on which car they should be put. Maybe, a couple logs not so good go in the same car, but this one good log is what the buyer wanted so he bought two logs that were a little bit less good to get a quality veneer log. We called it "merchandising." A railroad car held 20 to 25 tons and possibly more. Then, we had to arrange for these cars to go to Kentucky, these cars to Ohio, and these to Indiana, and so on to various mills. It was a lot of fun really, supervising and seeing that there wasn't a problem.

In Africa, particularly in Ghana, there were very well trained government officials, who weighed and classified each log before it was shipped. In Norfolk, we sold by the total net weight of a loaded railcar.

Sometimes you did get a few logs that were not mahogany. Some of the native chiefs were ruthless about that and you had to be there for constant supervision before shipment. There may have been a slight difference in weight, but that used to be minimal. Later on we sold by cubic foot contents. However, the chief was paid on the basis of weight. This was the official position. This is not absolutely correct in every instance. Many of the older chiefs couldn't care what the weight of the log was; they said: "You pay me so much per log."

The "modern" steel axe and cross-cut saw were apparently not introduced into many tropical areas until the 19th Century. One observer reported in 1908 that it was not uncommon as late as 1885 to see at New York and other ports large mahogany logs which had been burned-off into logs instead of being sawed..."the charred, ragged ends just in the same condition they left the unskilled hands of those tool-less woodsmen. [About 1905], a lot of logs [were seen] in the yard of Ichabod T. Williams, New York City, that had been gotten out of the woods in this way, no edged tool having been used on them" (Hamilton, 1908).

The American mahogany involvement in the Caribbean basin and the West Indies before the 1880s is sketchy at best. In any event, the Spanish were the first to exploit local mahogany resources soon after Columbus made his initial voyages to the Americas. Sir Walter Raleigh introduced the English to mahogany in 1595 and by 1665 "...considerable quantities of mahogany were [being] exported...during the early period of British occupation of Jamaica..." (McClure, 1985). For these reasons, the availability of West Indian mahogany was becoming scarce when American entrepreneurs became serious about extending their direct log procurement activities into the region. For over 300 years the easily accessible mahogany on the Caribbean islands had been heavily exploited. This was particularly true of the "figured" mahogany trees of Cuba and Santa Domingo. The texture and beauty of the figured wood were particularly prized in furniture manufacture. By 1913, "...the wood from Cuba [was] of comparatively small size, while from Santa Domingo supplies [were] practically unavailable" (Anon., 1913).

Mahoganies shipped from the West Indies tended to be "mostly of a plain character." West African mahogany, in contrast, was often figured and rapidly shifted the focus of American sawmill and veneer men before the turn-of-the-century to the European log auctions and the potentials of West African sources. Tropical American mahogany resources from Central and South America, nevertheless, remained valuable providers of mahogany logs for American mills. U. S. government statistics for 1908 reported that Mexico, Nicaragua, British and Spanish Honduras, Cuba and Santa Domingo furnished the bulk of mahogany used in the United States. According to government statistics for 1908, 65.5 percent of the mahogany imports came from these countries, 18.1 percent of unknown origin through European ports, and the balance (16.4 percent) from Africa and South America. Two-thirds of the volume was "laid down" in Atlantic and one-third in Gulf ports (Gifford, 1909).

In the 1980s, the romance and adventure associated with the quest for mahogany largely faded. What remains today are the instances of American entrepreneurs being hosted to cocktails and dinner by timber export development boards from West African countries in luxurious London hotels. Or an entrepreneur may be granted an audience with the King of Ashanti in Ghana. Johnson (1987) reported that particularly memorable was the "marvelous barbecue...on moonlit Busa beach outside Takoradi." Activities such as these appear to indicate that much continuing effort is going into the commercialization of the timber resource in these countries. Timber in Africa, as elsewhere in the American and Asian tropics, is increasingly being processed at both the primary and secondary manufacturing levels before being exported.

REFERENCES

Anonymous. 1913. Figured mahogany. The Wood Worker 32(7):24.

Gifford, John. 1909. Mahogany and its allies—statistics and character. Wood Craft Journal 11(5):135.

Hamilton, J. V. 1908. Handling mahogany in Africa. Veneers, June issue, p. 7.

Johnson, Paul. 1987. U. S. importers tour Ghana and Ivory Coast. Import/Export Wood Purchasing News, April/May issue, p. 3.

Maunder, Elwood R. 1961. Interview with T. R. Williams. Typescript on file, Forest History Society, Durham, North Carolina.

McClure, Paul L. 1985. Mahogany: The discovery of seafarers that made furniture makers famous. Wood (magazine), No. 5, p. 527.

Payson, William F. 1926. In: *Mahogany: Antique and Modern.* E. P. Dutton, New York. p. 3-12.

Appendix **C**

THE BLACK WALNUT EXPORT CONTROVERSY

1

U. S. Senate Hearing — March 31, 1965 — Washington, D. C.

The debate between Senator Vance Hartke (Indiana) and U. S. Secretary of Commerce John T. Connor over the Secretary's decision not to continue federal controls over the export of American walnut

Senator Hartke. Mr. Secretary, let's cut right to the heart of this thing. Is it not your opinion really, and is not this the basis of your decision, that you just do not believe that the Export Control Act really was intended to apply to this type of situation?

Secretary Connor. Senator, that is not the basis for my decision. The basis for my decision is that as applied in practice this fell more heavily on the export side than on the domestic, and I just do not know how we could have devised a way of having an even distribution of burden under the present legislation.

Senator Hartke. Under the authority of the legislation, very basically, you could have continued this, could you not?

Secretary Connor. We could have continued what seemed to us to be an unfair burden, yes, sir.

Senator Hartke. And it is not merely a matter of military significance but you are under the law required to make a finding about the economic significance which may adversely affect the United States? Is that not true?

Secretary Connor. Well, the legislative authority that applies here is that we are authorized to impose controls "to protect the domestic economy from the excessive drain of scarce materials and to reduce the inflationary impact of abnormal foreign demand."

I think this is a question of interpretation. I think it is a close question. It is not an open-and-shut case either way.

Senator Hartke. There is no real question that this item is in short supply?

Secretary Connor. It is not at the moment, sir. There are plenty of walnut logs and walnut veneer available to meet the demand this year and next year. But there is no question at all that at some time out here in the future, if we continue the present rate of domestic and export consumption, we are going to run out of logs, except for new growth.

Senator Hartke. Do you have an estimate as to when that D-day will be?

Secretary Connor. On the old estimate of the amount of reserves, the demand, making certain assumptions, would use up the supply in approximately 7 years.

Now, on the basis of the new estimate which is just available from the Department of Agriculture, the demand, if it kept up, would use up the supply in 10 years from the commercial growth.

There is some additional noncommercial growth, and by scrounging around this could be stretched out some. But there is a D-day.

Senator Hartke. In other words, there is no substantial disagreement upon this, whether you say 7 or 10 years? I mean there is sufficient indication of the day of reckoning which we are facing, and this is, of course, why there is a great concern.

Secretary Connor. Yes, and I do want to point out, just so it is clear, Mr. Chairman, that after that D-day, then the supply of this veneer quality is 8 to 10 million board feet per year, and that is far less than what we would anticipate the demand to be at that point. So that this at that time unquestionably is a very scarce item.

Senator Hartke. Now, this problem has gradually been coming upon us. I mean it is not necessarily an overnight development but could have been anticipated as we moved along. Is that not true?

Secretary Connor. Well, I do not have enough history on the question, Mr. Chairman, to answer that.

Senator Hartke. Let me quote to you from the Department of Commerce's statements. All right?

Secretary Connor. Fine.

Senator Hartke. In their chart which was prepared by them, from the years 1955 through 1964, in their estimated consumption growth of black walnut veneer quality timber, which is going to be continued in the testimony submitted by the Department of Agriculture and in which they refer to your statistics, there was a growth surplus in 1955 of 12 million board feet. At that time there was only an export of about 1.2 million board feet and domestic use of 7.3 million board feet.

Now, from 1955 on, that growth surplus of 12 million board feet went down progressively till 1959 when the growth surplus for the first time became a growth deficit of 1 million board feet.

That went on up until 1963, the year before the order was placed into effect, at which time that growth deficit had reached 19 million board feet.

However, when we come to 1964, the calendar year, not the order year, we have a deficit of 11 million board feet.

Now, that has to be due to something.

Secretary Connor. Mr. Chairman, we think it was due to the type of controls; yes, sir.

Senator Hartke. It was due to the controls?

Secretary Connor. But it did not get down to zero.

Senator Hartke. Well, I know, but here the year before you had, in 1962, a deficit of 13 million board feet, which increased by 6 million to 19 million board feet in 1964, and then a reduction back to below the 1962 deficit.

In other words, this was, in fact, within 3 million board feet of the 1960 deficit.

And the whole purpose of it—and we grant you, and I think everyone would say, you had no absolute 100 percent conservation measure, but you have placed us in the position where you have said, "Well, it didn't work 100 percent so we are going to scuttle the whole ship."

Secretary Connor. Mr. Chairman, getting back to this inequitable point, as a result of our program exports were cut back 51 percent while domestic consumption dropped 21 percent.

Senator Hartke. Yes, and there was a mistake in the Department of Commerce's own estimate. In other words, part of this—

Secretary Connor. These are on the corrected figures. And, in any event, the revised figures—any sharing of the mistake I think involves a lot of people, not just the Department of Commerce, because this is a very difficult situation to estimate.

Senator Hartke. I understand it is difficult, but somebody had to make that estimate, and the estimate was wrong, and the only agency making that estimate was the Department of Commerce.

Secretary Connor. It was agreed to by the industry, Mr. Chairman.

Senator Hartke. Well, it was agreed to, but the estimate was made by the Department of Commerce.

Secretary Connor. The figures came from the industry.

Senator Hartke. I understand that it is the best they could do, and they also did some other things. But the point about it was that, if they were wrong, when they were found to be wrong then this factor was never given any consideration in the determination as to whether or not there should be a reevaluation as to the extent of domestic cutback. Is that not true?

Secretary Connor. Well, as I understand, the revised figures came as the year went along, and at that point there would have to be a change not so much in the domestic target but in the export quota which had the force of law and which was what was really effective, completely 100 percent effective.

Senator Hartke. Let's come back to this. There was, in fact, a cutback of about how much on domestic use?

Secretary Connor. Pardon me, sir?

Senator Hartke. There was a cutback in domestic use of how much?

Secretary Connor. You mean in actual use?

Senator Hartke. Yes, during the control period.

Secretary Connor. The 1963 actual domestic figure was 23.5, and—

Senator Hartke. That is the calendar year.

Secretary Connor. Do you mean in percentages, sir?

Senator Hartke. I just want to get this straight. How much was it cut back under the export control order year, the year that order was in effect? How much cutback was there in the domestic use?

Secretary Connor. You mean in actual results?

Senator Hartke. That is right. Actual board feet.

Secretary Connor. The figure for 1964 was 19.5, and the figure for the year before was 23.5.

Senator Hartke. So that is 4 million board feet.

Secretary Connor. Yes, sir.

Senator Hartke. And the estimate of the cutback that was under the order was how much?

Secretary Connor. Somewhat higher than that but not appreciably higher.

Senator Hartke. Not appreciably higher? So the industry did come at least within 80 percent of the projected goal? Is that not true?

Secretary Connor. The industry did accomplish a fair percentage of the target. The target, however, was 15 million.

Senator Hartke. The cutback in use, though, that the industry was using in predicating its entire operation, was that they would cut back approximately 5 million board feet? Is that not right?

Secretary Connor. Mr. Chairman, I think the fairest way to express it was that the industry accepted a quota that would have resulted in a 16 percent decrease from the figures available for the base period. The actual accomplishment was 8 percent.

Now, the export quota was based on the same 16 percent figure, and, in fact, the export cutback was that amount.

Senator Hartke. Yes.

Secretary Connor. But we were working with figures available from time to time.

Senator Hartke. That is right.

Secretary Connor. I think that is the only reasonable thing we can do.

Senator Hartke. That is right. The point about it is, without regard now to the effect upon the exports, in fact and in truth the industry instituted its own voluntary program. All of us know in almost any order or legislation there has to be an allowance of lead time. Here there was about three months which necessitated changes in industry methods and in convincing of their customers that they could reduce the thickness of the veneer. They were really in operations for only three-quarters of a year. And they accomplished 80 percent compliance or more than three-quarters of the actual goal of the cutback.

Secretary Connor. Mr. Chairman, I just cannot accept that, because the target was 15 million. The actual accomplishment was 19.5. So the overrun was 4.5 over 15, which is a larger percentage.

Now, the industry—and I am fully sympathetic because I know what the problems are—but the industry knew of the problems of the changeover and how long it would take and the amount of persuasion to change an established practice.

Senator Hartke. Of course, the industry is operating here under the gun. And when they operate under the gun with the federal government they have a problem. In other words, they realize, sure, but this does not make it any easier for them.

If there is anything under the sun that harasses industry, it is government. And I think—

Secretary Connor. Mr. Chairman, this was a gun in a sense, but the export users were under a "Big Bertha," because the actual quota they accepted was 7.3 compared with their actual figure the year before of 14.3, and this is where the inequity comes in.

Senator Hartke. Mr. Secretary, I do not think there is any real disagreement about the actual statistics if we can establish these between yourself and myself. There is a disagreement on how to interpret them. I am not going to try to change your interpretation, but I do want to place this in the proper perspective.

There was under the order an anticipated cutback of 5 million board feet. Is that not right?

Secretary Connor. Yes, sir; that comes out to 16 percent target.

Senator Hartke. I am not going to try to dispute that.

Secretary Connor. I am not trying to be argumentative. I am trying to keep—

Senator Hartke. And we did cut back a little more than 4 million board feet actually?

Secretary Connor. Yes, sir; I think that is right.

Senator Hartke. And if you assume it was only 4 million—and 4 million out of 5 million is all I am saying—whether you want to say it was a 16-percent cutback of the target, I think it was an 80-percent accomplishment in regard to the actual amount of lumber which was going to be used.

Secretary Connor. Yes, sir; if we look at the domestic consumption alone, that is right. But we have to also look at exports. Let's go back to your own statistics. The truth of it was in 1955 when this report and chart were originally prepared by the Department of Commerce, the domestic consumption was 7.3, and at that time the export was 1.2.

Senator Hartke. Now, within a period of 10 years the domestic consumption had gone up, to 1963, just about 3 times what it was in 1955, to 23.5 in actuality, which was the estimate that was wrong. It was estimated to be lower. But the actual determination showed that the domestic consumption went up about 3 times in the period 1955 to the last before the control order was put into effect.

On the side of equity now, showing that the exporters certainly have no right to complain, their consumption and export went up from 1955 from 1.2 to 15 million, to the place where they were using almost three-fourths as much as the domestic industry alone.

In other words, this was a place where the domestic was using 7 times as much as export in 1955, and by the year of 1963 the export consumption had risen so rapidly that they had closed this gap to less than—well, to almost a 3-to-2 from a 7-to-1 average.

So, on the side of equity, they have certainly not been mistreated; is that true?

Secretary Connor. Well, Mr. Chairman, I accept your figures without necessarily accepting your conclusion.

Senator Hartke. You are not accepting "my" figures. I will give you a copy of this. We are going to accept your figures.

Secretary Connor. There is no question about the figures. The export demand in the free market increased proportionately more rapidly than the domestic demand.

Senator Hartke. Not alone proportionately, but excessively so.

Secretary Connor. Well, in any business we like export business as well as domestic business.

Senator Hartke. Well, not when we are going to deplete a natural resource as precious as walnut.

Secretary Connor. Mr. Chairman, I think that is the question. If there is to be a conservation program, what should it be?

Senator Hartke. All right. Let's come on back to this question of GATT. You do not feel that we could impose a complete embargo?

Secretary Connor. Not according to my understanding of the GATT requirements, no, sir. And I think it might be helpful there if you had testimony from the government officials who are responsible for the administration.

Senator Hartke. If in fact and in truth you are wrong on your interpretation of the General Agreements on Tariffs and Trade, commonly called GATT, would you then be willing to review this whole situation with the idea that if this is not a fact, that there could in fact be a complete embargo or any portion thereof, would you be willing in the interest of conservation of this natural resource to reinstitute the order?

Secretary Connor. Mr. Chairman, as I have said, the GATT factor was one factor. It was not the overriding factor. I think the imposition of these export controls, in the absence of any effective domestic control, was the decisive factor in my thinking.

I certainly am prepared and intend to read the record of these hearings and any report of your committee. But I sincerely believe that if Congress wants to conserve this natural resource it has to do it by legislation that differs from export control alone.

Senator Hartke. Well, this is why I asked the question in the first place, really, cutting through the heart of this whole matter, as I said it, the basis of your decision was the fact that you just did not believe this act really applies to walnut logs?

Secretary Connor. Mr. Chairman, I do not think export controls alone would accomplish the conservation purpose. I believe in export controls for the purposes that they can handle. But on a long-term conservation necessity, I do not think export controls were designed for this type of problem.

Senator Hartke. Let me ask you this: Now, we have a distinguished member here, the Senator from Kentucky, and he and I have been so congenial the last few days, and I am fearful that congeniality may wear off today.

Senator Morton. Never.

Senator Hartke. And he is an expert, a former member of the Department of State and is very familiar with GATT and all the treaty arrangements.

Are you prepared to state upon what authority and what provisions of the treaty that you have reference to did you base this decision?

Secretary Connor. Not at this moment, sir. I would appreciate, if you want to go into that in detail, the opportunity to consult further, and I will be prepared to come and testify further at a later date.

Senator Hartke. Let me read to you from article 20, and subsection (j) of the treaty.

> Nothing in this agreement shall be construed to prevent the adoption or enforcement by any contracting party of measures essential to the acquisition or distribution of products in general or local short supply provided that any such measures shall be consistent with the principle that all contracting parties are entitled to an equitable share of the international supply of such products and that any such measures which are inconsistent with other provisions of this agreement shall be discontinued as soon as the conditions giving rise to them cease to exist.

Now, under the authority of—

Secretary Connor. That is a provision we looked at, sir, and the word "equitable" plays a prominent role, and I cannot conclude that these controls were imposed equitably on the export users.

Senator Hartke. No, it just requires the equity be among those parties who are going to participate in the international—in other words, you cannot discriminate in favor of Canada against Mexico, for example, or against France or against some other country as long as the total of the international amount is—

Secretary Connor. I think the United States is a contracting part—

Senator Hartke. No, let's read that. Let's read the provision.

> *Provided,* That any such measures shall be consistent with the principle that all contracting parties are entitled to an equitable share of the international supply.

And if there would be a complete embargo there would not be any international supply, and all contracting parties would be treated equally in regard to that international quota.

Secretary Connor. My interpretation, Mr. Chairman, would be that the U. S. supply should not be inequitably large but I think on this perhaps the committee needs the opinion of people who are closer to these problems than I am.

Senator Hartke. Let me help you a little bit in this regard. There are some other countries who are involved in the question of hardwoods, and there are 21 contracting parties to GATT who have either direct prohibitions on the export of logs or quota systems. Are you familiar with that?

Secretary Connor. I am, sir; I am familiar—

Senator Hartke. Are they in violation of the treaty?

Secretary Connor. Perhaps they are, and this should be taken up through the GATT machinery. But two wrongs do not make a right.

Senator Hartke. Well, I would think that this was reaffirmed just two weeks ago at the treaty negotiations of GATT, where the revisions were under consideration. This provision of the treaty was reaffirmed at that time, and there was no question raised about any of these countries, contracting parties, violating the provisions of this section of the treaty.

Secretary Connor. Well, perhaps the question should be raised. But the point is, Mr. Chairman, that even if we have a complete embargo of this product so that none of it could move in export, we are going to run out of walnut log supply so long as the domestic consumption is in excess of the annual growth.

Senator Hartke. Well, let's go back. Under the control year—and this you must remember: Your estimates of consumption and growth in black walnut veneer quality timber as of 1964 are on a calendar year, whereas the actual order was on February—

Secretary Connor. Thirteenth.

Senator Hartke (continuing). February 13, 1964 to 1965. But taking the calendar year of 1964, there was a domestic consumption of 19.8 million board feet. There was an export consumption of 9.9 million board feet. There was a growth deficit of 11 million board feet, which means that you had reduced this deficit of 1963, if you would have a complete embargo—and which I am not arguing for completely but I am just pointing this out to you—that if you had a complete embargo upon export consumption of walnut logs you would have eliminated all but 1 million board feet of the deficit, which would have brought us back to the year of 1959. And that would be a substantial conservation, would it not?

Secretary Connor. Mr. Chairman, it would have postponed the inevitable H-day or D-day if domestic consumption continues to outpace the annual growth factor. But it does not eliminate it.

Senator Hartke. No. But the point about it is that the only thing that is going to eliminate it entirely is what you have asked for—a return to OPA.

2

Secretary Connor's letter to Senator Warren Magnuson, Chairman, Senate Committee on Commerce, dated September 22, 1965 (abstracted), rejecting controls on walnut exports

The basic arguments for and against these export controls which were known to me at the time of my initial decision last February have remained the same. In addition, I have given consideration to two other factors. First, the monthly volume of exports of veneer quality walnut logs has increased substantially since the export controls expired in February of this year. However, it is our judgment that a significant portion of this increased export activity during recent months has been prompted by the uncertainty as to whether export controls might be reimposed, particularly in view of the hearings of the Senate Commerce Committee as well as the announced reconsideration of my original decision.

Second, as a result of my current review, we have received confirmation from representatives of the domestic veneer and furniture industries that it is not possible in the foreseeable future to achieve any significant reduction in domestic consumption through voluntary use of one thirty-sixth-inch veneer. In short, the domestic furniture industry now makes it plain that it cannot or will not use this thickness of veneer nor will the furniture industry adopt this size as its standard. In view of this fact, I am convinced that no meaningful conservation can be achieved without controlling domestic consumption of walnut logs.

Concerning the factor of walnut log supply, it is quite clear that our long-term forecasts as to the rate of depletion of walnut resources are not and cannot be regarded as clearly reliable. The fact of the matter is that no one knows the number of walnut trees now standing in the United States, and no one can say with precision how long the present supply will last at the present rate of consumption. The reimposition of an annual export quota of 7.3 million board feet would not prevent ultimate exhaustion of walnut reserves since current domestic consumption alone exceeds estimated domestic growth on lands classified as commercial forest.

Taking into account the various conflicting economic interests, I have particularly considered the potential impact of export controls on: (1) the producers or growers of walnut logs, (2) the veneer producers, (3) the furniture manufacturers, and (4) the log exporters.

With respect to the growers or producers of walnut logs, it is clear that the use of export controls is contrary to their economic interests because of reduced demand and the depressant effect on the prices they can obtain for the logs.

With respect to veneer producers (the group most vigorous in support of export controls), control on the export of walnut logs is very clearly to their financial advantage. With exports of logs severely restricted, the principal purchasers in the market for walnut logs would be the domestic veneer producers. In a word, the foreign competition for walnut logs available would be significantly reduced.

With respect to the furniture industry—the user of the walnut veneer—it is evident that export controls are potentially and theoretically in their financial interest. This is true because the foreign demand would be limited.

Concerning the interest of the walnut log exporters, it is, of course, evident their financial interests are adversely affected by the imposition of export controls and for this reason they strongly oppose the reimposition of export controls on walnut logs.

Upon consideration of all these factors ...[I]t appears to me that reimposition of such controls would only assure a price advantage to one particular segment—the veneer producers, at the cost of an adverse financial impact on some other groups of American businessmen.

I am concerned about the pure conservation issue. I would prefer to be able to make an administrative decision—which would satisfy all interested groups of our economy—that would have the clear effect of insuring that the black walnut tree would never become extinct in America. But, as I have said before, no meaningful conservation measure can be achieved without controlling domestic consumption and any export control measure adopted under the misconception that export control alone is sufficient would only delay our squarely facing the issue. I do not have the legal authority to impose regulations on domestic consumption, and voluntary control of domestic consumption does not appear feasible in the foreseeable future.

If the Congress concludes that the commercial supply of black walnut logs is threatened with exhaustion and that this possibility is not likely to be prevented over the long term by the operation of market forces and technological advance, then the enactment of effective domestic conservation regulation should be considered.

The short supply control authority in the Export Control Act should, in my opinion, be used to protect the overall welfare of the domestic economy and not be used as a price control device to benefit one or two segments of an industry to the detriment of other segments in that industry.

3

J. Edward Day's response to two legal aspects in Secretary Connor's letter of September 22, 1965

As to the first point, the requirement for reduction of domestic requirement, the Secretary's letter states:

> On February 14, 1964, my predecessor, Secretary of Commerce Luther H. Hodges, placed "short supply" controls on exports of black walnut logs for a trial period of one year. An export quota was established as a part of an overall program, which included domestic conservation efforts, to bring annual consumption of veneer quality logs to balance with estimated new log growth.
>
> Because a curb on exports could not achieve any meaningful conservation without a concurrent reduction in domestic consumption, Secretary Hodges expressly conditioned the export controls on a voluntary reduction in domestic consumption to approximately 15 million board feet during the trial period, February 1964 to February 1965.

The Secretary's attitude, as shown by his latest letter and his earlier statements on this subject, is to penalize the domestic industry for errors which were in fact made by the Commerce Department itself.

First, the Export Control Act is an unusually clear and specific statue. It relates to "abnormal foreign demand," a condition which clearly exists here.

The act nowhere refers to or authorizes any other kind of controls than "export controls," and there is not the slightest suggestion in the act that the Department should or could condition export controls upon reduction in domestic consumption. The attempt of the Department to impose a domestic target of 15 million board feet was therefore without authority and should be considered of no legal effect.

Second, even if there had been authority to require some realistic reduction in domestic use, such a requirement would only have been valid if the domestic industry had some proper and legal way to meet that target. The domestic reduction the Department purported to impose here was more than the domestic users could meet by any lawful means. They did what they could by cutting thinner veneer. But any effort on their part to do more than that in an effort to reduce domestic use would have risked violation of the antitrust laws since action in concert would have been involved. The condition the Department attempted to impose was, therefore, an unlawful and invalid condition.

Third, the Department's annual domestic quota of 15 million board feet was much too low because of an error made by the Department of Commerce itself. The intention was to reduce domestic use by 5 million board feet a year. At the time the 15 million figure was arrived at in early 1964, the Department erroneously believed, due to incorrect advance calculations, that 1963 usage had been 20 million feet. The correct 1963 domestic usage figure turned out to be 23.5 million.

Only when the 1963 actual figures became available did it become apparent that a cut to the goal of 15 million would involve reducing domestic use not by 5 million but by 8.5 million feet. Careful checking reveals no previous case where any past decision by the Department under the Export Control Act has hinged on requiring a reduction in domestic consumption. Never before has the Department refused to apply the act because there was no authority for domestic controls.

Now, as to my second point which I outlined at the outset: The Secretary's statements about the price effects of controls ignore the clear language of the Export Control Act. In his September 22 letter the Secretary states:

> With respect to veneer producers (the group most vigorous in support of export controls), control on the export of walnut logs is very clearly to their financial advantage. With exports of logs severely restricted, the principal purchasers in the market for walnut logs would be the domestic veneer producers. In a word, the foreign competition for walnut logs available would be significantly reduced.
>
> It appears to me that reimposition of such controls would only assure a price advantage to one particular segment—the veneer producers, at the cost of an adverse financial impact on some other groups of American businessmen.

The Secretary thus refuses to impose controls because this would reduce the price pressure from foreign demand. He thus attacks the clear goal of the Export Control Act which is "to reduce the inflationary impact of abnormal foreign demand."

During the 1964-65 control period the average price of exports continued to go up—so sellers of logs for export did not suffer the "depressant effect" on prices which Secretary Connor, in his September 22 letter, says would take place with controls. Recently the value of walnut logs exported has been about twice what it was 5 years ago and three times what it was 10 years ago.

It is in my opinion absurd for the Secretary to refuse to apply the Export Control Act because a decision to do so would not please everybody. The act contemplates and requires a result which is bound to displease those who are gaining by abnormal foreign demand and its inflationary impact. That policy has been set by Congress in passing the act. The Secretary cannot properly ignore the standards of the act to suit his own preference as to whether there should be such a policy and such a result as Congress has prescribed in the act.

Appendix D

PAST PRESIDENTS AND EXECUTIVES OF TRADE ASSOCIATIONS RELATED TO THE FINE HARDWOODS

Presidents

VENEER ASSOCIATION

1933-1934	J. C. Rodahaffer	Penrod, Jurden & Clark Co.
1935-1936	Harry Hartke	Acme Veneer & Lumber Co.
1937-1938	Albert J. Heidt	Penrod, Jurden & Clark Co.
1939-1940	Thomas A. Dean	The Dean Company
1941-1942	Frank L. Montgomery	Penrod, Jurden & Clark Co.
1943-1944-1945	George A. Vehlow	Underwood Veneer Company
1946	Harry J. Hartke, Jr.	Acme Veneer & Lumber Co.
1947-1948	I. T. Williams	I. T. Williams & Sons
1949	Chester B. Stem	Chester B. Stem, Inc.
1950-1951	William C. Linn	Hoosier Veneer Company
1952	Joseph F. Mertes	R. S. Bacon Veneer Company

FINE HARDWOODS ASSOCIATION

1953	Joseph F. Mertes	R. S. Bacon Veneer Company
1954-1955	J. T. Mengel	The Mengel Company
1956-1957	William F. Ashby	Ashby Veneer & Lumber Co.
1958-1959	Bice A. Roth	Amos-Thompson Corporation
1960-1961	Edward A. Heidt, Sr.	Penrod, Jurden & Clark Co.
1962-1963	Robert E. Hollowell, Jr.	Pierson-Hollowell Company
1964-1965	Charles L. Guyer	Alexander Wood Products, Inc.
1966-1967	Richard K. Stem	Chester B. Stem, Inc.
1968-1969	Burton F. Swain III	National Veneer & Lumber Co.
1970-1971	George T. Kinley	William L. Marshall, Ltd.
1971-1985	(See Fine Hardwoods-American Walnut Association)	

FINE HARDWOOD VENEER ASSOCIATION

1986-1987	Werner Lorenz	Indiana Veneers Corporation
1988-1989	O. C. (Bo) Edwards	International Veneer Company
1990-1991	Kenneth J. Staugas	Norstam Veneers, Inc.

FINE HARDWOODS/AMERICAN WALNUT ASSN.

In 1971, the American Walnut Manufacturers Association and the Fine Hardwoods Association merged and started doing business as the Fine Hardwoods/American Walnut Association.

1971	Bice A. Roth	Amos-Thompson Corporation
1972-1973	George W. Hartzell	Hartzell Industries, Inc.
1974-1975	Daniel W. Duncan	Iowa-Missouri Walnut Co.
1976-1977	G. Karl Bordner	Penrod, Jurden & Clark Co.
1978-1979	Albert B. Cox	Amos-Hill Veneer & Lumber Co.
1980-1981	John A. Grunwald	David R. Webb Company
1982	J. Robert Bolon	American Walnut Company
1983-1984	Burton F. Swain III	National Veneer & Lumber Co.
1985	David M. Dubrule	Interforest Ltd.

In 1986, members of the combined association agreed to separate into their original groups. The Fine Hardwood Veneer Association and the American Walnut Manufacturers Association now maintain their own programs and membership lists although they share office space and staff personnel.

Association Executives

American Walnut Manufacturers Association

1913-1929	George N. Lamb
1929-1953	Burdett L. Green
1953-1956	Charles H. White
1956-1961	George C. Romeiser
1961-1971	Donald H. Gott

Veneer Association

1933-1935	Burdett L. Green
1935-1936	L. W. Anderson
1936-1953	Charles E. Close

Fine Hardwoods Association

1953-1957	Burdett L. Green
1957-1971	E. Howard Gatewood

Fine Hardwoods/American Walnut Association

1971-1978	Donald H. Gott
1978-1985	Larry R. Frye

American Walnut Manufacturers Association

1986-	Larry R. Frye

Fine Hardwood Veneer Association

1986-	Larry R. Frye

AMERICAN WALNUT MANUFACTURERS ASSOCIATION

— Presidents —

1912-1914	Steering Committee	
1914-1915	H. A. McCowen	A. McCowen Company
1916	Unknown	
1917	Unknown	
1918	John N. Penrod	Penrod Walnut & Veneer Company
1919	Ray E. Pickrel	Pickrel Walnut Company
1920	W. W. Knight	Long-Knight Lumber Company
1921	Louis Moschel	Langston Lumber Company
1922	Max Kosse	Kosse, Shoe & Schleyer
1923	W. H. Day	Wood-Mosaic Company
1924	J. C. Rodahaffer	Penrod Walnut & Veneer Company
1925-1926	Charles H. Barnaby	Charles H. Barnaby Company
1927	Alex Schmidt	Kosse, Shoe & Schleyer
1928-1929	Robert E. Hollowell, Sr.	Pierson-Hollowell Company
1930	Joseph J. Egan	Wood-Mosaic Company
1931	George W. Hartzell	Hartzell Industries, Inc.
1932	Chester B. Stem	Pickrell Walnut Company
1933	Roy Amos	Amos-Thompson Corporation
1934	Frank J. Purcell	Frank Purcell Walnut Lumber Co.
1935	Robert N. Hartzell	Hartzell Industries, Inc.
1936	R. E. Pickrel	Pickrel Walnut Company
1937	John K. Shaw	Ozark Hardwood Lumber Company
1938-1939	J. C. Rodahaffer	Penrod, Jurden & Clark Company
1940	Burton F. Swain, Jr.	National Veneer & Lumber Co.
1941-1942	William F. Franket	Wood-Mosaic Company
1943	J. B. Petrus, Sr.	Midwest Walnut Company
1944	J. P. Mikesell	Hartzell Industries, Inc.
1945	H. B. Nickell	Pierson-Hollowell Company
1946	Robert Mathers	Thiesing Veneer Company
1947	William C. Linn	Hoosier Veneer Company
1948-1949	Bice A. Roth	Amos-Thompson Corporation
1950	J. R. Rodahaffer	Penrod, Jurden & Clark Company
1951	R. T. (Bud) Miller	Curry-Miller Veneers, Inc.
1952	Charles R. Crouch	Campbell Lumber Company
1953	Richard K. Stem	Chester B. Stem, Inc.
1954	Louis Vonderbrink	P. V. Shoe Walnut Company
1955-1956	Robert E. Hollowell, Jr.	Pierson-Hollowell Company, Inc.
1957-1958	J. Carl Daugherty	Hoosier Veneer Company
1959-1960	J. B. Petrus, Jr.	Midwest Walnut Company
1961	Clarence Swords	Swords-Morton Veneer & Lbr. Co.
1962-1963	George W. Hartzell	Hartzell Industries, Inc.

1964-1965	Burton F. Swain III	National Veneer & Lumber Co.
1966-1967	William F. Beckwith	Thiesing Veneer Company
1968-1969	David O. Holmes, Jr.	Holmes & Company, Inc.
1970-1971	Kenneth H. Christgen, Sr.	Missouri Valley Walnut Company
1971-1985	(See Fine Hardwoods-American Walnut Association)	
1986-1987	Robert E. Hollowell, Jr.	Pierson-Hollowell Company
1988-1989	Robert E. Hollowell, Jr.	Pierson-Hollowell Company
1990-1991	Robert E. Hollowell, Jr.	Pierson-Hollowell Company

HARDWOOD PLYWOOD MANUFACTURERS ASSOCIATION & PREDECESSOR ORGANIZATIONS

National Veneer and Panel Association (Hardwood Plywood Institute)

Year	President	Year	President
1921	Thomas D. Perry	1946	George A. Vehlow
1922	H. Brooks Sale	1947	Glenn Willson
1938	?	1948	M. H. Speltz
1939	?	1949	Walter L. Jolin
1940	?	1950	A. D. Patchen
1941	A. D. Patchen	1951	William F. Durban
1942	C. G. Yerkes	1952	M. C. McIver
1943-1945	Harold M. Zang	1953	F. S. Kuhns

(In 1954, the Southern Plywood Manufacturers Association merged with the Hardwood Plywood Institute.)

Southern Plywood Manufacturers Association

(Atlanta, Georgia)

Year	President
1946	Bonner Thomason
1947	E. B. Meyercord
1948	E. B. Meyercord
1949	D. E. Nichols
1950	Julius J. Stern
1951	Phil Adams
1952	Verne L. Toussaint
1953	B. E. Bryan

Hardwood Plywood Institute

Year	President	Year	President
1954	George Lester	1959	Sam Nickey
1955	Don C. Nichols	1960	C. D. Wheeler
1956	C. V. Gamelspacher	1961	R. J. Weber
1957	Eric Larson	1962	F. R. McKnight
1958	Donald Gott	1963	J. M. Prince

In 1964, the name was changed to:

Hardwood Plywood Manufacturers Association

Year	President	Year	President
1964	D. W. Tyler	1973	W. D. Kornegay, Jr.
1965	J. J. Stern	1974	James H. Stough
1966	E. M. Magarian	1975	Richard K. Weber
1967	W. Waldo Bradley	1976	J. F. DePledge
1968	B. E. Bryan	1977	C. Herb Warwick
1969	J. R. Berger	1978	J. Tait Hardaway
1970	J. R. Berger	1979	Gail Overgard
1971	R. A. Jones	1980	James B. Lester
1972	Robert O'Donoghue	1981	Robert T. Renfro

(The position of President became "Chairman of the Board" in 1982.)

Year	Chairman	Year	Chairman
1982	Michael Speltz	1987	Donald MacLean Bell
1983	Clint Ludeman	1988	Gene Nelson
1984	R. L. "Dick" Crutchfield	1989	Clay Howell
1985	Clifton R. Jones	1990	Herb McClaugherty
1986	Brad Mills	1991	Diane Montoya

— Association Executives —

1958(?)-81	Clark E. McDonald	Secretary/Treasurer & Managing Director
1982-1986	Clark E. McDonald	President
1987-	E. T. "Bill" Altman	President

NATIONAL HARDWOOD LUMBER ASSOCIATION

— Presidents —

1898-1901	William A. Bennett	Cincinnati, Ohio
1901-1902	Frank H. Smith	St. Louis, Missouri
1902-1904	Burton F. Swain	Shelbyville, Indiana
1904-1906	Earl Palmer	Paducah, Kentucky

1906-1908	William H. Russe	Memphis, Tennessee
1908-1910	Oliver O. Agler	Chicago, Illinois
1910-1912	Fred A. Diggins	Cadillac, Michigan
1912-1914	Charles H. Barnaby	Greencastle, Indiana
1914-1916	Edward Vose Babcock	Pittsburgh, Pennsylvania
1916-1918	John M. Woods	Boston, Massachusetts
1918-1920	Charles A. Goodman	Marinette, Wisconsin
1920-1922	Horace F. Taylor	Buffalo, New York
1922-1924	John W. McClure	Memphis, Tennessee
1924-1926	Harry B. Curtin	Clarksburg, West Virginia
1926-1928	Ben C. Currie	Philadelphia, Pennsylvania
1928-1930	Robert C. Stimson	Memphis, Tennessee
1930-1932	John I. Shafer	Logansport, Indiana
1932-1934	John R. Thistlethwaite	Opelousas, Louisiana
1934-1936	George N. Harder	Wells, Michigan
1936-1938	Wilson H. Lear	Philadelphia, Pennsylvania
1938-1940	Joseph J. Linehan	Cincinnati, Ohio
1940-1942	Gordon E. Reynolds	Albany, Georgia
1942-1944	Abbott M. Fox	Iron Mountain, Michigan
1944-1946	D. C. MacLea	Baltimore, Maryland
1946-1948	George H. Henderson	Lufkin, Texas
1948-1950	Frank W. Hutcheson	Huntsville, Ontario
1950-1952	Lee Robinson	Mt. Vernon, Alabama
1952-1954	Harry D. Gaines	St. Louis, Missouri
1954-1956	T. M. Millett	Louisville, Kentucky
1956-1958	Howard A. Hanlon	Odessa, New York
1958-1960	Bernard Bock	Montreal, Quebec
1960-1962	Mark Townsend	Stuttgart, Arkansas
1962-1964	James P. Hamer	Kenova, West Virginia
1964-1966	Donald F. White	Oakland, California
1966-1968	Lawrence D. Kellogg	Alexandria, Louisiana
1968-1970	H. F. Staniforth	Montreal, Quebec
1970-1972	Gordon R. Connor	Wausau, Wisconsin
1972-1974	Richard S. Elliott	Potsdam, New York
1974-1976	Robert E. Hollowell, Jr.	Indianapolis, Indiana
1976-1978	Leonard J. Barrett	Montreal, Quebec
1978-1980	Henry W. Jones, Jr.	Memphis, Tennessee
1980-1982	Paul D. Webster	Wayzata, Minnesota
1982-1984	John B. Veach, Jr.	Alcoa, Tennessee
1984-1986	E. R. (Ted) Thomas, Jr.	St. Louis, Missouri
1986-1988	Walter M. Fields, Jr.	Memphis, Tennessee
1988-1990	John G. Thomson	Alliston, Ontario

— Association Executives —

(Secretaries/Executive Managers)

1898–1904	Allen R. Vinnedge
1904–1905	Charles D. Strode
1905–1930	Frank Fish
1930–1934	Louis Beale
1934–1951	John W. McClure
1951–1956	Joseph L. Muller
1956–1971	Maurice B. Pendleton
1971–1978	E. Howard Gatewood
1978–1988	S. Carroll White
1988–	Ernest J. Stebbins

INDEX OF INDIVIDUALS

INDEX OF FIRMS, ORGANIZATIONS, AND AGENCIES